QA2.71
M362

AN
INTRODUCTION
TO THE
THEORY OF

LINEAR SPACES

GEORGI E. SHILOV

Professor of Mathematics
Moscow State University

Translated from the Russian by
Richard A. Silverman

DOVER PUBLICATIONS, INC.

New York

Published in Canada by General Publishing Company, Ltd., 30 Lesmill Road, Don Mills, Toronto, Ontario.
Published in the United Kingdom by Constable and Company, Ltd., 10 Orange Street, London WC 2.

This Dover edition, first published in 1974, is an unabridged and corrected republication of the work originally published by Prentice-Hall, Inc., in 1961.

International Standard Book Number: 0-486-63070-6
Library of Congress Catalog Card Number: 73-94345

Manufactured in the United States of America
Dover Publications, Inc.
180 Varick Street
New York, N.Y. 10014

AUTHOR'S PREFACE

This book came about as a result of reworking material presented by the author in lecture courses and seminars given in recent years at Moscow and Kiev Universities. Stated briefly, the basic idea of the book is the following: It is more useful to treat algebra, geometry, and analysis as parts of a connected whole than as separate subjects. Thus, geometric notions help to clarify and often anticipate facts from algebra or analysis, just as algebraic methods often suggest the proper approach to be taken in a geometric or analytic context. Of course, this is not a new idea, and in fact, we can trace its influence on many generations of mathematicians from Descartes to Hilbert.

In the present book, the idea of the unity of algebra, geometry, and analysis is pursued in connection with elementary topics, accessible to students of mathematics and physics, even on the undergraduate level. Moreover, the book contains considerable material which these students are required to study in one form or another, and, in my opinion, the approach adopted here is the most accessible, pleasant, and useful way to master this material. In the U.S.S.R., this book is also used by students and candidates in technical institutes, by staff-workers in computation centers (despite the book's lack of explicit computational procedures) and by engineers in various professional-advancement courses. I also feel that the book is suitable for self-study.

To some extent, the problems given here are intended to help the student acquire technique, but most of them serve to illustrate and develop further the basic subject matter of the text. Some of the problems stem from elementary seminars, others from more advanced seminars, in which a certain degree of enthusiasm on the part of the participants is taken for granted.

I am happy to have the opportunity to express my appreciation to M. A. Krasnosyelski and M. G. Krein for numerous valuable suggestions, and to N. V. Efimov and D. A. Raikov for their careful reading of the manuscript before its publication; in many instances, their critical remarks allowed me to improve the presentation.

I am especially grateful to the Prentice-Hall Publishing Company, and in particular to Dr. Richard A. Silverman, for undertaking to make this book available in an English-language edition.

G. E. S.

TRANSLATOR'S PREFACE

The present volume is the first in a new series of translations of outstanding Russian textbooks and monographs in the fields of mathematics, physics, and engineering. It is my privilege to serve as Editor of the series. It is hoped that this book by Professor G. E. Shilov will set the standard for the volumes to follow.

The translation is a faithful one, to the extent that this is compatible with the syntactic and stylistic differences between Russian and English. However, I have occasionally made slight changes, and I have attempted to detect and correct all typographical errors. I have also added a Bibliography, containing suggestions for collateral and supplementary reading. Finally, it should be noted that sections marked with asterisks contain material of a more advanced nature, which can be omitted without loss of continuity.

R. A. S.

CONTENTS

1

DETERMINANTS

I. Systems of Linear Equations

In this and the next two chapters, we shall study systems of linear equations. In the most general case, such a system has the form

$$
\begin{aligned}
a_{11}x_1 + a_{12}x_2 + \cdots + a_{1n}x_n &= b_1, \\
a_{21}x_1 + a_{22}x_2 + \cdots + a_{2n}x_n &= b_2, \\
\cdots \qquad \cdots \qquad \cdots \qquad \cdots & \\
a_{k1}x_1 + a_{k2}x_2 + \cdots + a_{kn}x_n &= b_k.
\end{aligned}
\tag{1}
$$

Here $x_1, x_2, \ldots, x_n$ denote the unknowns which are to be determined. (Note that we do not necessarily assume that the number of unknowns equals the number of equations.) The quantities $a_{11}, a_{12}, \ldots, a_{kn}$ are called the *coefficients* of the system. The first index of a coefficient indicates the number of the equation in which the coefficient appears, while the second index indicates the number of the unknown with which the coefficient is associated.[1] The quantities $b_1, b_2, \ldots, b_k$ appearing in the right-hand side of (1) are called the *constant terms* of the system; like the coefficients, they are assumed to be known. By a *solution* of the system (1) we mean any set of numbers $c_1, c_2, \ldots, c_n$ which when substituted for the unknowns $x_1, x_2, \ldots, x_n$, turns all the equations of the system into identities.[2]

---

[1] Thus, for example, the symbol a_{34} should be read as "*a* three four" and not as "*a* thirty-four."

[2] We emphasize that the set of numbers $c_1, c_2, \ldots, c_n$ represents *one* solution of the system and not *n* solutions.

Not every system of linear equations of the form (1) has a solution. For example, the system

$$2x_1 + 3x_2 = 5,$$
$$2x_1 + 3x_2 = 6 \tag{2}$$

obviously has no solution at all. Indeed, whatever numbers c_1, c_2 we substitute in place of the unknowns x_1, x_2, the left-hand sides of the equations of the system (2) are the same, while the right-hand sides are different. Therefore, no such substitution can simultaneously convert both equations of the system into identities.

A system of equations of the form (1) which has at least one solution is called *compatible*; a system which does not have solutions is called *incompatible*. A compatible system can have one solution or several solutions. In the latter case, we distinguish the solutions by indicating the number of the solution by a superscript in parentheses; for example, the first solution will be denoted by $c_1^{(1)}, c_2^{(1)}, \ldots, c_n^{(1)}$, the second solution by $c_1^{(2)}, c_2^{(2)}, \ldots, c_n^{(2)}$, and so on. The solutions $c_1^{(1)}, c_2^{(1)}, \ldots, c_n^{(1)}$ and $c_1^{(2)}, c_2^{(2)}, \ldots, c_n^{(2)}$ are considered to be *distinct* if at least one of the numbers $c_i^{(1)}$ does not coincide with the corresponding numbers $c_i^{(2)}$ ($i = 1, 2, \ldots, n$). For example, the system

$$2x_1 + 3x_2 = 0,$$
$$4x_1 + 6x_2 = 0 \tag{3}$$

has the distinct solutions

$$c_1^{(1)} = c_2^{(1)} = 0 \quad \text{and} \quad c_1^{(2)} = 3, c_2^{(2)} = -2$$

(and also infinitely many other solutions). If a compatible system has a unique solution, the system is called *determinate*; if a compatible system has at least two different solutions, it is called *indeterminate*.

We can now formulate the basic problems which arise in studying the system (1):

1. *To ascertain whether the system* (1) *is compatible or incompatible;*

2. *If the system* (1) *is compatible, to ascertain whether it is determinate;*

3. *If the system* (1) *is compatible and determinate, to find its unique solution;*

4. *If the system* (1) *is compatible and indeterminate, to describe the set of all its solutions.*

The basic mathematical tool for studying linear systems is *the theory of determinants*, which we consider next.

2. Determinants of Order n

2.1 Suppose that we are given a *square matrix*, i.e., an array of n^2 numbers a_{ij} $(i, j = 1, 2, \ldots, n)$:

$$
\begin{Vmatrix}
a_{11} & a_{12} & \cdots & a_{1n} \\
a_{21} & a_{22} & \cdots & a_{2n} \\
\cdot & \cdot & \cdots & \cdot \\
\cdot & \cdot & \cdots & \cdot \\
a_{n1} & a_{n2} & \cdots & a_{nn}
\end{Vmatrix}
\tag{4}
$$

The number of rows and columns of the matrix (4) is called its *order*. The numbers a_{ij} are called the *elements* of the matrix. The first index indicates the row and the second index the column in which a_{ij} appears.

Consider any product of n elements which appear in different rows and different columns of the matrix (4), i.e., a product containing *just one element from each row and each column*. Such a product can be written in the form

$$
a_{\alpha_1 1} a_{\alpha_2 2} \ldots a_{\alpha_n n}.
\tag{5}
$$

Actually, for the first factor we can always choose the element appearing in the first column of the matrix (4); then, if we denote by α_1 the number of the row in which the element appears, the indices of the element will be $\alpha_1, 1$. Similarly, for the second factor we can choose the element appearing in the second column; then its indices will be $\alpha_2, 2$, where α_2 is the number of the row in which the element appears, and so on. Thus, the indices $\alpha_1, \alpha_2, \ldots, \alpha_n$ are the numbers of the rows in which the factors of the product (5) appear, when we agree to write the column indices in increasing order. Since, by hypothesis, the elements $a_{\alpha_1 1}, a_{\alpha_2 2}, \ldots, a_{\alpha_n n}$ appear in *different* rows of the matrix (4), one from each row, then the numbers $\alpha_1, \alpha_2, \ldots, \alpha_n$ are all different and represent some permutation of the numbers $1, 2, \ldots, n$.

By an *inversion* in the sequence $\alpha_1, \alpha_2, \ldots, \alpha_n$, we mean an arrangement of two indices such that the larger index comes before the smaller index. The total number of inversions will be denoted by $N(\alpha_1, \alpha_2, \ldots, \alpha_n)$. For example, in the permutation 2, 1, 4, 3, there are two inversions (2 before 1, 4 before 3), so that

$$
N(2, 1, 4, 3) = 2.
$$

In the permutation 4, 3, 1, 2, there are five inversions (4 before 3, 4 before 1, 4 before 2, 3 before 1, 3 before 2), so that

$$
N(4, 3, 1, 2) = 5.
$$

If the number of inversions in the sequence $\alpha_1, \alpha_2, \ldots, \alpha_n$ is even, we put a plus sign before the product (5); if the number is odd, we put a minus sign before the product. In other words, we agree to write in front of each product of the form (5) the sign determined by the expression

$$(-1)^{N(\alpha_1, \alpha_2, \ldots, \alpha_n)}.$$

The total number of products of the form (5) which can be formed from the elements of a given matrix of order n is equal to the total number of permutations of the numbers $1, 2, \ldots, n$. As is well known, this number is equal to $n!$.

We now introduce the following definition:

By the determinant D of the matrix (4) *is meant the algebraic sum of the n! products of the form* (5), *each of which is preceded by the sign determined by the rule just given, i.e.,*

$$D = \sum (-1)^{N(\alpha_1, \alpha_2, \ldots, \alpha_n)} a_{\alpha_1 1} a_{\alpha_2 2} \ldots a_{\alpha_n n}. \tag{6}$$

Henceforth, the products of the form (5) will be called the *terms* of the determinant. The elements a_{ij} of the matrix (4) will be called the *elements* of the determinant. We denote the determinant corresponding to the matrix (4) by one of the following symbols:

$$D = \begin{vmatrix} a_{11} & a_{12} & \ldots & a_{1n} \\ a_{21} & a_{22} & \ldots & a_{2n} \\ \cdot & & \ldots & \cdot \\ a_{n1} & a_{n2} & \ldots & a_{nn} \end{vmatrix} = \det \|a_{ij}\|. \tag{7}$$

For example, we obtain the following expressions for the determinants of orders two and three:

$$\begin{vmatrix} a_{11} & a_{12} \\ a_{21} & a_{22} \end{vmatrix} = a_{11}a_{22} - a_{21}a_{12},$$

$$\begin{vmatrix} a_{11} & a_{12} & a_{13} \\ a_{21} & a_{22} & a_{23} \\ a_{31} & a_{32} & a_{33} \end{vmatrix} = a_{11}a_{22}a_{33} + a_{21}a_{32}a_{13} + a_{31}a_{12}a_{23} \\ - a_{31}a_{22}a_{13} - a_{21}a_{12}a_{33} - a_{11}a_{32}a_{23}.$$

We now indicate the role of determinants in solving systems of linear equations, by considering the example of a system of two equations in two unknowns:

$$a_{11}x_1 + a_{12}x_2 = b_1,$$
$$a_{21}x_1 + a_{22}x_2 = b_2.$$

Eliminating one of the unknowns in the usual way, we can easily obtain the formulas

$$x_1 = \frac{b_1 a_{22} - b_2 a_{12}}{a_{11} a_{22} - a_{21} a_{12}}, \qquad x_2 = \frac{a_{11} b_2 - a_{21} b_1}{a_{11} a_{22} - a_{21} a_{12}},$$

assuming that these ratios have nonvanishing denominators. The numerators and denominators of the ratios can be represented by the second-order determinants

$$a_{11}a_{22} - a_{21}a_{12} = \begin{vmatrix} a_{11} & a_{12} \\ a_{21} & a_{22} \end{vmatrix},$$

$$b_1 a_{22} - b_2 a_{12} = \begin{vmatrix} b_1 & a_{12} \\ b_2 & a_{22} \end{vmatrix},$$

$$a_{11}b_2 - a_{21}b_1 = \begin{vmatrix} a_{11} & b_1 \\ a_{21} & b_2 \end{vmatrix}.$$

It turns out that similar formulas hold for the solutions of systems with an arbitrary number of unknowns (Sec. 7).

2.2. The rule for determining the sign of a given term of a determinant can be formulated somewhat differently, in geometric terms. Corresponding to the enumeration of elements in the matrix (4), we can distinguish two natural positive directions: from left to right along the rows, and from top to bottom along the columns. Moreover, the slanting lines joining any two elements of the matrix can be furnished with a direction: we shall say that the line segment joining the element a_{ij} with the element a_{kl} has *positive slope* if its right endpoint lies lower than its left endpoint, and that it has *negative slope* if its right endpoint lies higher than its left endpoint.[3] Now imagine that in the matrix (4) we draw all the segments with *negative* slope joining pairs of elements $a_{\alpha_1 1}, a_{\alpha_2 2}, \ldots, a_{\alpha_n n}$ of the product (5). Then we put a plus sign before the product (5) if the number of all such segments is even, and a minus sign if the number is odd.

For example, in the case of a fourth-order matrix, a plus sign must be put before the product $a_{21}a_{12}a_{43}a_{34}$, since there are two segments of negative slope joining the elements of this product:

$$\begin{Vmatrix} a_{11} & \boxed{a_{12}} & a_{13} & a_{14} \\ \boxed{a_{21}} & a_{22} & a_{23} & a_{24} \\ a_{31} & a_{32} & a_{33} & \boxed{a_{34}} \\ a_{41} & a_{42} & \boxed{a_{43}} & a_{44} \end{Vmatrix}.$$

However, a minus sign must be put before the product $a_{41}a_{32}a_{13}a_{24}$, since in the matrix there are five segments of negative slope joining these elements:

$$\begin{Vmatrix} a_{11} & a_{12} & \boxed{a_{13}} & a_{14} \\ a_{21} & a_{22} & a_{23} & \boxed{a_{24}} \\ a_{31} & \boxed{a_{32}} & a_{33} & a_{34} \\ \boxed{a_{41}} & a_{42} & a_{43} & a_{44} \end{Vmatrix}.$$

---

[3] This definition of "slope" is not to be confused with the geometric notion with the same name. In fact, the sign convention adopted here is the opposite of that used in geometry (*Translator*).

In these examples, the number of segments of negative slope joining the elements of a given term equals the number of *inversions* in the order of the first indices of the elements appearing in the term. In the first example, the sequence 2, 1, 4, 3 of first indices has two inversions; in the second example, the sequence 4, 3, 1, 2 of first indices has five inversions.

We now show that *the second definition of the sign of a term in a determinant is equivalent to the first.* To show this, it suffices to prove that the number of inversions in the sequence of first indices of a given term (with the second indices in natural order) is always equal to the number of segments of negative slope joining the elements of the given term in the matrix. But this is almost obvious, since the presence of a segment of negative slope joining the elements $a_{\alpha_i i}$ and $a_{\alpha_j j}$ means that $\alpha_i > \alpha_j$ for $i < j$, i.e., there is an inversion in the order of the first indices.

Problem 1. With what sign do the terms

   (a) $a_{23}a_{31}a_{42}a_{56}a_{14}a_{65}$,

   (b) $a_{32}a_{43}a_{14}a_{51}a_{66}a_{25}$

appear in the determinant of order 6?   *Ans.* (a) $+$, (b) $+$.

Problem 2. Write down all the terms appearing in the determinant of order 4 which have a minus sign and contain the factor a_{23}.

         Ans. $a_{11}a_{32}a_{23}a_{44}$, $a_{41}a_{12}a_{23}a_{34}$, $a_{31}a_{42}a_{23}a_{14}$.

Problem 3. With what sign does the term $a_{1n}a_{2,n-1}\ldots a_{n1}$ appear in the determinant of order n?   *Ans.* $(-1)^{n(n-1)/2}$.

3. Properties of Determinants of Order n

3.1. The transposition operation. The determinant

$$\begin{vmatrix} a_{11} & a_{21} & \cdots & a_{n1} \\ a_{12} & a_{22} & \cdots & a_{n2} \\ \cdot & \cdot & \cdots & \cdot \\ a_{1n} & a_{2n} & \cdots & a_{nn} \end{vmatrix} \tag{8}$$

obtained from the determinant (7) by interchanging rows and columns with the same indices is said to be the *transpose* of the determinant (7). We now show that *the transpose of a determinant has the same value as the original determinant.* In fact, the determinants (7) and (8) obviously consist of the same terms; therefore, it is enough for us to show that identical terms in the determinants (7) and (8) have identical signs. Transposition of the matrix of a determinant is clearly the result of rotating it (in space) through 180° about the diagonal $a_{11}a_{22}\ldots a_{nn}$. As a result of this rotation, every segment with negative slope (e.g., making an angle $\alpha < 90°$ with the rows of the matrix) again becomes a segment with negative slope (i.e., making the angle

$90° - \alpha$ with the rows of the matrix). Therefore, the number of segments with negative slope joining the elements of a given term does not change after transposition. Consequently, the sign of the term does not change either. Thus, the signs of all the terms are preserved, which means that the value of the determinant remains unchanged.

The property just proved establishes the equivalence of the rows and columns of a determinant. Therefore, further properties of determinants will be stated and proved only for columns.

3.2. The antisymmetry property. By the property of being *antisymmetric with respect to columns*, we mean the fact that a determinant changes sign when two of its columns are interchanged. We consider first the case where two adjacent columns are interchanged, for example columns j and $j + 1$. The determinant which is obtained after these columns are interchanged obviously still consists of the same terms as the original determinant. Consider any of the terms of the original determinant. Such a term contains an element of the jth column and an element of the $(j + 1)$th column. If the segment joining these two elements originally had negative slope, then after the interchange of columns, its slope becomes positive, and conversely. As for the other segments joining pairs of elements of the term in question, each of these segments does not change the character of its slope after the column interchange. Consequently, the number of segments with negative slope joining the elements of the given term changes by one when the two columns are interchanged; therefore, each term of the determinant, and hence the determinant itself, changes sign when the columns are interchanged.

Suppose now that two nonadjacent columns are interchanged, e.g., column j and column k $(j < k)$, where there are m other columns between them. This interchange can be accomplished by successive interchanges of adjacent columns as follows: First column j is interchanged with column $j + 1$, then with columns $j + 2, j + 3, \ldots, k$. Then the column $k - 1$ so obtained (which was formerly column k) is interchanged with columns $k - 2$, $k - 3, \ldots, j$. In all, $m + 1 + m = 2m + 1$ interchanges of adjacent columns are required, each of which, according to what has been proved, changes the sign of the determinant. Therefore, at the end of the process, the determinant will have a sign opposite to its original sign (since for any integer m, the number $2m + 1$ is odd).

COROLLARY. *A determinant with two identical columns vanishes.*

Proof. Interchanging the columns, we do not change the determinant; on the other hand, by what has been proved, the determinant must change its sign. Thus $D = -D$, which implies that $D = 0$.

Problem. Show that of the $n!$ terms of a determinant, exactly half $(n!/2)$

have a plus sign according to the definition of Sec. 2, while the other half have a minus sign.

Hint. Consider the determinant all of whose elements equal 1.

3.3. The linear property of determinants. This property can be formulated as follows:

If all the elements of the j'th column of a determinant D are "linear combinations" of two columns of numbers, i.e.,

$$a_{ij} = \lambda b_i + \mu c_i \qquad (i = 1, 2, \ldots, n),$$

where λ and μ are fixed numbers, then the determinant D is equal to a linear combination of two determinants, i.e.,

$$D = \lambda D_1 + \mu D_2. \tag{9}$$

Here both determinants have the same columns as the determinant D except for the j'th column; the j'th column of D_1 consists of the numbers b_i, while the j'th column of D_2 consists of the numbers c_i.

Proof. Every term of the determinant D can be represented in the form

$$a_{\alpha_1 1} a_{\alpha_2 2} \ldots a_{\alpha_j j} \ldots a_{\alpha_n n} = a_{\alpha_1 1} a_{\alpha_2 2} \ldots (\lambda b_{\alpha_j} + \mu c_{\alpha_j}) \ldots a_{\alpha_n n}$$

$$= \lambda a_{\alpha_1 1} a_{\alpha_2 2} \ldots b_{\alpha_j} \ldots a_{\alpha_n n} + \mu a_{\alpha_1 1} a_{\alpha_2 2} \ldots c_{\alpha_j} \ldots a_{\alpha_n n}.$$

Adding up all the first terms (with the signs which the corresponding terms have in the original determinant), we clearly obtain the determinant D_1, multiplied by the number λ. Similarly, adding up all the second terms, we obtain the determinant D_2, multiplied by the number μ. This proves formula (9).

It is convenient to write this formula in a somewhat different form. Let D be an arbitrary fixed determinant. Denote by $D_j(p_i)$ the determinant which is obtained by replacing the elements of the jth column of D by the numbers p_i $(i = 1, 2, \ldots, n)$. Then (9) takes the form

$$D_j(\lambda b_i + \mu c_i) = \lambda D_j(b_i) + \mu D_j(c_i).$$

This linear property of determinants can easily be extended to the case where every element of the jth column is a linear combination not of two terms but of any other number of terms, i.e.

$$a_{ij} = \lambda b_i + \mu c_i + \cdots + \tau f_i.$$

In this case

$$D_j(a_{ij}) = D_j(\lambda b_i + \mu c_i + \cdots + \tau f_i)$$
$$= \lambda D_j(b_i) + \mu D_j(c_i) + \cdots + \tau D_j(f_i). \tag{10}$$

COROLLARY 1. *We can factor out of a determinant any common factor of a column of the determinant.*

Proof. If $a_{ij} = \lambda b_i$, then by (10) we have

$$D_j(a_{ij}) = D_j(\lambda b_i) = \lambda D_j(b_i),$$

as required.

COROLLARY 2. *If a column of a determinant consists entirely of zeros, then the determinant vanishes.*

Proof. Since 0 is a common factor of the elements of one of the columns, we can factor it out of the determinant, obtaining

$$D_j(0) = D_j(0 \cdot 1) = 0 \cdot D_j(1) = 0.$$

Problem. By making a termwise expansion, calculate the determinant

$$\Delta = \begin{vmatrix} am + bp & an + bq \\ cm + dp & cn + dq \end{vmatrix}.$$

Ans. $\Delta = (mq - np)(ad - bc)$.

3.4. Addition of an arbitrary multiple of one column to another column:
We do not change the value of a determinant by adding the elements of one column multiplied by an arbitrary number to the corresponding elements of another column.

Suppose that we add the kth column multiplied by the number λ to the jth column ($k \neq j$). The jth column of the resulting determinant consists of elements of the form $a_{ij} + \lambda a_{ik}$ ($i = 1, 2, \ldots, n$). By (10) we have

$$D_j(a_{ij} + \lambda a_{ik}) = D_j(a_{ij}) + \lambda D_j(a_{ik}).$$

The jth column of the second determinant consists of the elements a_{ik}, and hence is identical with the kth column. It follows from the corollary on p. 7 that $D_j(a_{ik}) = 0$, so that

$$D_j(a_{ij} + \lambda a_{ik}) = D_j(a_{ij}),$$

as required.

This property can also be formulated more generally:

Suppose we add to the elements of the j'th column of a determinant first the corresponding elements of the k'th column multiplied by λ, next the elements of the l'th column multiplied by μ, etc., and finally the elements of the p'th column multiplied by τ ($k \neq j, l \neq j, \ldots, p \neq j$). Then the value of the determinant remains unchanged.

Problem. The numbers 20604, 53227, 25755, 20927 and 78421 are divisible by 17. Show that the determinant

$$\begin{vmatrix} 2 & 0 & 6 & 0 & 4 \\ 5 & 3 & 2 & 2 & 7 \\ 2 & 5 & 7 & 5 & 5 \\ 2 & 0 & 9 & 2 & 7 \\ 7 & 8 & 4 & 2 & 1 \end{vmatrix}$$

is also divisible by 17.

Hint. Multiply the first column by 10^4, the second by 10^3, the third by 10^2, the fourth by 10^1, and add them to the last column; then use Corollary 1 to Property 3.3.

Because of the invariance of determinants under transposition (Property 3.1), all the properties of determinants proved in this section for columns remain valid for rows as well.

4. Expansion of a Determinant with Respect to a Row or a Column. Cofactors

Consider any column, the jth say, of the determinant D. Let a_{ij} be any element of this column. Add up all the terms containing the element a_{ij} appearing in the right-hand side of the equality

$$D = \sum (-1)^{N(\alpha_1, \alpha_2, \ldots, \alpha_n)} a_{\alpha_1 1} a_{\alpha_2 2} \cdots a_{\alpha_n n}$$

(cf. equation (6) of Sec. 2), and then factor out the element a_{ij}. The quantity which remains will be denoted by A_{ij}. It is called the *cofactor* (or *algebraic complement*) of the element a_{ij} of the determinant D.

Since every term of the determinant D contains an element from the jth column, (6) can now be given the form

$$D = a_{1j} A_{1j} + a_{2j} A_{2j} + \cdots + a_{nj} A_{nj}; \tag{11}$$

(11) is called *the expansion of the determinant D with respect to the elements of the j'th column.* Of course, we can write a similar formula for any *row* of the determinant D; for example, for the ith row we have the formula

$$D = a_{i1} A_{i1} + a_{i2} A_{i2} + \cdots + a_{in} A_{in}. \tag{12}$$

Equations (11) and (12) can be used to calculate determinants, but first we must know how to calculate cofactors. In the next section, we shall show how this is done.

To conclude this section, we note a consequence of (11) and (12) which will be useful later. Equation (11) is an identity in the quantities $a_{1j}, a_{2j}, \ldots, a_{nj}$. Therefore, it remains valid if we replace a_{ij} ($i = 1, 2, \ldots, n$) by any other quantities. The quantities $A_{1j}, A_{2j}, \ldots, A_{nj}$ remain unchanged when such a replacement is made, since they do not depend on the elements a_{ij}. Suppose that in the right- and left-hand sides of the equality (11) we replace the elements $a_{1j}, a_{2j}, \ldots, a_{nj}$ by the corresponding elements of any other column, say the kth. Then the determinant in the left-hand side of (11) will have two identical columns and will therefore vanish, according to Property 3.2 of Sec. 3. Thus, we obtain the relation

$$a_{1k} A_{1j} + a_{2k} A_{2j} + \cdots + a_{nk} A_{nj} = 0 \tag{13}$$

for $k \neq j$. Similarly, from (12) we obtain

$$a_{l1}A_{i1} + a_{l2}A_{i2} + \cdots + a_{ln}A_{in} = 0 \tag{14}$$

for $l \neq i$. We formulate the results just obtained in the form of two theorems:

THEOREM 1. *The sum of all the products of the elements of a column (or row) of the determinant D with the corresponding cofactors is equal to the determinant D itself.*

THEOREM 2. *The sum of all the products of the elements of a column (or row) of the determinant D with the cofactors of the corresponding elements of another column (row) is equal to zero.*

5. Minors. Expression of Cofactors in Terms of Minors

If we delete a row and a column from a matrix of order n, then, of course, the remaining elements form a matrix of order $n - 1$. The determinant of this matrix is called a *minor* of the original nth-order matrix (and also a minor of its determinant D). If we delete the ith row and the jth column of D, then the minor so obtained is denoted by M_{ij} or $M_{ij}(D)$.

We now show that the relation

$$A_{ij} = (-1)^{i+j}M_{ij} \tag{15}$$

holds, so that the calculation of cofactors reduces to the calculation of the corresponding minors. First, we prove (15) for the case $i = 1, j = 1$. We add up all the terms in the right-hand side of (6) which contain the element a_{11}, and consider one of these terms. It is clear that the product of all the elements of this term except a_{11} gives a term c of the minor M_{11}. Since in the matrix of the determinant D, there are no segments of negative slope joining the element a_{11} with the other elements of the term selected, the sign ascribed to the term $a_{11}c$ of the determinant D is the same as the sign ascribed to the term c in the minor M_{11}. Moreover, by suitably choosing a term of the determinant D containing a_{11} and then deleting a_{11}, we can obtain any term of the minor M_{11}. Thus, the algebraic sum of all the terms of the determinant D containing a_{11}, with a_{11} deleted, equals the product M_{11}. But according to Sec. 4, this sum is equal to the product A_{11}. Therefore, $A_{11} = M_{11}$ as required.

Now we prove (15) for arbitrary i and j, making essential use of the fact that the formula is valid for $i = j = 1$. Consider the element $a_{ij} = a$, located at the intersection of the ith row and the jth column of the determinant D. By successively interchanging adjacent rows and columns, we can move the element a over to the upper left-hand corner of the matrix; to do this, we need

$$i - 1 + j - 1 = i + j - 2$$

interchanges. As a result, we obtain the determinant D_1 with the same terms as those of the original determinant D multiplied by

$$(-1)^{i+j-2} = (-1)^{i+j}.$$

The minor $M_{11}(D_1)$ of the determinant D_1 is clearly identical with the minor $M_{ij}(D)$ of the determinant D. By what has been proved already, the sum of the terms of the determinant D_1 which contain the element a, with a deleted, is equal to $M_{11}(D_1)$. Therefore, the sum of the terms of the original determinant D which contain the element $a_{ij} = a$, with a deleted, is equal to

$$(-1)^{i+j}M_{11}(D_1) = (-1)^{i+j}M_{ij}(D).$$

According to Sec. 4, this same sum is equal to A_{ij}. Consequently,

$$A_{ij} = (-1)^{i+j}M_{ij},$$

which completes the proof of (15).

Example. The determinant of the form

$$D_n = \begin{vmatrix} a_{11} & 0 & 0 & \ldots & 0 \\ a_{21} & a_{22} & 0 & \ldots & 0 \\ a_{31} & a_{32} & a_{33} & \ldots & 0 \\ \cdot & \cdot & \cdot & \ldots & \cdot \\ a_{n1} & a_{n2} & a_{n3} & \ldots & a_{nn} \end{vmatrix}$$

is called *triangular*. Expanding D_n with respect to the first row, we find that D_n equals the product of the element a_{11} with the triangular determinant

$$D_{n-1} = \begin{vmatrix} a_{22} & 0 & \ldots & 0 \\ a_{32} & a_{33} & \ldots & 0 \\ \cdot & \cdot & \ldots & \cdot \\ a_{n2} & a_{n3} & \ldots & a_{nn} \end{vmatrix}$$

of order $n - 1$. Again expanding D_{n-1} with respect to the first row, we find that

$$D_{n-1} = a_{22}D_{n-2},$$

where D_{n-2} is a triangular determinant of order $n - 2$. Continuing in this way, we finally obtain

$$D_n = a_{11}a_{22} \ldots a_{nn}.$$

6. Practical Evaluation of Determinants

Equation (12) takes a particularly simple form when all the elements of the ith row vanish except one element, say a_{ik}. In this case

$$D = a_{ik}A_{ik}, \tag{16}$$

and the calculation of the determinant D of order n reduces at once to the calculation of a determinant of order $n - 1$. If in addition to a_{ik}, there is another nonzero element a_{ij} in the ith row, then multiplying the kth column by $\lambda = a_{ij}/a_{ik}$ and subtracting it from the ith column, we obtain a determinant which is equal to the original one (cf. Sec. 3.4) but which now has a zero in the ith row and jth column. By a sequence of similar operations, we change any determinant with a nonzero element a_{ik} in the ith row into a determinant in which all the elements of the ith row equal zero except a_{ik}. This new determinant can then be evaluated by (16). Of course, similar operations can also be performed on the *columns* of a determinant.

Example. We calculate the following determinant of order five:

$$D = \begin{vmatrix} -2 & 5 & 0 & -1 & 3 \\ 1 & 0 & 3 & 7 & -2 \\ 3 & -1 & 0 & 5 & -5 \\ 2 & 6 & -4 & 1 & 2 \\ 0 & -3 & -1 & 2 & 3 \end{vmatrix}.$$

There are already two zeros in the third column of this determinant. In order to obtain two more zeros in this column, we multiply the fifth row by 3 and add it to the second row and then multiply the fifth row by 4 and subtract it from the fourth row. After performing these operations and expanding the determinant with respect to the third column, we obtain

$$D = \begin{vmatrix} -2 & 5 & 0 & -1 & 3 \\ 1 & -9 & 0 & 13 & 7 \\ 3 & -1 & 0 & 5 & -5 \\ 2 & 18 & 0 & -7 & -10 \\ 0 & -3 & -1 & 2 & 3 \end{vmatrix} = (-1)^{3+5}(-1) \begin{vmatrix} -2 & 5 & -1 & 3 \\ 1 & -9 & 13 & 3 \\ 3 & -1 & 5 & -5 \\ 2 & 18 & -7 & -10 \end{vmatrix}$$

$$= - \begin{vmatrix} -2 & 5 & -1 & 3 \\ 1 & -9 & 13 & 7 \\ 3 & -1 & 5 & -5 \\ 2 & 18 & -7 & -10 \end{vmatrix}.$$

The simplest thing to do now is to produce three zeros in the first column; to do this, we add twice the second row to the first row, subtract three times the second row from the third row and subtract twice the second row from the fourth row:

$$D = - \begin{vmatrix} -2 & 5 & -1 & 3 \\ 1 & -9 & 13 & 7 \\ 3 & -1 & 5 & -5 \\ 2 & 18 & -7 & -10 \end{vmatrix} = - \begin{vmatrix} 0 & -13 & 25 & 17 \\ 1 & -9 & 13 & 7 \\ 0 & 26 & -34 & -26 \\ 0 & 36 & -33 & -24 \end{vmatrix}$$

$$= -(-1)^{1+2} \begin{vmatrix} -13 & 25 & 17 \\ 26 & -34 & -26 \\ 36 & -33 & -24 \end{vmatrix}.$$

To simplify the calculation of the third-order determinant just obtained, we try to decrease the absolute values of its elements.   To do this, we factor the common factor 2 out of the second row, add the second row to the first and subtract twice the second row from the third row:

$$D = 2 \begin{vmatrix} -13 & 25 & 17 \\ 13 & -17 & -13 \\ 36 & -33 & -24 \end{vmatrix} = 2 \begin{vmatrix} 0 & 8 & 4 \\ 13 & -17 & -13 \\ 10 & 1 & 2 \end{vmatrix} = 2 \cdot 4 \begin{vmatrix} 0 & 2 & 1 \\ 13 & -17 & -13 \\ 10 & 1 & 2 \end{vmatrix}.$$

There is already one zero in the first row.   To obtain still another zero, we subtract twice the third column from the second column.   After this, the evaluation of the determinant is easily completed.

$$D = 8 \begin{vmatrix} 0 & 2 & 1 \\ 13 & -17 & -13 \\ 10 & 1 & 2 \end{vmatrix} = 8 \begin{vmatrix} 0 & 0 & 1 \\ 13 & 9 & -13 \\ 10 & -3 & 2 \end{vmatrix} = 8(-1)^{1+3} \begin{vmatrix} 13 & 9 \\ 10 & -3 \end{vmatrix}$$

$$= 8 \cdot 3 \begin{vmatrix} 13 & 3 \\ 10 & -1 \end{vmatrix} = 8 \cdot 3(-13 - 30) = -8 \cdot 3 \cdot 43 = -1032.$$

Problem 1. Calculate the determinants

$$\Delta_1 = \begin{vmatrix} 246 & 427 & 327 \\ 1014 & 543 & 443 \\ -342 & 721 & 621 \end{vmatrix}, \quad \Delta_2 = \begin{vmatrix} 2 & 1 & 1 & 1 & 1 \\ 1 & 3 & 1 & 1 & 1 \\ 1 & 1 & 4 & 1 & 1 \\ 1 & 1 & 1 & 5 & 1 \\ 1 & 1 & 1 & 1 & 6 \end{vmatrix}.$$

Ans. $\Delta_1 = -29,400,000$; $\Delta_2 = 394$.

Problem 2. Calculate the determinant

$$P(x) = \begin{vmatrix} 1 & 1 & 2 & 3 \\ 1 & 2-x^2 & 2 & 3 \\ 2 & 3 & 1 & 5 \\ 2 & 3 & 1 & 9-x^2 \end{vmatrix}.$$

Hint. $P(x)$ is obviously a polynomial of degree 4.   We first find its leading coefficient, and then determine its roots by making rows of the determinant coincide.

Problem 3. Calculate the nth-order determinant

$$\Delta = \begin{vmatrix} x & a & a & \dots & a \\ a & x & a & \dots & a \\ a & a & x & \dots & a \\ . & . & . & \dots & . \\ a & a & a & \dots & x \end{vmatrix}.$$

Hint.   Add all the columns to the first.

Ans. $\Delta = [x + (n-1)a](x-a)^{n-1}$.

Problem 4. Calculate the *Vandermonde determinant*

$$\Delta(x_1, x_2, ..., x_n) = \begin{vmatrix} 1 & 1 & \cdots & 1 \\ x_1 & x_2 & \cdots & x_n \\ x_1^2 & x_2^2 & \cdots & x_n^2 \\ \cdot & \cdot & \cdots & \\ x_1^{n-1} & x_2^{n-1} & \cdots & x_n^{n-1} \end{vmatrix}.$$

Hint. Subtract the first column from all the others, expand with respect to the first row, and afterwards subtract from each row the preceding row multiplied by x_1. Then use induction.

Ans. $\Delta(x_1, x_2, ..., x_n) = (x_2 - x_1)(x_3 - x_1) ... (x_n - x_1)$
$$\times (x_3 - x_2) ... (x_n - x_2)$$
$$\cdot \quad \cdot \quad \cdot \quad \cdot \quad \cdot$$
$$\cdot \quad \cdot \quad \cdot \quad \cdot \quad \cdot$$
$$\times (x_n - x_{n-1})$$

7. Cramer's Rule

We are now in a position to solve systems of linear equations. First, we consider a system of the special form

$$\begin{aligned} a_{11}x_1 + a_{12}x_2 + \cdots + a_{1n}x_n &= b_1, \\ a_{21}x_1 + a_{22}x_2 + \cdots + a_{2n}x_n &= b_2, \\ \cdots \quad \cdots \quad \cdots \quad \cdots & \\ a_{n1}x_1 + a_{n2}x_2 + \cdots + a_{nn}x_n &= b_n, \end{aligned} \tag{17}$$

i.e., a system which has the same number of unknowns and equations. The coefficients a_{ij} $(i, j = 1, 2, ..., n)$ form the *coefficient matrix* of the system; we assume that the determinant of this matrix is different from zero. We now show that such a system is always *compatible* and *determinate*, and we obtain a formula which gives the unique solution of the system.

We begin by assuming that $c_1, c_2, ..., c_n$ is a solution of (17), so that

$$\begin{aligned} a_{11}c_1 + a_{12}c_2 + \cdots + a_{1n}c_n &= b_1, \\ a_{21}c_1 + a_{22}c_2 + \cdots + a_{2n}c_n &= b_2, \\ \cdots \quad \cdots \quad \cdots \quad \cdots & \\ a_{n1}c_1 + a_{n2}c_2 + \cdots + a_{nn}c_n &= b_n. \end{aligned} \tag{18}$$

We multiply the first of the equations (18) by the cofactor A_{11} of the element a_{11} in the coefficient matrix, then we multiply the second equation by A_{21}, the third by A_{31}, and so on, and finally the last equation by A_{n1}. Then we add all the equations so obtained. The result is

$$(a_{11}A_{11} + a_{21}A_{21} + \cdots + a_{n1}A_{n1})c_1$$
$$+ (a_{12}A_{11} + a_{22}A_{21} + \cdots + a_{n2}A_{n1})c_2 + \cdots \tag{19}$$
$$+ (a_{1n}A_{11} + a_{2n}A_{21} + \cdots + a_{nn}A_{n1})c_n = b_1A_{11} + b_2A_{21} + \cdots + b_nA_{n1}.$$

By Theorem 1, the coefficient of c_1 in (19) equals the determinant D itself. By Theorem 2, the coefficients of all the other c_j $(j \neq 1)$ vanish. The expression in the right-hand side of (19) is the expansion of the determinant

$$D_1 = \begin{vmatrix} b_1 & a_{12} & \ldots & a_{1n} \\ b_2 & a_{22} & \ldots & a_{2n} \\ . & . & \ldots & . \\ b_n & a_{n2} & \ldots & a_{nn} \end{vmatrix}$$

with respect to its first column. Therefore, (19) can now be written in the form

$$Dc_1 = D_1,$$

so that

$$c_1 = \frac{D_1}{D}.$$

In a completely analogous way, we can obtain the expression

$$c_j = \frac{D_j}{D} \qquad (j = 1, 2, \ldots, n), \tag{20}$$

where

$$D_j = \begin{vmatrix} a_{11} & a_{12} & \ldots & a_{1,j-1} & b_1 & a_{1,j+1} & \ldots & a_{1n} \\ a_{21} & a_{22} & \ldots & a_{2,j-1} & b_2 & a_{2,j+1} & \ldots & a_{2n} \\ . & . & \ldots & \ldots & & \ldots & \ldots & . \\ a_{n1} & a_{n2} & \ldots & a_{n,j-1} & b_n & a_{n,j+1} & \ldots & a_{nn} \end{vmatrix}$$

is the determinant obtained from the determinant D by replacing its jth column by the numbers $b_1, b_2, \ldots, b_n$. Thus we obtain the following result:

If a solution of the system (17) exists, then (20) expresses the solution in terms of the coefficients of the system and the numbers in the right-hand side of (17). In particular, we find that if a solution of the system (17) exists, it is unique.

It now remains to show that a solution of the system (17) always exists. Consider the quantities

$$c_j = \frac{D_j}{D} \qquad (j = 1, 2, \ldots, n),$$

and substitute them into the system (17) in place of the unknowns $x_1, x_2, \ldots, x_n$. Then this reduces all the equations of the system (17) to identities. In fact, for the ith equation we obtain

$$a_{i1}c_1 + a_{i2}c_2 + \cdots + a_{in}c_n = a_{i1}\frac{D_1}{D} + /a_{i2}\frac{D_2}{D} + \cdots + a_{in}\frac{D_n}{D}$$

$$= \frac{1}{D} [a_{i1}(b_1 A_{11} + b_2 A_{21} + \cdots + b_n A_{n1})$$
$$+ a_{i2}(b_1 A_{12} + b_2 A_{22} + \cdots + b_n A_{n2}) + \cdots$$
$$+ a_{in}(b_1 A_{1n} + b_2 A_{2n} + \cdots + b_n A_{nn})]$$
$$= \frac{1}{D} [b_1(a_{i1} A_{11} + a_{i2} A_{12} + \cdots + a_{in} A_{1n}) + \cdots$$
$$+ b_2(a_{i1} A_{21} + a_{i2} A_{22} + \cdots + a_{in} A_{2n}) + \cdots$$
$$+ b_n(a_{i1} A_{n1} + a_{i2} A_{n2} + \cdots + a_{in} A_{nn})].$$

By Theorems 1 and 2, only one of the coefficients of the quantities $b_1, b_2, \ldots, b_n$ is different from zero, namely the coefficient of b_1, which is equal to the determinant D itself. Consequently, the above expression reduces to

$$\frac{1}{D} b_i D = b_i,$$

i.e., is identical with the right-hand side of the ith equation of the system. Thus, the quantities c_j actually constitute a solution of the system (17), and we have found the following rule for obtaining solutions of the system (17) (*Cramer's rule*):

If the determinant of the system (17) *is different from zero, then its unique solution is obtained by taking for the value of the unknown x_j the fraction whose denominator is the determinant D of the system* (17) *and whose numerator is the determinant obtained by replacing the j'th column of D by the column consisting of the terms on the right-hand side of the system.*

Thus, finding the solution of the system (17) reduces to calculating determinants.

    Problem. Solve the system of equations

$$x_1 + 2x_2 + 3x_3 + 4x_4 + 5x_5 = 13,$$
$$2x_1 + x_2 + 2x_3 + 3x_4 + 4x_5 = 10,$$
$$2x_1 + 2x_2 + x_3 + 2x_4 + 3x_5 = 11,$$
$$2x_1 + 2x_2 + 2x_3 + x_4 + 2x_5 = 6,$$
$$2x_1 + 2x_2 + 2x_3 + 2x_4 + x_5 = 3.$$

                 Ans. $c_1 = 0, c_2 = 2, c_3 = -2, c_4 = 0, c_5 = 3.$

Ways of solving more general systems (with vanishing determinants, or with a number of equations different from the number of unknowns) will be given in the next two chapters.

    Remark. One sometimes encounters systems of linear equations whose constant terms are not numbers but vectors (e.g., in analytic geometry or in mechanics). Cramer's rule and its proof remain valid in this case as well; one must only bear in mind that the values of the unknowns $x_1, x_2, \ldots, x_n$ will be vectors rather than numbers. For example, the system

$$x_1 + x_2 = \mathbf{i} - 3\mathbf{j},$$
$$x_1 - x_2 = \mathbf{i} + 5\mathbf{j}$$

has the unique solution

$$c_1 = \mathbf{i} + \mathbf{j}, \qquad c_2 = -4\mathbf{j}.$$

*8. Minors of Any Order. Laplace's Theorem

The theorem in Sec. 4 on the expansion of a determinant with respect to a row or a column is a special case of a more general theorem on the expansion of a determinant with respect to a whole set of rows or columns. Before formulating this general theorem (Laplace's theorem), we introduce some new notation.

Suppose that in a square matrix of order n we specify any $k \leqslant n$ different rows and the same number of different columns. The elements appearing at the intersections of these rows and columns form a square matrix of order k. The determinant of this matrix is called a *minor of order k* of the original matrix of order n (also a minor of order k of the determinant D); it is denoted by

$$M = M_{j_1, j_2, \ldots, j_k}^{i_1, i_2, \ldots, i_k},$$

where $i_1, i_2, \ldots, i_k$ are the numbers of the deleted rows, and $j_1, j_2, \ldots, j_k$ are the numbers of the deleted columns.

If in the original matrix we delete the rows and columns which make up the minor M, then the remaining elements again form a square matrix, this time of order $n - k$. The determinant of this matrix is called the *complementary minor* of the minor M, and is denoted by the symbol

$$\bar{M} = \bar{M}_{j_1, j_2, \ldots, j_k}^{i_1, i_2, \ldots, i_k}.$$

In particular, if the original minor is of order 1, i.e., is just some element a_{ij} of the determinant D, then the complementary minor is the same as the minor M_{ij} discussed in Sec. 5.

Consider now the minor

$$M_1 = M_{1, 2, \ldots, k}^{1, 2, \ldots, k}$$

formed from the first k rows and the first k columns of the determinant D; its complementary minor is

$$M_2 = \bar{M}_1 = \bar{M}_{1, 2, \ldots, k}^{1, 2, \ldots, k}.$$

In the right-hand side of equation (6) of Sec. 2, group together all the terms of the determinant whose first k elements belong to the minor M_1 (and thus whose remaining $n - k$ elements belong to the minor M_2). Let one of these terms be denoted by c; we now wish to determine the sign which must be ascribed to c. The first k elements of c belong to a term c_1 of the minor M_1. If we denote by N_1 the number of segments of negative slope corresponding to these elements, then the sign which must be put in front of the term c_1 in the minor M_1 is $(-1)^{N_1}$. The remaining $n - k$ elements of c belong to a term c_2 of the minor M_2; the sign which must be put in front of this term in the minor M_2 is $(-1)^{N_2}$, where N_2 is the number of segments of negative slope corresponding to the $n - k$ elements of c_2. Since in the matrix of the determinant D there is not a single segment with negative slope joining an element of the minor M_1 with an element of the minor M_2, the total number of segments of negative slope joining elements of the term c equals the sum $N_1 + N_2$. Therefore, the sign which must be put in front of the term c is given by the expression $(-1)^{N_1+N_2}$, and hence is equal to the product of the signs of the terms c_1 and c_2 in the minors M_1 and M_2. Moreover, we note that the product of any term of the minor M_1 and any term of the minor M_2 gives us one of the terms of the determinant D that have been grouped together. It follows that the sum of all the terms that we have grouped together from the expression for the determinant D (i.e., equation (6) of Sec. 2) is equal to the *product of the minors M_1 and M_2.*

Next we solve the analogous problem for an arbitrary minor

$$M_1 = M_{j_1, j_2, \ldots, j_k}^{i_1, i_2, \ldots, i_k},$$

with complementary minor M_2. By successively interchanging adjacent rows and columns, we can move the minor M_1 over to the upper left-hand corner of the determinant D; to do so we need a total of

$$(i_1 - 1) + (i_2 - 2) + \cdots + (i_k - k)$$
$$+ (j_1 - 1) + (j_2 - 2) + \cdots + (j_k - k)$$

interchanges. As a result, we obtain a determinant D_1 with the same terms as in the original determinant but multiplied by $(-1)^{i+j}$, where

$$i = i_1 + i_2 + \cdots + i_k, \qquad j = j_1 + j_2 + \cdots + j_k.$$

By what has just been proved, the sum of all the terms in the determinant D_1 whose first k elements appear in the minor M_1 is equal to the product $M_1 M_2$. It follows from this that the sum of the corresponding terms of the determinant D is equal to the product

$$(-1)^{i+j} M_1 M_2 = M_1 A_2,$$

where the quantity

$$A_2 = (-1)^{i+j} M_2$$

is called the *cofactor* (or *algebraic complement*) of the minor M_1 in the determinant D. Sometimes one uses the notation

$$A_2 = \bar{A}^{i_1, i_2, \ldots, i_k}_{j_1, j_2, \ldots, j_k},$$

where the indices indicate the numbers of the deleted rows and columns.

Finally, let the rows of the determinant D with indices $i_1, i_2, \ldots, i_k$ be fixed; some elements from these rows appear in every term of D. We group together all the terms of D such that the elements from the fixed rows $i_1, i_2, \ldots, i_k$ belong to the columns with indices $j_1, j_2, \ldots, j_k$. Then, by what has just been proved, the sum of all these terms equals the product of the minor

$$M^{i_1, i_2, \ldots, i_k}_{j_1, j_2, \ldots, j_k}$$

with the corresponding cofactor. In this way, all the terms of D can be divided into groups, each of which is characterized by specifying k columns. The sum of the terms in each group is equal to the product of the corresponding minor and its cofactor. Therefore, the entire determinant can be represented as the sum

$$D = \sum M^{i_1, i_2, \ldots, i_k}_{j_1, j_2, \ldots, j_k} \bar{A}^{i_1, i_2, \ldots, i_k}_{j_1, j_2, \ldots, j_k}, \tag{21}$$

where the indices $i_1, i_2, \ldots, i_k$ (the indices selected above) are fixed, and the sum is over all possible values of the column indices $j_1, j_2, \ldots, j_k$ ($1 \leqslant j_1 < j_2 < \cdots < j_k \leqslant n$). The expansion of D given by (21) is called *Laplace's theorem*; it is clear that Laplace's theorem constitutes a generalization of the formula for expanding a determinant with respect to one of its rows (derived in Sec. 4).

Example. The determinant of the form

$$D = \begin{vmatrix} a_{11} & \cdots & a_{1k} & 0 & \cdots & 0 \\ a_{21} & \cdots & a_{2k} & 0 & \cdots & 0 \\ & \cdots & \cdot & \cdot & \cdots & \cdot \\ a_{kk} & \cdots & a_{kk} & 0 & \cdots & 0 \\ a_{k+1, 1} & \cdots & a_{k+1,k} & a_{k+1,k+1} & \cdots & a_{k+1,n} \\ \cdot & \cdots & \cdot & \cdot & \cdots & \cdot \\ a_{n1} & \cdots & a_{nk} & a_{n,k+1} & \cdots & a_{nn} \end{vmatrix}$$

such that all the elements appearing in both the first k rows and the last $n - k$ columns vanish, is called *quasi-triangular*. To calculate the determinant, we expand it with respect to the first k rows by using Laplace's theorem. Only one term survives in the sum (21), and we obtain

$$D = \begin{vmatrix} a_{11} & \cdots & a_{1k} \\ \cdot & \cdots & \cdot \\ a_{k1} & \cdots & a_{kk} \end{vmatrix} \times \begin{vmatrix} a_{k+1,k+1} & \cdots & a_{k+1,n} \\ \cdot & \cdots & \cdot \\ a_{n,k+1} & \cdots & a_{nn} \end{vmatrix}.$$

Problem. Formulate and prove the theorem which bears the same relation to Laplace's theorem as Theorem 2 bears to Theorem 1 (see Sec. 4).

$$\textit{Ans.} \quad \sum M^{i_1, i_2, \ldots, i_k}_{j_1, j_2, \ldots, j_k} \bar{A}^{i'_1, i'_2, \ldots, i'_k}_{j_1, j_2, \ldots, j_k} = 0,$$

where $i_1 < i_2 < \cdots < i_k$ and $i'_1 < i'_2 < \cdots < i'_k$ are fixed, and at least one of the i_α differs from the corresponding i'_α (Cauchy).

9. Linear Dependence between Columns of a Determinant

9.1. Suppose we are given m columns of numbers with n numbers in each:

$$A_1 = \begin{Vmatrix} a_{11} \\ a_{21} \\ \vdots \\ a_{n1} \end{Vmatrix}, \qquad A_2 = \begin{Vmatrix} a_{12} \\ a_{22} \\ \vdots \\ a_{n2} \end{Vmatrix}, \qquad \ldots, \qquad A_m = \begin{Vmatrix} a_{1m} \\ a_{2m} \\ \vdots \\ a_{nm} \end{Vmatrix}.$$

We multiply every element of the first column by some number λ_1, every element of the second column by λ_2, etc., and finally every element of the last (mth) column by λ_m; we then add corresponding elements of the columns. As a result, we get a new column of numbers, whose elements we denote by $c_1, c_2, \ldots, c_n$. We can represent all these operations schematically as follows:

$$\lambda_1 \begin{Vmatrix} a_{11} \\ a_{21} \\ \vdots \\ a_{n1} \end{Vmatrix} + \lambda_2 \begin{Vmatrix} a_{21} \\ a_{22} \\ \vdots \\ a_{n2} \end{Vmatrix} + \cdots + \lambda_m \begin{Vmatrix} a_{1m} \\ a_{2m} \\ \vdots \\ a_{nm} \end{Vmatrix} = \begin{Vmatrix} c_1 \\ c_2 \\ \vdots \\ c_n \end{Vmatrix},$$

or more briefly as

$$\lambda_1 A_1 + \lambda_2 A_2 + \cdots + \lambda_m A_m = C,$$

where C denotes the column whose elements are $c_1, c_2, \ldots, c_n$. The column C is called a *linear combination* of the columns $A_1, A_2, \ldots, A_m$, and the numbers $\lambda_1, \lambda_2, \ldots, \lambda_m$ are called the *coefficients* of the linear combination.

Suppose now that our columns are not chosen independently, but rather make up a determinant D of order n. Then the following theorem holds:

THEOREM 3. *If one of the columns of the determinant D is a linear combination of the other columns, then $D = 0$.*

Proof. Suppose, for example, that the qth column of the determinant D is a linear combination of the jth, kth, ..., pth columns of D, with coefficients $\lambda_j, \lambda_k, \ldots, \lambda_p$, respectively. Then, according to Property 3.4 of Sec. 3, by subtracting from the qth column first the jth column multiplied by λ_j, then the kth column multiplied by λ_k, etc., and finally the pth column multiplied by λ_p, we do not change the value of the determinant D. However, as a result, the qth column consists of zeros only, from which it follows that $D = 0$.

It is remarkable that the converse theorem is also true:

If a given determinant D is equal to zero, then at least one of its columns is a linear combination of the other columns.

The proof of this theorem requires some preliminary considerations, to which we now turn.

9.2. Again suppose that we have m columns of numbers with n elements in each. We can write them in the form of a matrix

$$A = \begin{Vmatrix} a_{11} & a_{12} & \cdots & a_{1m} \\ a_{21} & a_{22} & \cdots & a_{2m} \\ \cdot & \cdot & \cdots & \cdot \\ a_{n1} & a_{n2} & \cdots & a_{nm} \end{Vmatrix}$$

with n rows and m columns. If k columns and k rows of this matrix are held fixed, then the elements appearing at the intersections of these columns and rows form a square matrix of order k, whose determinant is called a *minor of order* k of the original matrix A; this determinant may either be vanishing or nonvanishing. If, as we shall always assume, not all of the a_{ik} are zero, then we can always find an integer r which has the following two properties:

1. The matrix A has a minor of order r which does not vanish;
2. Every minor of the matrix A of order $r + 1$ and higher (if such actually exist) vanishes.

The number r which has these properties is called the *rank* of the matrix A. If all the a_{ik} vanish, then the rank of the matrix A is considered to be zero ($r = 0$). Henceforth, we shall assume that $r > 0$. The minor of order r which is different from zero is called the *basis minor* of the matrix A. (Of course, A can have several basis minors, but they all have the same order r.) The columns which contain the basis minor are called the *basis columns*. We have the following important theorem concerning the basis columns:

THEOREM 4. ("*Basis minor theorem*") *Any column of the matrix A is a linear combination of its basis columns.*

Proof. To be explicit, we assume that the basis minor of the matrix A is located in the first r rows and first r columns of A. Let s be any integer from 1 to m, let k be any integer from 1 to n, and consider the determinant

$$D = \begin{vmatrix} a_{11} & a_{12} & \cdots & a_{1r} & a_{1s} \\ a_{21} & a_{22} & \cdots & a_{2r} & a_{2s} \\ \cdot & \cdot & \cdots & \cdot & \cdot \\ \cdot & \cdot & \cdots & \cdot & \cdot \\ a_{r1} & a_{r2} & \cdots & a_{rr} & a_{rs} \\ a_{k1} & a_{k2} & \cdots & a_{kr} & a_{ks} \end{vmatrix}$$

of order $r + 1$. If $k \leqslant r$, the determinant D is obviously zero, since it then has two identical rows. Similarly, $D = 0$ for $s \leqslant r$. If $k > r$ and $s > r$, then the determinant D is also equal to zero, since it is then a minor of order $r + 1$ of a matrix of rank r. Consequently, $D = 0$ for any values of k and s.

We now expand D with respect to its last row, obtaining the relation

$$a_{k1}A_{k1} + a_{k2}A_{k2} + \cdots + a_{kr}A_{kr} + a_{ks}A_{ks} = 0, \tag{22}$$

where the numbers $A_{k1}, A_{k2}, \ldots, A_{kr}, A_{ks}$ denote the cofactors of the elements $a_{k1}, a_{k2}, \ldots, a_{kr}, a_{ks}$ appearing in the last row of D. These cofactors do not depend on the number k, since they are formed by using elements a_{ij} with $i \leqslant r$. Therefore, we can introduce the notation

$$A_{k1} = C_1, \quad A_{k2} = C_2, \quad \ldots, \quad A_{kr} = C_r, \quad A_{ks} = C_s.$$

Substituting the values $k = 1, 2, \ldots, n$ in turn into (22), we obtain the system of equations

$$\begin{aligned} C_1 a_{11} + C_2 a_{12} + \cdots + C_r a_{1r} + C_s a_{1s} &= 0, \\ C_1 a_{21} + C_2 a_{22} + \cdots + C_r a_{2r} + C_s a_{2s} &= 0, \\ \cdots \quad \cdots \quad \cdots \quad \cdots \quad \cdots & \\ C_1 a_{n1} + C_2 a_{n2} + \cdots + C_r a_{nr} + C_s a_{ns} &= 0. \end{aligned} \tag{23}$$

The number $C_s = A_{ks}$ is different from zero, since A_{ks} is a basis minor of the matrix A. Dividing each of the equations (23) by C_s, transposing all the terms except the last to the right-hand side, and denoting $-(C_j/C_s)$ by λ_j $(j = 1, 2, \ldots, r)$, we obtain

$$\begin{aligned} a_{1s} &= \lambda_1 a_{11} + \lambda_2 a_{12} + \cdots + \lambda_r a_{1r}, \\ a_{2s} &= \lambda_1 a_{21} + \lambda_2 a_{22} + \cdots + \lambda_r a_{2r}, \\ \cdot \quad \cdots \quad \cdots \quad \cdots \quad \cdots & \\ a_{ns} &= \lambda_1 a_{n1} + \lambda_2 a_{n2} + \cdots + \lambda_r a_{nr}. \end{aligned} \tag{24}$$

These equations show that the sth column of the matrix A is a linear combination of the first r columns of the matrix (with coefficients $\lambda_1, \lambda_2, \ldots, \lambda_r$). Since s can be any number from 1 to m, this completes the proof of Theorem 4.

9.3. We can now prove the converse of Theorem 3 (stated at the end of Sec. 9.1).

THEOREM 5. *If the determinant D vanishes, then it has at least one column which is a linear combination of the other columns.*

Proof. Consider the matrix of the determinant D. Since $D = 0$, the basis minor of this matrix is of order $r < n$. Therefore, after specifying the r basis columns, we can still find at least one column which is not one of the basis columns. By the basis minor theorem, this column is a linear combination of the basis columns. Thus, we have found a column of the determinant D which is a linear combination of the other columns, Q.E.D. (We note that we can include all the remaining columns of the determinant D in this linear combination by assigning them zero coefficients.)

9.4. The results just obtained can be formulated in a somewhat more symmetric way. If the coefficients $\lambda_1, \lambda_2, \ldots, \lambda_m$ of a linear combination of m columns $A_1, A_2, \ldots, A_m$ (see Sec. 9.1) are equal to zero, then obviously the linear combination is just the *zero column*, i.e., the column consisting entirely of zeros. But it may also be possible to obtain the zero column from the given columns by using coefficients $\lambda_1, \lambda_2, \ldots, \lambda_m$ which are not all equal to zero. In this case, the given columns $A_1, A_2, \ldots, A_m$ are called *linearly dependent*. For example, the columns

$$A_1 = \begin{Vmatrix} 1 \\ 2 \\ 3 \\ 4 \end{Vmatrix}, \qquad A_2 = \begin{Vmatrix} 2 \\ 4 \\ 6 \\ 8 \end{Vmatrix}, \qquad A_3 = \begin{Vmatrix} 1 \\ 1 \\ 1 \\ 1 \end{Vmatrix}$$

are linearly dependent, since the zero column can be obtained as the linear combination

$$2 \cdot A_1 - 1 \cdot A_2 + 0 \cdot A_3.$$

A more detailed statement of the definition of linear dependence is the following: The columns

$$A_1 = \begin{Vmatrix} a_{11} \\ a_{21} \\ \vdots \\ a_{n1} \end{Vmatrix}, \qquad A_2 = \begin{Vmatrix} a_{12} \\ a_{22} \\ \vdots \\ a_{n2} \end{Vmatrix}, \qquad \ldots, \qquad A_m = \begin{Vmatrix} a_{m1} \\ a_{m2} \\ \vdots \\ a_{nm} \end{Vmatrix}$$

are called *linearly dependent*, if there exist numbers $\lambda_1, \lambda_2, \ldots, \lambda_m$, not all equal to zero, such that the system of equation

$$\lambda_1 a_{11} + \lambda_2 a_{12} + \cdots + \lambda_m a_{1m} = 0,$$
$$\lambda_1 a_{21} + \lambda_2 a_{22} + \cdots + \lambda_m a_{2m} = 0,$$
$$\cdots \qquad \cdots \qquad \cdots \qquad \cdots \qquad .$$
$$\lambda_1 a_{n1} + \lambda_2 a_{n2} + \cdots + \lambda_m a_{nm} = 0$$

is satisfied, or equivalently such that

$$\lambda_1 A_1 + \lambda_2 A_2 + \cdots + \lambda_m A_m = 0,$$

where the symbol 0 on the right-hand side denotes the zero column.   If one of the columns $A_1, A_2, \ldots, A_m$, e.g., the last column, is a linear combination of the others, i.e.,

$$A_m = \lambda_1 A_1 + \lambda_2 A_2 + \cdots + \lambda_{m-1} A_{m-1}, \tag{25}$$

then the columns $A_1, A_2, \ldots, A_m$ are linearly dependent.   In fact, (25) is equivalent to the relation

$$\lambda_1 A_1 + \lambda_2 A_2 + \cdots + \lambda_{m-1} A_{m-1} - A_m = 0.$$

Consequently, there exists a linear combination of the columns $A_1, A_2, \ldots, A_m$, whose coefficients are not equal to zero (e.g., with the last coefficient equal to -1) whose sum is the zero column; this just means that the columns $A_1, A_2, \ldots, A_m$ are linearly dependent.

Conversely, if the columns $A_1, A_2, \ldots, A_m$ are linearly dependent, then at least one of the columns is a linear combination of the other columns.   In fact, suppose that in the relation

$$\lambda_1 A_1 + \lambda_2 A_2 + \cdots + \lambda_{m-1} A_{m-1} + \lambda_m A_m = 0 \tag{26}$$

expressing the linear dependence of the columns $A_1, A_2, \ldots, A_m$, the coefficient λ_m, say, is nonzero.   Then (26) is equivalent to the relation

$$A_m = -\frac{\lambda_1}{\lambda_m} A_1 - \frac{\lambda_2}{\lambda_m} A_2 - \cdots - \frac{\lambda_{m-1}}{\lambda_m} A_{m-1},$$

which shows that the column A_m is a linear combination of the columns $A_1, A_2, \ldots, A_{m-1}$.   Thus, finally, *the columns $A_1, A_2, \ldots, A_m$ are linearly dependent if and only if one of the columns is a linear combination of the other columns.*

Theorems 3 and 5 show that the determinant D vanishes if and only if one of its columns is a linear combination of the other columns.   Using the property just proved, we can state the following theorem:

THEOREM 6. *The determinant D vanishes if and only if there is linear dependence between its columns.*

9.5. Since the value of a determinant does not change when it is transposed (Sec. 3.1), and since transposition changes columns to rows, we can change columns to rows in all the statements made in this section.   In particular, the following result holds: *The determinant D vanishes if and only if there is linear dependence between its rows.*

Problem 1. Construct four linearly independent columns of four numbers each.

Hint. It is sufficient for the corresponding fourth-order determinant to be nonzero.

Problem 2. Show that if the rows of a determinant of order n are linearly dependent, its columns are also linearly dependent.

Hint. Use the results of Secs. 9.4 and 9.5.

2

LINEAR SPACES

---

10. Introduction

A system of linear equations with all its constant terms equal to zero, i.e.,

$$\begin{array}{cccc}
a_{11}x_1 + a_{12}x_2 + \cdots + a_{1n}x_n = 0, \\
a_{21}x_1 + a_{22}x_2 + \cdots + a_{2n}x_n = 0, \\
\cdots \qquad \cdots \qquad \cdots \qquad \cdots \\
a_{k1}x_1 + a_{k2}x_2 + \cdots + a_{kn}x_n = 0
\end{array} \qquad (1)$$

is called a *homogeneous* linear system. Such a system is always compatible, since it obviously has the *trivial solution*

$$x_1 = x_2 = \cdots = x_n = 0.$$

The solutions of this system have the following remarkable property: Let

$$c_1^{(1)}, c_2^{(1)}, \ldots, c_n^{(1)} \quad \text{and} \quad c_1^{(2)}, c_2^{(2)}, \ldots, c_n^{(2)}$$

be two solutions of the system, and form the numbers

$$c_1 = c_1^{(1)} + c_1^{(2)}, \quad c_2 = c_2^{(1)} + c_2^{(2)}, \quad \ldots, \quad c_n = c_n^{(1)} + c_n^{(2)}. \qquad (2)$$

It is clear that $c_1, c_2, \ldots, c_n$ is again a *solution of the system* (1). In fact, substituting these numbers in the ith equation of the system, we obtain

$$\begin{aligned}
a_{i1}c_1 &+ a_{i2}c_2 + \cdots + a_{in}c_n \\
&= a_{i1}(c_1^{(1)} + c_1^{(2)}) + a_{i2}(c_2^{(1)} + c_2^{(2)}) + \cdots + a_{in}(c_n^{(1)} + c_n^{(2)}) \\
&= (a_{i1}c_1^{(1)} + a_{i2}c_2^{(1)} + \cdots + a_{in}c_n^{(1)}) \\
&\quad + (a_{i1}c_1^{(2)} + a_{i2}c_2^{(2)} + \cdots + a_{in}c_n^{(2)}) = 0,
\end{aligned}$$

as required. We shall call this solution the *sum* of the solutions

$$c_1^{(1)}, c_2^{(1)}, \ldots, c_n^{(1)} \quad \text{and} \quad c_1^{(2)}, c_2^{(2)}, \ldots, c_n^{(2)}.$$

Similarly, if $c_1, c_2, \ldots, c_n$ is an arbitrary solution of the system (1), then, for any fixed real λ, the numbers $\lambda c_1, \lambda c_2, \ldots, \lambda c_n$ also form a solution of the system (1); we shall call this solution the *product of the solution* $c_1, c_2, \ldots, c_n$ *and the number* λ. *Thus, the solutions of a homogeneous system can be added to one another and multiplied by real numbers.*

Objects on which we can perform the operations of addition and multiplication by real numbers are encountered in various branches of algebra, analysis and geometry. First and foremost, the real numbers themselves (and the complex numbers as well) are examples of such objects. Other examples are the two-dimensional and three-dimensional free vectors used in analytic geometry and mechanics; for them there also exist perfectly definite rules for addition and multiplication by real numbers. As we have just seen, the operations of addition and multiplication by numbers can also be introduced for the solutions of a homogeneous system of linear equations, i.e., for algebraic objects. At the end of the first chapter, we added together and multiplied by real numbers columns consisting of n elements; these columns might be called arithmetical objects. In the analysis of functions defined on some interval, one can also add and multiply by real numbers, in accordance with known rules.

One might raise the objection that comparisons of this type are useless, since the objects described above are of completely different kinds, and therefore the operations of addition and multiplication by real numbers which we have defined on them have nothing in common other than perhaps their names. However, to conclude this would be premature. If we examine more carefully the addition operations applied to the different kinds of objects enumerated above, we see that the operations have many properties in common. For example, in every case the associative law of addition is satisfied, as well as the distributive law of multiplication (with respect to addition). The first of these laws is expressed by the equality

$$(x + y) + z = x + (y + z),$$

and the second by the equality

$$\lambda(x + y) = \lambda x + \lambda y,$$

where x, y, and z are arbitrary objects from the set under consideration, and λ is an arbitrary real number. Other arithmetical laws are also satisfied, and moreover, all facts derived solely from these laws are valid for every one of the sets of objects under consideration. If we are interested only in facts of this kind, we do not have to take into account the concrete nature of the objects of the given set; indeed, the only effect of introducing the concrete nature of the objects is to make the investigation more complicated by involving it with spurious facts which have no bearing on the problem.[1]

We now proceed to introduce the concept of a *linear space*, by which we mean a set of elements of any kind, on which certain operations, provisionally called "addition" and "multiplication by real numbers," can be performed. Since the nature of the elements is not specified, we cannot tell just how to perform the operations on them. However, we shall assume that the operations obey definite arithmetical laws, which will be suitably formulated in the form of axioms. We shall call the elements of a linear space *vectors*, regardless of the fact that their concrete nature may be quite unlike the directed line segments usually encountered. The geometric notions associated with the name "vectors" will help us to explain and often anticipate important results, as well as to find a direct geometric interpretation (which would otherwise not be obvious) of various facts from algebra and analysis. In particular, in the next chapter, we shall obtain a simple geometric characterization of all the solutions of a homogeneous or inhomogeneous system of linear equations. In the last chapter of this book, we shall become acquainted with the geometric treatment of Fourier series and other objects from the field of analysis.

11. Definition of a Linear Space

11.1. A set R (whose elements are subsequently denoted by the letters $x, y, z, \ldots$ and referred to as "vectors") is called a *linear* (or *affine*) *space*, if (1) there is a rule which allows us to construct for every two elements x and y in R a third element $z \in R$,[2] called the "sum" of the elements x and y and denoted by $x + y$, (2) there is a rule which allows us to construct for every element $x \in R$ and every real number λ an element $u \in R$, called the "product of the element x and the number λ" and denoted by λx, and (3) the rules for forming sums of elements and products of elements with numbers have the following properties:

1. a) $x + y = y + x$ for any x and y in R;
   b) $(x + y) + z = x + (y + z)$ for any x, y, z in R;

---

[1] Of course, this depends on the level of generality at which the problem is posed. In Sec. 90 we shall encounter a different situation; the problem solved there is such that general considerations are not sufficient, and we have to introduce the concrete nature of the objects involved.

[2] Here and subsequently, we use some notation from set theory. By $a \in A$, we mean that the element a belongs to the set A; by $B \subset A$, we mean that the set B is a part of the set A (B may also coincide with A). The two relations $B \subset A$ and $A \subset B$ are equivalent to the assertion that the sets A and B coincide. The symbols $\in$ and $\subset$ are called *inclusion relations*.

   c) there exists an element $0 \in R$ (the "zero vector") such that $x + 0 = x$ for any $x \in R$;

   d) for any $x \in R$, there exists an element $y \in R$ (the "inverse element") such that $x + y = 0$;

2. e) $1 \cdot x = x$ for any $x \in R$;

   f) $\alpha(\beta x) = (\alpha\beta)x$ for any $x \in R$ and any real α and β;

3. g) $(\alpha + \beta)x = \alpha x + \beta x$ for any $x \in R$ and any real α and β;

   h) $\alpha(x + y) = \alpha x + \alpha y$ for any x and y in R and any real α.

We now derive the following theorems from these axioms:

THEOREM 7. *In any linear space there exists a unique zero vector.*

Proof. The existence of at least one zero vector is asserted in Axiom 1(c). Suppose that there are two zero vectors 0_1 and 0_2 in the space R. Setting $x = 0_1$, $0 = 0_2$ in Axiom 1(c), we obtain

$$0_1 + 0_2 = 0_1.$$

Setting $x = 0_2$, $0 = 0_1$ in the same axiom, we obtain

$$0_2 + 0_1 = 0_2.$$

Comparing the first of these relations with the second and using Axiom 1(a), we find that $0_1 = 0_2$, as required.

THEOREM 8. *In any linear space, there exists a unique inverse for every element.*

Proof. The existence of at least one inverse element is asserted in Axiom 1(d). Suppose that an element x has two inverses y_1 and y_2. Adding the element y_2 to both sides of the equation $x + y_1 = 0$ and using Axioms 1(b) and 1(c), we obtain

$$y_2 + (x + y_1) = (y_2 + x) + y_1 = 0 + y_1 = y_1,$$
$$y_2 + (x + y_1) = y_2 + 0 = y_2,$$

whence $y_1 = y_2$, as required.

THEOREM 9. *In any linear space, the relation*

$$0 \cdot x = 0$$

holds for every element x. (In the right-hand side of the equation, 0 denotes the zero vector, and in the left-hand side the number 0.)

Proof. Consider the element $0 \cdot x + 1 \cdot x$. Using Axioms 3(g) and 2(e), we obtain

$$0 \cdot x + 1 \cdot x = (0 + 1)x = 1 \cdot x = x, \qquad 0 \cdot x + 1 \cdot x = 0 \cdot x + x,$$

whence

$$x = 0 \cdot x + x.$$

Adding to both sides of this equation the element y which is the inverse of x, we find that

$$0 = x + y = (0 \cdot x + x) + y = 0 \cdot x + (x + y) = 0 \cdot x + 0 = 0 \cdot x,$$

whence

$$0 = 0 \cdot x,$$

as required.

THEOREM 10. *Given any element x of a linear space, the element*

$$y = (-1) \cdot x$$

serves as the inverse of x.

Proof. Form the sum $x + y$. Using the axioms and Theorem 9, we find that

$$x + y = 1 \cdot x + (-1) \cdot x = (1 - 1) \cdot x = 0 \cdot x = 0,$$

as required.

The inverse of a given element x will now be denoted by $-x$, since Theorem 10 makes this a natural notation. The presence of an inverse allows us to introduce the operation of subtraction, i.e., the difference $x - y$ is defined as the sum of x and $-y$. This definition agrees with the definition of subtraction in arithmetic.

11.2. If the nature of the elements $x, y, z, \ldots$ and the rules for operating on them are specified (where the axioms of Sec. 11.1 must be satisfied), then we call the linear space *concrete*. The following three kinds of concrete spaces will be of particular importance:

The space V_3. The elements of this space are the free vectors studied in three-dimensional analytic geometry. Each vector is characterized by a length and a direction (with the exception of the zero vector, whose length is zero and whose direction is arbitrary). The addition of vectors is defined in the usual way by the parallelogram rule. The multiplication of a vector by a number λ is also defined in the usual way, i.e., the length of the vector is multiplied by $|\lambda|$, while its direction remains unchanged if $\lambda > 0$ and is reversed if $\lambda < 0$. It is easily verified that in this case, all the axioms of Sec. 11.1 are satisfied. We denote the analogous sets of two-dimensional and one-dimensional vectors, which are also linear spaces, by V_2 and V_1, respectively.

The space T_n. An element of this space is any ordered set of n real numbers, i.e.,

$$x = (\xi_1, \xi_2, \ldots, \xi_n).$$

The numbers $\xi_1, \xi_2, \ldots, \xi_n$ are called the *components* of the element x.

The operations of addition and multiplication by a number are specified by the following rules:

$$(\xi_1, \xi_2, \ldots, \xi_n) + (\eta_1, \eta_2, \ldots, \eta_n) = (\xi_1 + \eta_1, \xi_2 + \eta_2, \ldots, \xi_n + \eta_n),$$

$$\lambda(\xi_1, \xi_2, \ldots, \xi_n) = (\lambda\xi_1, \lambda\xi_2, \ldots, \lambda\xi_n).$$

It is easily verified that all the axioms of Sec. 11.1 are satisfied. In particular, the element 0 is the set consisting of n zeros, i.e. $0 = (0, 0, \ldots, 0)$. Actually, we dealt with the elements of this space in Sec. 9, except that we wrote them there in the form of a column of numbers rather than a row of numbers.

The space $C(a, b)$. An element of this space is any continuous function $x = x(t)$, defined on the interval $a \leqslant t \leqslant b$. The operations of addition of functions and multiplication of functions by real numbers are defined by the usual rules of analysis; it is obvious that the axioms of Sec. 11.1 are satisfied. In this case, the element 0 is the function which is identically zero.

We note that all the properties of elements of concrete spaces (e.g., the vectors of the space V_3) which are based only on the axioms of Sec. 11.1 are also valid for the elements of an arbitrary linear space. For example, analyzing the proof of Cramer's rule for solving the system of linear equations

$$a_{11}x_1 + a_{12}x_2 + \cdots + a_{1n}x_n = b_1,$$
$$a_{21}x_1 + a_{22}x_2 + \cdots + a_{2n}x_n = b_2,$$
$$\cdots \qquad \cdots \qquad \cdots \qquad \cdots$$
$$a_{n1}x_1 + a_{n2}x_2 + \cdots + a_{nn}x_n = b_n,$$

we observe that insofar as the quantities $b_1, b_2, \ldots, b_n$ are concerned, the proof is based only on the fact that these quantities can be added and multiplied by real numbers and the fact that the axioms of Sec. 11.1 are valid. As has already been pointed out in Sec. 7, this permits us to generalize Cramer's rule to systems in which the quantities $b_1, b_2, \ldots, b_n$ are vectors (elements of the space V_3). Furthermore, this permits us to assert that Cramer's rule is also valid for systems in which the elements $b_1, b_2, \ldots, b_n$ are elements of any linear space R; we note only that then the values of the unknowns $x_1, x_2, \ldots, x_n$ are also elements of the space R, and in fact can be expressed linearly in terms of the quantities $b_1, b_2, \ldots, b_n$.

Problem 1. Consider the set of vectors in the plane whose initial points are located at the origin of coordinates and whose final points lie within the first quadrant. Does this set form a linear space (with the usual operations)?

Ans. No, since we cannot multiply by -1 and stay within the set.

Problem 2. Consider the set of all vectors in a plane with the exception of the vectors which are parallel to a given straight line. Does this set form a linear space?

Ans. No, since we cannot add two vectors which are symmetric with respect to the given line and still stay within the set.

Problem 3. Consider the set P consisting of the *positive* real numbers only. We introduce operations according to the following rules: By the "sum" of two numbers, we mean their product (in the usual sense), and by the "product of an element $r \in P$ with a real number λ," we mean r raised to the power λ (in the usual sense). With these operations, is P a linear space?

Ans. Yes. In particular, the number $1 \in P$ serves as the "zero vector" of the space P.

11.3. In analytic geometry, it is sometimes convenient to consider vectors which are not free but have their initial points attached to the origin of coordinates; then the endpoint of every vector is associated with a point of space, and every point of space can be specified by giving the corresponding vector, the so-called *radius vector* of the point. With this picture in mind, we sometimes call the elements of a linear space *points* instead of vectors. Of course, this change in terminology is not accompanied by any change whatsoever in the definitions, and merely appeals to our geometric intuition.

12. Linear Dependence

Let $x_1, x_2, \ldots, x_k$ be vectors of the linear space R, and let $C_1, C_2, \ldots, C_k$ be real numbers. The vector

$$y = C_1 x_1 + C_2 x_2 + \cdots + C_k x_k$$

is called a *linear combination* of the vectors $x_1, x_2, \ldots, x_k$; the numbers $C_1, C_2, \ldots, C_k$ are called the *coefficients* of the linear combination. If $C_1 = C_2 = \cdots = C_k = 0$, then by Theorem 9 and Axiom 1(c), we find that $y = 0$. However, there may exist a linear combination of the vectors $x_1, x_2, \ldots, x_k$, with coefficients that are not all equal to zero, which is nevertheless equal to the zero vector; in this case, the vectors $x_1, x_2, \ldots, x_k$ are called *linearly dependent*. In other words, the vectors $x_1, x_2, \ldots, x_k$ are said to be *linearly dependent* if there exist numbers $C_1, C_2, \ldots, C_k$, not all equal to zero, such that

$$C_1 x_1 + C_2 x_2 + \cdots + C_k x_k = 0. \tag{3}$$

If (3) is possible only in the case where

$$C_1 = C_2 = \cdots = C_k = 0,$$

then the vectors $x_1, x_2, \ldots, x_k$ are called *linearly independent*.

Example 1. In the linear space V_3, linear dependence of two vectors means that they are parallel to the same straight line. Linear dependence of three vectors means that they are parallel to the same plane. Any four vectors are linearly dependent.

Example 2. We now explain what is meant by linear dependence of the vectors $x_1, x_2, \ldots, x_k$ of the linear space T_n. Let the vector x_i have components $\xi_1^{(i)}, \xi_2^{(i)}, \ldots, \xi_n^{(i)}$ ($i = 1, 2, \ldots, k$); then the linear dependence expressed by

$$C_1 x_1 + C_2 x_2 + \cdots + C_k x_k = 0$$

means that the n equations

$$
\begin{aligned}
C_1 \xi_1^{(1)} + C_2 \xi_1^{(2)} + \cdots + C_k \xi_1^{(k)} &= 0, \\
C_1 \xi_2^{(1)} + C_2 \xi_2^{(2)} + \cdots + C_k \xi_2^{(k)} &= 0, \\
\cdots \quad\quad \cdots \quad\quad \cdots \quad\quad \cdots & \\
C_1 \xi_n^{(1)} + C_2 \xi_n^{(2)} + \cdots + C_k \xi_n^{(k)} &= 0
\end{aligned}
\tag{4}
$$

hold, where not all the constants $C_1, C_2, \ldots, C_k$ are equal to zero. This is the same definition of linear dependence which we gave in Sec. 9 for columns of numbers. Thus, the question of whether the vectors $x_1, x_2, \ldots, x_k$ are linearly dependent reduces in the general case to the question of whether there exists a nontrivial solution of the homogeneous system of equations with coefficients equal to the corresponding components of the given vectors. In the next chapter (Sec. 20) this question will be solved completely, and we shall thereby obtain the rule which will allow us to decide whether or not given vectors of the space T_n are linearly dependent by examining their components. However, in some cases we can even now decide whether or not a given system of vectors is linearly dependent. For example, suppose we take the n vectors

$$
\begin{aligned}
e_1 &= (1, 0, 0, \ldots, 0) \\
e_2 &= (0, 1, 0, \ldots, 0) \\
\cdot\quad & \quad\quad \cdots\cdots \\
e_n &= (0, 0, 0, \ldots, 1)
\end{aligned}
$$

of the space T_n. For these vectors, the system (4) has the form

$$
\begin{aligned}
C_1 \cdot 1 + C_2 \cdot 0 + C_3 \cdot 0 + \cdots + C_n \cdot 0 &= 0, \\
C_1 \cdot 0 + C_2 \cdot 1 + C_3 \cdot 0 + \cdots + C_n \cdot 0 &= 0, \\
\cdots \quad\quad \cdots \quad\quad \cdots \quad\quad \cdots \quad\quad \cdots & \\
C_1 \cdot 0 + C_2 \cdot 0 + C_3 \cdot 0 + \cdots + C_n \cdot 1 &= 0,
\end{aligned}
$$

and obviously has the unique solution

$$C_1 = C_2 = \cdots = C_n = 0.$$

Thus, the vectors $e_1, e_2, \ldots, e_n$ in the space T_n are linearly independent.

Problem. Show that a criterion for the linear independence of n given vectors in the space T_n is that the determinant formed from the coordinates of the vectors does not vanish.

Hint. Use Theorem 6 of Sec. 9.

Example 3. Linear dependence of the vectors

$$x_1 = x_1(t), \quad x_2 = x_2(t), \quad \ldots, \quad x_k = x_k(t)$$

of the space $C(a, b)$ means that a relation

$$C_1 x_1(t) + C_2 x_2(t) + \cdots + C_k x_k(t) \equiv 0$$

holds for the functions $x_1(t), x_2(t), \ldots, x_k(t)$, where not all the constants $C_1, C_2, \ldots, C_k$ are equal to zero. For example, the functions

$$x_1(t) = \cos^2 t, \quad x_2(t) = \sin^2 t, \quad x_3(t) = 1$$

are linearly dependent, since the relation

$$x_1(t) + x_2(t) - x_3(t) \equiv 0$$

holds. On the other hand, we shall verify that the functions $1, t, t^2, \ldots, t^k$ are linearly independent. Assume that there exists a relation

$$C_0 \cdot 1 + C_1 t + \cdots + C_k t^k \equiv 0. \tag{5}$$

Then, by successively differentiating (5) k times, we obtain a system of $k + 1$ equations in the quantities $C_0, C_1, \ldots, C_k$, with a determinant which is clearly different from zero (see the example given in Sec. 5). Solving the system by Cramer's rule (Sec. 7) we find that

$$C_0 = C_1 = \cdots = C_k = 0.$$

Consequently, the functions $1, t, t^2, \ldots, t^k$ are linearly independent in the space $C(a, b)$, as asserted.

Problem. Show that the functions $t^{\alpha_1}, t^{\alpha_2}, \ldots, t^{\alpha_k}$ are linearly independent in the space $C(a, b)$, where $\alpha_1, \alpha_2, \ldots, \alpha_k$ are different real numbers.

We now note two simple properties of systems of vectors, properties which are connected with linear dependence.

LEMMA 1. *If some of the vectors $x_1, x_2, \ldots, x_k$ are linearly dependent, then the whole system $x_1, x_2, \ldots, x_k$ is also linearly dependent.*

Proof. Without loss of generality, we can assume that the vectors $x_1, x_2, \ldots, x_j$ $(j \leqslant k)$ are linearly dependent. Thus, a relation

$$C_1 x_1 + C_2 x_2 + \cdots + C_j x_j = 0$$

holds, where at least one of the constants $C_1, C_2, \ldots, C_j$ is different from zero. By Theorem 9 (Sec. 11) and Axiom 1(c), the relation

$$C_1 x_1 + C_2 x_2 + \cdots + C_j x_j + 0 \cdot x_{j+1} + \cdots + 0 \cdot x_k = 0$$

is valid, which shows that the vectors $x_1, x_2, \ldots, x_k$ are also linearly dependent, since at least one of the constants $C_1, C_2, \ldots, C_j, 0, \ldots, 0$ is different from zero.

LEMMA 2. *The vectors $x_1, x_2, \ldots, x_k$ are linearly dependent if and only if one of the vectors can be expressed as a linear combination of the others.*

A statement similar to this one has already been encountered previously; in fact, it was proved for columns of numbers in Sec. 9.4. If we look over this proof, we see that it is based only on the possibility of performing on columns the operations of addition and multiplication by real numbers, i.e., the proof can be carried through for the elements of any linear space. Thus, our lemma is valid for any linear space, as was to be shown.

13. Bases and Components

13.1. By definition, a system of linearly independent vectors $e_1, e_2, \ldots, e_n$ in a linear space R is a *basis* for R if there exists an expansion

$$x = \xi_1 e_1 + \xi_2 e_2 + \cdots + \xi_n e_n \tag{6}$$

for every vector $x \in R$. It is easy to see that under these conditions *the coefficients in the expansion* (6) *are uniquely determined.* In fact, if we can write two expansions

$$\begin{aligned} x &= \xi_1 e_1 + \xi_2 e_2 + \cdots + \xi_n e_n \\ &= \eta_1 e_1 + \eta_2 e_2 + \cdots + \eta_n e_n \end{aligned}$$

for a vector x, then, subtracting them term by term, we obtain the relation

$$0 = (\xi_1 - \eta_1)e_1 + (\xi_2 - \eta_2)e_2 + \cdots + (\xi_n - \eta_n)e_n,$$

from which, by the assumption that the vectors $e_1, e_2, \ldots, e_n$ are linearly independent, we find that

$$\xi_1 = \eta_1, \ \xi_2 = \eta_2, \ldots, \xi_n = \eta_n.$$

These uniquely defined numbers $\xi_1, \xi_2, \ldots, \xi_n$ are called the *components of the vector x with respect to the basis* $e_1, e_2, \ldots, e_n$.

Example 1. A familiar basis in the space V_3 is formed by the three orthogonal unit vectors $\mathbf{i}, \mathbf{j}, \mathbf{k}$. The components ξ_1, ξ_2, ξ_3 of the vector $\mathbf{x}$ with respect to this basis are the projections of the vector $\mathbf{x}$ along the coordinate axes.

Example 2. An example of a basis in the space T_n is the system of vectors

$$e_1 = (1, 0, \ldots, 0), \ e_2 = (0, 1, \ldots, 0), \ \ldots, e_n = (0, 0, \ldots, 1),$$

which we have already considered in Sec. 12. Indeed, it is obvious that the relation

$$x = \xi_1(1, 0, \ldots, 0) + \xi_2(0, 1, \ldots, 0) + \cdots + \xi_n(0, 0, \ldots, 1)$$

holds for any vector

$$x = (\xi_1, \xi_2, \ldots, \xi_n) \in T_n.$$

This fact, together with the linear independence of the vectors $e_1, e_2, \ldots, e_n$ already proved, shows that these vectors form a basis in the space T_n. In particular, we see that the numbers $\xi_1, \xi_2, \ldots, \xi_n$ are just the components of the vector x with respect to the basis $e_1, e_2, \ldots, e_n$.

Example 3. In the space $C(a, b)$, there does not exist a basis in the sense in which it is defined here. The proof of this statement will be given in the next section.

13.2. The fundamental significance of the concept of a basis for a linear space consists in the fact that when a basis is specified, the linear operations in the space which were originally abstract, become ordinary linear operations with numbers, i.e., the components of the vectors with respect to the given basis. Thus, the following theorem holds:

THEOREM 11. *When two vectors of a space R are added, their components (with respect to any basis) are added. When a vector is multiplied by a number λ, all its components are multiplied by λ.*

Proof. Let

$$x = \xi_1 e_1 + \xi_2 e_2 + \cdots + \xi_n e_n, \qquad y = \eta_1 e_1 + \eta_2 e_2 + \cdots + \eta_n e_n.$$

Then by the axioms of Sec. 11.1, we have

$$x + y = (\xi_1 + \eta_1)e_1 + (\xi_2 + \eta_2)e_2 + \cdots + (\xi_n + \eta_n)e_n,$$
$$\lambda x = \lambda \xi_1 e_1 + \lambda \xi_2 e_2 + \cdots + \lambda \xi_n e_n,$$

as required.

Problem 1. The following is known about a system of vectors $e_1, e_2, \ldots, e_n$ in a linear space R:

a) Every vector $x \in R$ can be expanded as

$$x = \xi_1 e_1 + \xi_2 e_2 + \cdots + \xi_n e_n;$$

b) This expansion is *unique* for some fixed vector $x_0 \in R$.
Show that the system $e_1, e_2, \ldots, e_n$ forms a basis in the space R.

Hint. Show that the zero vector also has a unique expansion with respect to the system $e_1, e_2, \ldots, e_n$. From this deduce the linear independence of the vectors of the system.

Problem 2. Does there exist a basis in the space P (Problem 3 of Sec. 11)?
Ans. Yes, consisting of a single vector, i.e., any element $x \in P$ different
from 1.

14. Dimension

If in a linear space R we can find n linearly independent vectors, while
every $n + 1$ vectors of the space are linearly dependent, then the number n
is called the *dimension* of the space R, and the space R itself is called
n-dimensional. A linear space in which we can find an arbitrarily large
number of linearly independent vectors is called *infinite-dimensional*.

THEOREM 12. *In a space R of dimension n there exists a basis consisting
of n vectors; moreover, any set of n linearly independent vectors of the
space R is a basis for the space.*

Proof. Let $e_1, e_2, \ldots, e_n$ be a system of n linearly independent vectors
of the given n-dimensional space R. If x is any vector of the space, then
the set of $n + 1$ vectors

$$x, e_1, e_2, \ldots, e_n$$

is linearly dependent, i.e., there exists a relation of the form

$$C_0 x + C_1 e_1 + \cdots + C_n e_n = 0, \tag{7}$$

where at least one of the coefficients $C_0, C_1, \ldots, C_n$ is different from
zero. We can assert that the coefficient C_0 is different from zero; in
fact, if this were not the case, then we would obtain linear dependence
between the vectors $e_1, e_2, \ldots, e_n$, which by hypothesis cannot occur.
Thus, in the usual way, i.e., by dividing the equation by C_0 and transposing all the other terms to the other side, we find that x can be expressed
as a linear combination of the vectors $e_1, e_2, \ldots, e_n$. Since x is an
arbitrary vector of the space R, we have shown that the vectors
$e_1, e_2, \ldots, e_n$ form a basis for the space, Q.E.D.

The following theorem is the converse of Theorem 12:

THEOREM 13. *If there is a basis in the space R, then the dimension of
R equals the number of basis vectors.*

Proof. Let the vectors $e_1, e_2, \ldots, e_n$ be a basis for R. By the
definition of a basis, the vectors $e_1, e_2, \ldots, e_n$ are linearly independent;
thus, we already have n linearly independent vectors. We now show
that any $n + 1$ vectors of the space R are linearly dependent.

Suppose we are given $n + 1$ vectors of the space R:

$$x_1 = \xi_1^{(1)}e_1 + \xi_2^{(1)}e_2 + \cdots + \xi_n^{(1)}e_n,$$
$$x_2 = \xi_1^{(2)}e_1 + \xi_2^{(2)}e_2 + \cdots + \xi_n^{(2)}e_n,$$
$$\cdots \qquad \cdots \qquad \cdots \qquad \cdots$$
$$x_{n+1} = \xi_1^{(n+1)}e_1 + \xi_2^{(n+1)}e_2 + \cdots + \xi_n^{(n+1)}e_n.$$

Writing the components of each of these vectors as a column of numbers, we form the matrix

$$A = \left\| \begin{matrix} \xi_1^{(1)} & \xi_1^{(2)} & \cdots & \xi_1^{(n+1)} \\ \xi_2^{(1)} & \xi_2^{(2)} & \cdots & \xi_2^{(n+1)} \\ \cdot & \cdot & \cdots & \cdot \\ \xi_n^{(1)} & \xi_n^{(2)} & \cdots & \xi_n^{(n+1)} \end{matrix} \right\|$$

with n rows and $n + 1$ columns. The basis minor of the matrix A (see Sec. 9) is of order $r \leqslant n$. If $r = 0$, the linear dependence is obvious. Let $r > 0$. After specifying the r basis columns, we can still find at least one column which is not one of the basis columns. But then, according to the basis minor theorem, this column is a linear combination of the basis columns. Thus, the corresponding vector of the space R is a linear combination of some other vectors among the given $x_1, x_2, \ldots, x_{n+1}$. But in this case, according to Lemma 2 of Sec. 12, the vectors $x_1, x_2, \ldots, x_{n+1}$ are linearly dependent, Q.E.D.

Example 1. The space V_3 is three-dimensional, since it has a basis consisting of the three vectors **i**, **j**, **k** (see Sec. 13). Similarly, V_2 is two-dimensional and V_1 is one-dimensional.

Example 2. The space T_n is n-dimensional, since it contains a basis consisting of the n vectors $e_1, e_2, \ldots, e_n$ (see Sec. 13).

Example 3. In the space $C(a, b)$, there is an arbitrarily large number of linearly independent vectors (see Sec. 12), and hence this space is infinite-dimensional. Therefore, it does not have a basis, for the presence of a basis would contradict Theorem 13. The generalization of the concept of a basis to the case of infinite-dimensional spaces will be deferred until Chapter 12.

Problem. What is the dimension of the space P (Problem 3 of Sec. 11)?

Ans. 1.

15. Subspaces

15.1. Suppose that a set L of elements of a linear space R has the following properties:

I. If $x \in L$, $y \in L$, then $x + y \in L$;
II. If $x \in L$, and λ is a real number, then $\lambda x \in L$.

Thus, we are given a set of elements with linear operations defined on them. We now show that this set is also a *linear space*. To do so, we have to verify that the set L with the operations I and II satisfies the axioms of Sec. 12. Axioms 1(a), 1(b) and 2(e) through 3(h) are satisfied, since they are satisfied quite generally for all elements of the space R. It remains to verify Axioms 1(c) and 1(d). Let x be any element of L. Then by hypothesis, $\lambda x \in L$ for any real λ. First we choose $\lambda = 0$; then, since $0 \cdot x = 0$ by Theorem 3 (Sec. 11), the zero vector belongs to the set L. This means that Axiom 1(c) is satisfied. Next we choose $\lambda = -1$; then, by Theorem 4 (Sec. 11), the element $(-1)x$ is the inverse of the element x. Thus, if an element x belongs to the set L, so does the inverse of x. This means that Axiom 1(d) is also satisfied, so that the proof of our assertion is now complete. Consequently, every set $L \subset R$ which satisfies the conditions I and II is called a *linear subspace* (or simply a *subspace*) of the space R.

Example 1. The zero vector of a space R is obviously the smallest possible subspace of the space R.

Example 2. The whole space R is the largest possible subspace of the space R.

Example 3. Let L_1 and L_2 be two subspaces of the same linear space R. The set of all vectors $x \in R$ belonging both to L_1 and L_2 forms a subspace called the *intersection* of the subspaces L_1 and L_2. The set of all vectors of the form $y + z$ where $y \in L_1$, $z \in L_2$ forms a subspace, called the *sum* of the subspaces L_1 and L_2.

Example 4. All the vectors in the space V_3 parallel to a plane (or to a line) form a subspace. If we talk about points rather than about vectors (see Sec. 11.3), then the set of points lying on a plane (or on a line) passing through the origin of coordinates is a subspace of the space V_3.

Example 5. In the space T_n we consider the set L of vectors $x = (\xi_1, \xi_2, \ldots, \xi_n)$ whose coordinates satisfy a homogeneous system of linear equations

$$
\begin{aligned}
a_{11}x_1 + a_{12}x_2 + \cdots + a_{1n}x_n &= 0, \\
a_{21}x_1 + a_{22}x_2 + \cdots + a_{2n}x_n &= 0, \\
\cdots \qquad \cdots \qquad \cdots \qquad \cdots & \\
a_{k1}x_1 + a_{k2}x_2 + \cdots + a_{kn}x_n &= 0.
\end{aligned}
\tag{8}
$$

We saw in Sec. 10 that if we add together solutions of this system or multiply them by real numbers, then we again obtain solutions of the system. Thus, the set L is a subspace of the space T_n, and is therefore a linear space in its own right. We shall call L the *solution space of the system* (8). In Sec. 23 we shall calculate the dimension of this space and construct a basis for it.

Problem. Consider two different two-dimensional subspaces L_1 and L_2 of the space V_3 (two different planes passing through the origin of coordinates). What is their intersection and sum?

Ans. Their intersection is the line of intersection (in the usual sense) of the two planes. Their sum is the whole space.

15.2. We now consider some properties of subspaces which are related to the definitions of Secs. 12–14. First of all, we note that every linear relation which connects the vectors $x, y, \ldots, z$ in the subspace L is also valid in the whole space R, and conversely; in particular, the fact that the vectors $x, y, \ldots, z \in L$ are linearly dependent holds true simultaneously in the subspace L and in the space R. For example, if every set of $n + 1$ vectors is linear dependent in the space R, then this fact is true *a fortiori* in the subspace L. It follows that *the dimension of any subspace L in an n-dimensional space R does not exceed the number n.* According to Theorem 12, *in any subspace $L \subset R$ we can construct a basis with the same number of vectors as the dimension of L.* Of course, if a basis $e_1, e_2, \ldots, e_n$ is chosen in R, then in the general case we cannot choose the basis vectors of the subspace L from the vectors $e_1, e_2, \ldots, e_n$, because none of these vectors may belong to L. However, we can assert the converse:

If a basis $f_1, f_2, \ldots, f_l$ is chosen in the subspace L (which, to be explicit, we assume has the dimension $l < n$), then additional vectors $f_{l+1}, \ldots, f_n$ can always be chosen in the whole space R, such that the system $f_1, f_2, \ldots, f_l, \ldots, f_n$ is a basis for all of R.

To prove this, we argue as follows: *In the space R there are vectors which cannot be expressed as linear combinations of $f_1, f_2, \ldots, f_l$.* Indeed, if there were no such vectors, then the vectors $f_1, f_2, \ldots, f_l$, which are linearly independent by hypothesis, would constitute a basis for the space R, and then by Theorem 13 the dimension of R would be l rather than n. We denote by f_{l+1} any of the vectors that cannot be expressed as a linear combination of $f_1, f_2, \ldots, f_l$. *The system $f_1, f_2, \ldots, f_l, f_{l+1}$ is linearly independent.* Indeed, if there exists a relation of the form

$$C_1 f_1 + C_2 f_2 + \cdots + C_l f_l + C_{l+1} f_{l+1} = 0,$$

then if $C_{l+1} \neq 0$, the vector f_{l+1} could be expressed as a linear combination of $f_1, f_2, \ldots, f_l$, while if $C_{l+1} = 0$, the vectors $f_1, f_2, \ldots, f_l$ would be linearly dependent; both these results contradict the construction. If now every vector of the space R can be expressed as a linear combination of $f_1, f_2, \ldots, f_l, f_{l+1}$, then the system $f_1, f_2, \ldots, f_l, f_{l+1}$ forms a basis for R (and $l + 1 = n$), and our construction is finished. If $l + 1 < n$, then there is a vector f_{l+2} which cannot be expressed as a linear combination of

$f_1, f_2, \ldots, f_l, f_{l+1}$, and therefore we can continue the construction. Eventually, after $n - l$ steps, we obtain a basis for the space R.

Problem. Prove the following theorem: If the dimension of the space $L \subset R$ is the same as the dimension of the space R, then $L \equiv R$.

*15.3. We now find the dimension of the sum of two finite-dimensional subspaces L and M of a linear space R. Let l and m denote the dimensions of these subspaces. We denote by K the intersection of the subspaces L and M, and we denote the dimension of K by k. We choose a basis $e_1, e_2, \ldots, e_k$ in K. Then, by using the argument of Sec. 13.2, we augment the basis $e_1, e_2, \ldots, e_k$ by the vectors $f_{k+1}, f_{k+2}, \ldots, f_l$ to make a basis for the whole subspace L and by the vectors $g_{k+1}, g_{k+2}, \ldots, g_m$ to make a basis for the whole subspace M. By definition, every vector in the sum $L + M$ is the sum of a vector from L and a vector from M, and therefore can be expressed as a linear combination of the vectors

$$e_1, e_2, \ldots, e_k, f_{k+1}, \ldots, f_l, g_{k+1}, \ldots, g_m.$$

We now show that these vectors *form a basis for the subspace $L + M$.* To show this, it remains to verify their linear independence. Assume that there exists a linear relation of the form

$$\alpha_1 e_1 + \cdots + \alpha_k e_k + \beta_{k+1} f_{k+1} + \cdots \\ + \beta_l f_l + \gamma_{k+1} g_{k+1} + \cdots + \gamma_m g_m = 0, \tag{9}$$

where at least one of the coefficients $\alpha_1, \ldots, \gamma_m$ is different from zero. We can then assert that at least one of the numbers $\gamma_{k+1}, \ldots, \gamma_m$ is different from zero, since otherwise the vectors

$$e_1, e_2, \ldots, e_k, f_{k+1}, \ldots, f_l$$

would be linearly dependent, which is impossible in view of the fact that they form a basis for the subspace L. Consequently, the vector

$$x = \gamma_{k+1} g_{k+1} + \cdots + \gamma_m g_m \neq 0, \tag{10}$$

for otherwise the vectors $g_{k+1}, \ldots, g_m$ would be linearly dependent. But it follows from (9) that

$$-x = \alpha_1 e_1 + \cdots + \beta_l f_l \in L,$$

while (10) shows that $x \in M$. Thus, x belongs to both L and M, and therefore belongs to the subspace K. But then

$$x = \gamma_{k+1} g_{k+1} + \cdots + \gamma_m g_m = \lambda_1 e_1 + \lambda_2 e_2 + \cdots + \lambda_k e_k,$$

and since the vectors

$$e_1, e_2, \ldots, e_k, g_{k+1}, \ldots, g_m$$

are linearly independent, we have

$$\gamma_{k+1} = \cdots = \gamma_m = 0.$$

This contradiction shows that the vectors

$$e_1, e_2, \ldots, e_k, f_{k+1}, \ldots, f_l, g_{k+1}, \ldots, g_m$$

are actually linearly independent. Now, according to Theorem 13, the dimension of the subspace $L + M$ is equal to the number of basis vectors

$$e_1, \ldots, e_k, f_{k+1}, \ldots, f_l, g_{k+1}, \ldots, g_m,$$

and this number is equal to $l + m - k$. Thus, finally, *the dimension of the sum of two subspaces is equal to the sum of their dimensions minus the dimension of their intersection.*

16. Linear Manifolds

16.1. An important way of constructing subspaces is to form the *linear manifold spanned by a given system of vectors.* Let $x, y, z, \ldots$ be a system of vectors of a linear space R; by the *linear manifold* spanned by the system $x, y, z, \ldots$ is meant the set of all (finite) linear combinations

$$\alpha x + \beta y + \gamma z + \cdots, \tag{11}$$

with real coefficients $\alpha, \beta, \gamma, \ldots$. It is easily verified that this set satisfies conditions I and II of Sec. 15.1; therefore, the linear manifold spanned by a system $x, y, z, \ldots$ is a subspace of the space R. Obviously, every subspace containing the vectors $x, y, z, \ldots$ also contains all their linear combinations (11); consequently, *the linear manifold spanned by the vectors $x, y, z, \ldots$ is the smallest subspace containing these vectors.* The linear manifold spanned by the vectors $x, y, z, \ldots$ is denoted by $L(x, y, z, \ldots)$.

Example 1. The linear manifold spanned by the basis vectors $e_1, e_2, \ldots, e_n$ of a space R is obviously the whole space R.

Example 2. The linear manifold spanned by a pair of (noncollinear) vectors of the space V_3 consists of all the vectors parallel to the plane of the pair of vectors.

Example 3. The linear manifold spanned by the system of functions $1, t, t^2, \ldots, t^k$ of the space $C(a, b)$ consists of the set of all polynomials in t of degree no higher than k. The linear manifold spanned by the infinite system of functions $1, t, t^2, \ldots$ consists of all polynomials (of any degree) in the variable t.

16.2. We now note two simple properties of linear manifolds.

LEMMA 1. *If the vectors x', y', ... belong to the linear manifold spanned by the vectors x, y, ..., then the linear manifold $L(x, y, ...)$ contains the whole linear manifold $L(x', y', ...)$.*

Proof. Since the vectors x', y', ... belong to the subspace $L(x, y, ...)$, then all their linear combinations (whose totality constitutes the linear manifold $L(x', y', ...)$)) also belong to the subspace $L(x, y, ...)$.

LEMMA 2. *We can eliminate every vector of the system x, y, ... which is linearly dependent on the other vectors of the system, without changing the linear manifold spanned by x, y, ...*

Proof. If the vector x, say, is linearly dependent on the vectors $y, z, ...$, this means that $x \in L(y, z, ...)$. It follows by Lemma 1 that

$$L(x, y, z, ...) \subset L(y, z, ...).$$

On the other hand, obviously

$$L(y, z, ...) \subset L(x, y, z, ...).$$

This pair of relations shows that

$$L(y, z, ...) = L(x, y, z, ...), \quad \text{Q.E.D.}$$

16.3. We now ask how to construct a basis for a linear manifold and how to determine the dimension of a linear manifold. In answering these questions, we shall assume that the number of vectors x, y, ... spanning the linear manifold $L(x, y, ...)$ is *finite* (although some of the conclusions actually do not require this assumption). Suppose that among the vectors x, y, ... spanning the linear manifold $L(x, y, ...)$ we can find r linearly independent vectors denoted by $x_1, x_2, ..., x_r$, in terms of which any vector of the system x, y, ... can be expressed linearly. If this can be done, we can assert that *the vectors $x_1, x_2, ..., x_r$ form a basis for the space $L(x, y, ...)$*. Indeed, by the definition of a linear manifold, every vector $z \in L(x, y, ...)$ can be expressed as a linear combination of a finite number of vectors of the system x, y, ... But by hypothesis, each of these vectors can be expressed as a linear combination of $x_1, x_2, ..., x_r$. Thus, eventually the vector z can also be expressed directly as a linear combination of the vectors $x_1, x_2, ..., x_r$. This, together with the assumption that the vectors $x_1, x_2, ..., x_r$ are linearly independent, shows that both of the conditions for a basis (Sec. 13) are satisfied, Q.E.D.

According to Theorem 13, the dimension of the space $L(x, y, ...)$ is equal to the number r. Since in an r-dimensional space there can be no more than r linearly independent vectors, we can draw the following conclusions:

1. *If the number of vectors $x, y, \ldots$ spanning $L(x, y, \ldots)$ is larger than the number r, then the vectors $x, y, \ldots$ are linearly dependent; if the number of these vectors equals r, then the vectors are linearly independent.*

2. *Every set of $r + 1$ vectors from the system $x, y, \ldots$ is linearly dependent.*

3. *The dimension of the space $L(x, y, \ldots)$ can be defined as the maximum number of linearly independent vectors in the system $x, y, \ldots$.*

17. Hyperplanes

17.1. As we have seen in Sec. 15, if we adopt the "point" rather than the "vector" interpretation in the space V_3, then the geometric entity corresponding to the notion of a subspace is a plane (or a straight line) passing through the origin of coordinates. But it is also desirable to include in our scheme of things planes and straight lines which do not pass through the origin of coordinates. Noting that such planes and straight lines are obtained from planes and straight lines passing through the origin of coordinates by means of parallel displacement in space, i.e., by a *shift*, we are led in a natural way to the following general construction.

Let L be a subspace of the linear space R and let x_0 be a fixed vector which in general does not belong to L. Consider the set H of all vectors x obtained by the formula

$$x = x_0 + y,$$

where the vector y ranges over the whole subspace L. The set H which is *the result of shifting the subspace L by the vector x_0 is called a hyperplane.* We note that a hyperplane is generally not a subspace itself.

Example 1. In the space V_3, the set of all vectors starting from the origin of coordinates and terminating on a plane γ forms a hyperplane. It is easily verified that this hyperplane is a subspace if and only if the plane γ passes through the origin of coordinates.

Example 2. In the space T_n, consider the set H consisting of the vectors

$$x = (\xi_1, \xi_2, \ldots, \xi_n)$$

whose components satisfy the compatible inhomogeneous system of linear equations

$$
\begin{aligned}
a_{11}x_1 + a_{12}x_2 + \cdots + a_{1n}x_n &= b_1, \\
a_{21}x_1 + a_{22}x_2 + \cdots + a_{2n}x_n &= b_2, \\
\cdots \quad\quad \cdots \quad\quad \cdots \quad\quad \cdots \quad & \quad . \\
a_{k1}x_1 + a_{k2}x_2 + \cdots + a_{kn}x_n &= b_k,
\end{aligned}
\tag{12}
$$

and consider the set L consisting of the vectors $y = (\eta_1, \eta_2, \ldots, \eta_n)$ whose components satisfy the homogeneous system of linear equations with the same coefficients, i.e.

$$
\begin{aligned}
a_{11}y_1 + a_{12}y_2 + \cdots + a_{1n}y_n &= 0, \\
a_{21}y_1 + a_{22}y_2 + \cdots + a_{2n}y_n &= 0, \\
\cdots \quad\quad \cdots \quad\quad\quad \cdots \quad\quad\; . \\
a_{k1}y_1 + a_{k2}y_2 + \cdots + a_{kn}y_n &= 0.
\end{aligned}
\tag{13}
$$

As we already know, the set L is a subspace of the space T_n.   Let

$$
x_0 = (\xi_1^{(0)}, \xi_2^{(0)}, \ldots, \xi_n^{(0)})
$$

be a solution of the system (12).   We now show that the set H is identical with the set of all sums $x_0 + y$, where y ranges over the whole subspace L. In fact, if

$$
y = (\eta_1, \eta_2, \ldots, \eta_n)
$$

is a solution of the system (13), then the vector

$$
x = x_0 + y = (\xi_1^{(0)} + \eta_1, \xi_2^{(0)} + \eta_2, \ldots, \xi_n^{(0)} + \eta_n)
$$

is obviously a solution of the system (12), i.e., belongs to the set H.   Conversely, if x is any vector of the set H, then the difference $y = x - x_0$ certainly satisfies the system (13), i.e., the vector y belongs to the subspace L.   In view of the definition which we have given above, the set H is a hyperplane, namely, the result of shifting the space L by the vector x_0.

17.2. We can assign a dimension to every hyperplane, even if it is not a subspace, i.e., we consider the dimension of the hyperplane H to be equal to the dimension of the subspace L from which H was obtained by shifting. For this definition to be suitable, we must show that the given hyperplane H can be obtained as a shift of *only one* subspace.   To prove this statement, suppose that the hyperplane H is both the result of shifting the subspace L by the vector x_0 and the result of shifting the subspace L' by the vector x_0'. Then, for any $z \in H$, we have both $z = x_0 + y$, where $y \in L$, and $z = x_0' + y'$, where $y' \in L'$.   It follows that L' is the set of vectors given by the formula $y' = (x_0 - x_0') + y$, where y is an arbitrary vector in L, i.e., the subspace L' is the result of shifting the subspace L by the vector $x_1 = x_0 - x_0'$.   We now show that the vector x_1 belongs to the subspace L.   The zero vector, just like any other element of the space L', can be represented in the form $x_1 + y_1$, where $y_1 \in L$ (since L' is the subspace L shifted by the vector x_1). Hence, $x_1 = -y_1$ and therefore x_1 belongs to L, as asserted.   But in this case, every vector $y' \in L'$ also belongs to the subspace L, since y' is the sum of a vector $x_1 \in L$ and a vector $y \in L$.   Consequently, we have the relation $L' \subset L$.   Because of the complete symmetry of the hypothesis, we can prove similarly that $L \subset L'$, from which it follows that $L \equiv L'$, as required.

In what follows, hyperplanes of dimension 1 will be called *straight lines*, and hyperplanes of dimension 2 will be called *planes*.

Problem 1. Is the shift vector x_0 figuring in the construction of a hyperplane uniquely determined by the hyperplane itself?

Ans. No. It can be replaced by any other vector of the hyperplane.

Problem 2. Show that every hyperplane $H \subset R$ has the following property: If $x \in H$, $y \in H$, then $\alpha x + (1 - \alpha)y \in H$ for any real α. Conversely, if a *subset* $H \subset R$ has the stated property, then H is a hyperplane. What geometric characteristic of a hyperplane is expressed by this property?

Ans. With the "point" interpretation, the property means that every hyperplane contains the line passing through any two of its points.

Problem 3. The hyperplanes H_1 and H_2 have dimensions p and q, respectively. What is the least dimension which the hyperplane H_3 must have in order to be sure to contain H_1 and H_2?

Ans. In general $p + q + 1$, if this number does not exceed the dimension of the whole space.

Problem 4. Solve the problem for three hyperplanes H_1, H_2, and H_3 with dimensions p, q, and r, respectively.

Ans. In general $p + q + r + 2$, if this number does not exceed the dimension of the whole space.

18. Isomorphism of Linear Spaces

Theorem 11 proved in Sec. 13.2 shows that any linear space R with a basis consisting of n vectors (in particular, any n-dimensional space), does not differ essentially from any other space with a basis consisting of n vectors (e.g., the space T_n). In every such space, a vector is specified by n numbers, its components $\xi_1, \xi_2, \ldots, \xi_n$, and linear operations on vectors reduce to entirely identical operations on their components. This idea is expressed precisely in Theorem 14 below. First we define the concept of isomorphism.

Two linear spaces R' and R'' are called *isomorphic* if we can establish a one-to-one correspondence[3] between the vectors $x' \in R'$ and $x'' \in R''$, such that

1) If the vector $x'' \in R''$ corresponds to the vector $x' \in R'$ and the vector $y'' \in R''$ corresponds to the vector $y' \in R'$, then the vector $x'' + y'' \in R''$ corresponds to the vector $x' + y' \in R'$;

2) If the vector $x'' \in R''$ corresponds to the vector $x' \in R'$ and λ is a real number, then the vector $\lambda x'' \in R''$ corresponds to the vector $\lambda x' \in R'$.

---

[3] A correspondence between elements $x' \in R'$ and $x'' \in R''$ is called *one-to-one* if:

   1) To each element $x' \in R'$ corresponds one and only one element $x'' \in R''$;

   2) To each element $x'' \in R''$ corresponds one and only one element $x' \in R'$.

THEOREM 14. *Any two n-dimensional spaces R' and R" are isomorphic.*

Proof. Let $e'_1, e'_2, \ldots, e'_n$ be a basis in R' and let $e''_1, e''_2, \ldots, e''_n$ be a basis in R''. (By Theorem 12, both these bases exist.) Furthermore, let

$$x' = \sum_{k=1}^{n} \xi_k e'_k$$

be an arbitrary vector in R', and let the vector

$$x'' = \sum_{k=1}^{n} \xi_k e''_k \in R''$$

correspond to x' (x'' has the same components with respect to the basis $e''_1, e''_2, \ldots, e''_n$ as x' has with respect to the basis $e'_1, e'_2, \ldots, e'_n$). It is obvious that this correspondence is one-to-one.

We now verify that the conditions for an isomorphism are satisfied. If we choose two vectors

$$x' = \sum_{k=1}^{n} \xi_k e'_k, \qquad y' = \sum_{k=1}^{n} \eta_k e'_k$$

in the space R', then by Theorem 11

$$z' \equiv x' + y' = \sum_{k=1}^{n} (\xi_k + \eta_k) e'_k.$$

The corresponding vectors in the space R'' are

$$x'' = \sum_{k=1}^{n} \xi_k e''_k, \qquad y'' = \sum_{k=1}^{n} \eta_k e''_k, \qquad z'' = \sum_{k=1}^{n} (\xi_k + \eta_k) e''_k.$$

According to Theorem 11, $z'' = x'' + y''$, whence it follows that condition 1 is satisfied.

Similarly, we verify that condition 2 is satisfied. If

$$x' = \sum_{k=1}^{n} \xi_k e'_k,$$

then by Theorem 11

$$z' \equiv \lambda x' = \sum_{k=1}^{n} \lambda \xi_k e'_k,$$

and the corresponding elements in the space R'' are

$$x'' = \sum_{k=1}^{n} \xi_k e''_k, \qquad z'' = \sum_{k=1}^{n} \lambda \xi_k e''_k.$$

By Theorem 11, again, $z'' = \lambda x''$, which implies that condition 2 is met. This completes the proof of Theorem 14.

COROLLARY.   *Every n-dimensional space is isomorphic to the space T_n.*

Problem 1. Show that if the spaces R' and R'' are isomorphic, and the vectors $e'_1, e'_2, \ldots, e'_n$ are a basis in R', then the corresponding vectors $e''_1, e''_2, \ldots, e''_n \in R''$ are a basis in R''.

Problem 2. Show that if the spaces R' and R'' are isomorphic and R' has dimension n, then R'' also has dimension n.

Problem 3. According to Theorem 14, the one-dimensional spaces V_1 and P (see Problem 3 of Sec. 11 and Problem 1 of Sec. 14) are isomorphic. How can one establish this isomorphism in practice?

Ans.   With each positive number associate its logarithm.

3

SYSTEMS OF
LINEAR EQUATIONS

19. More on the Rank of Matrices

We have already touched upon the subject of matrices several times. In this section, we shall study in more detail those properties of matrices which are connected with the concept of rank (see Sec. 9). This will allow us to solve quite generally the problem of systems of linear equations (formulated in Sec. 1).

We begin by recalling some basic definitions from Sec. 9. Suppose we have a matrix

$$\begin{Vmatrix} a_{11} & a_{12} & \cdots & a_{1k} \\ a_{21} & a_{22} & \cdots & a_{2k} \\ \cdot & \cdot & \cdots & \cdot \\ a_{n1} & a_{n2} & \cdots & a_{nk} \end{Vmatrix} \quad (1)$$

with n rows and k columns, consisting of the numbers a_{ij},[1] where i is the row index, ranging from 1 to n, and j is the column index, ranging from 1 to k. If we choose any m rows and m columns of this matrix, then the elements which appear at the intersections of these rows and columns form a square matrix of order m. The determinant of this matrix is called a *minor of order m of the matrix A*. The integer r is said to be the *rank* of the matrix A, if A has a nonvanishing minor of order r and all its minors of

---

[1] Sometimes the indices of an element of the matrix A will be written differently, i.e., sometimes we shall denote the element appearing in the ith row and jth column by the symbol $a_i^{(j)}$.

order $r + 1$ and higher vanish. If the matrix A has rank $r > 0$, then each of its nonvanishing minors of order r is called a *basis minor*. The columns and rows of the matrix which intersect at the elements of the basis minor are called the *basis columns* and *basis rows*.

The considerations which follow are based on the possibility of regarding any column of numbers as a geometric object, i.e., as a vector in an n-dimensional space R (Sec. 14). With this geometric interpretation, the matrix A itself corresponds to a definite set of k vectors of the space R; we denote by x_j $(j = 1, 2, \ldots, k)$ the vector corresponding to the jth column of A. Then, any linear relation between the columns of A can be interpreted as a linear relation (the same one) between the corresponding vectors (Sec. 12).

Let $L(x_1, x_2, \ldots, x_k)$ be the linear manifold spanned by the vectors $x_1, x_2, \ldots, x_k$ of R (Sec. 16). We now prove that *the vectors corresponding to the basis columns of the matrix A form a basis for this linear manifold.* To be explicit, suppose that the first r columns of A are basis columns. Then, to prove our assertion, it suffices to show (a) that the vectors $x_1, x_2, \ldots, x_r$ are linearly independent, and (b) that any of the other vectors $x_{r+1}, \ldots, x_n$ is a linear combination of the first r vectors. To prove (a), suppose that the vectors $x_1, x_2, \ldots, x_r$ are linearly dependent. This means that the first r columns of A are linearly dependent; therefore, by Theorem 6 of Sec. 9, any determinant of order r constructed from these columns and any r rows of A vanishes. In particular, the basis minor of A vanishes. This contradiction establishes (a). The assertion (b), as applied to columns of the matrix A, has already been proved in Sec. 9, under the guise of the "basis minor theorem." This completes our proof of the fact that the vectors $x_1, x_2, \ldots, x_r$ form a basis for the space $L(x_1, x_2, \ldots, x_k)$. According to Theorem 13, the dimension of this space equals the number r, i.e., the rank of the matrix A. Thus, we have obtained the following important result:

THEOREM 15. *The dimension of the linear manifold spanned by the vectors corresponding to the columns of the matrix A equals the rank of A. Moreover, this linear manifold has for a basis the vectors corresponding to the basis columns of A.*

The following theorems are obvious consequences of Remarks 1, 2, and 3 of Sec. 16.3:

THEOREM 16. *If the rank of the matrix A is less than the number of columns in A ($r < k$), then the columns of A are linearly dependent. If the rank of A equals the number of columns in A ($r = k$), then the columns of A are linearly independent.*

THEOREM 17. *Any r + 1 columns of the matrix A are linearly dependent.*

THEOREM 18. *The rank of any matrix A equals the maximum number of linearly independent columns in A.*

This last theorem is of fundamental importance, since it constitutes a *new definition of the rank of a matrix.*

Consider the transpose of the matrix A, i.e., the matrix A' whose rows are the columns of A. Clearly, the rank of the transposed matrix A' is the same as the rank of A. But according to Theorem 18, the rank of A' equals the maximum number of linearly independent columns in A', or equivalently, the maximum number of linearly independent rows in A. Thus, we arrive at the following somewhat unexpected conclusion:

THEOREM 19. *The maximum number of linearly independent rows in a matrix A is the same as the maximum number of linearly independent columns in A.*

We note that Theorem 19 is not trivial. In fact, any direct proof of this theorem would require a chain of reasoning equivalent to the proof of Theorems 4 and 15. Finally, we note the following result, which is a consequence of Theorem 15 and Lemma 2 of Sec. 16.3:

THEOREM 20. *We can delete any column of the matrix A which is a linear combination of the other columns without changing the rank of A.*

Problem 1. Prove the following theorem: A necessary and sufficient condition for the matrix $\|a_{ij}\|$ of order m to have rank $r \leqslant 1$ is that there exist numbers $a_1, a_2, \ldots, a_m$ and $b_1, b_2, \ldots, b_m$ such that

$$a_{ij} = a_i b_j \qquad (i, j = 1, 2, \ldots, m).$$

Problem 2. Let $x_1, x_2, \ldots, x_k$ be k linearly independent vectors of the n-dimensional space R, and let $A = \|a_i^{(j)}\|$ be the matrix consisting of the components of the vectors $x_1, x_2, \ldots, x_k$ with respect to some basis $e_1, e_2, \ldots, e_n$. Show that the linear manifold $L = L(x_1, x_2, \ldots, x_k)$ is uniquely determined, if one knows the values of all the minors of A of order k.

Hint. We have to write the conditions for a vector y to belong to the subspace L in such a way that they contain only minors of A of order k. Now $y \in L$ if and only if the matrix B obtained by adding to A the column consisting of the components of the vector y has rank k, or, equivalently, if and only if every minor of B of order $k + 1$ vanishes. Expanding every minor of B of order $k + 1$ with respect to elements of the last column, we obtain a system of equations in the components of y, with coefficients which are minors of A of order k.

20. Nontrivial Compatibility of a Homogeneous Linear System

Suppose we have a homogeneous linear system

$$
\begin{aligned}
a_{11}x_1 + a_{12}x_2 + \cdots + a_{1n}x_n &= 0, \\
a_{21}x_1 + a_{22}x_2 + \cdots + a_{2n}x_n &= 0, \\
\cdots \qquad \cdots \qquad \cdots \qquad \cdots & \qquad . \\
a_{k1}x_1 + a_{k2}x_2 + \cdots + a_{kn}x_n &= 0.
\end{aligned} \tag{2}
$$

As we know, this system is always compatible, since it has the trivial solution

$$ x_1 = x_2 = \cdots = x_n = 0. $$

The basic problem encountered in studying homogeneous linear systems is the following: *Under what conditions is a homogeneous system "nontrivially compatible," i.e., under what conditions does a homogeneous system have other solutions in addition to the trivial solution?* The results of Sec. 19 allow us to solve this problem immediately. In fact, as we have seen in Sec. 12, the existence of a nontrivial solution of the system (2) is equivalent to the columns of the matrix

$$
A = \begin{Vmatrix}
a_{11} & a_{12} & \cdots & a_{1n} \\
a_{21} & a_{22} & \cdots & a_{2n} \\
\cdot & \cdot & \cdots & \cdot \\
a_{k1} & a_{k2} & \cdots & a_{kn}
\end{Vmatrix}
$$

being linearly dependent, which, according to Theorem 16, occurs if and only if the rank of the matrix A is less than the number of columns in A. Thus, we obtain the following theorem:

THEOREM 21. *The system* (2) *is nontrivially compatible, i.e., has nontrivial solutions, if and only if the rank of the matrix A is less than n. If the rank of the matrix A equals n, the system* (2) *does not have nontrivial solutions.*

In particular, if the number of equations in the system (2) is less than the number of unknowns ($k < n$), the rank of the matrix A is certainly less than n, and in this case nontrivial solutions will always exist. If $k = n$, the question of whether or not nontrivial solutions exist depends on the value of $\det A$. If $\det A \neq 0$, there are no nontrivial solutions ($r = n$), while if $\det A = 0$, there are nontrivial solutions ($r < n$). If $k > n$, we have to examine all possible determinants of order n which are obtained by fixing any n rows of the matrix A. If all these determinants vanish, then $r < n$, and nontrivial solutions exist. If at least one of these determinants is nonvanishing, then $r = n$, and there is only the trivial solution.

Problem. Show that when $k = n$, the system (2) has the solution

$$ c_1 = A_{i1}, c_2 = A_{i2}, \ldots, c_n = A_{in} \qquad (1 \leqslant i \leqslant n), $$

where A_{ik} is the cofactor of the element a_{ik} (i fixed), provided that the rank of the matrix A is less than n.

Comment. This fact makes it easy to construct nontrivial solutions of the system (2) in the case where the rank of the matrix of the system is $n - 1$.

21. The Compatibility Condition for an Inhomogeneous Linear System

Suppose we have a system of inhomogeneous linear equations

$$
\begin{aligned}
a_{11}x_1 + a_{12}x_2 + \cdots + a_{1n}x_n &= b_1, \\
a_{21}x_1 + a_{22}x_2 + \cdots + a_{2n}x_n &= b_2, \\
\cdots \quad \cdots \quad \cdots \quad \cdots \quad \cdot \\
a_{k1}x_1 + a_{k2}x_2 + \cdots + a_{kn}x_n &= b_k.
\end{aligned}
\tag{3}
$$

We associate two matrices with this system, the matrix

$$
A = \left\|
\begin{matrix}
a_{11} & a_{12} & \cdots & a_{1n} \\
a_{21} & a_{22} & \cdots & a_{2n} \\
\cdot & \cdot & \cdots & \cdot \\
a_{k1} & a_{k2} & \cdots & a_{kn}
\end{matrix}
\right\|,
$$

called the *coefficient matrix of the system* (3), and the matrix

$$
A_1 = \left\|
\begin{matrix}
a_{11} & a_{12} & \cdots & a_{1n} & b_1 \\
a_{21} & a_{22} & \cdots & a_{2n} & b_2 \\
\cdot & \cdot & \cdots & \cdot & \cdot \\
a_{k1} & a_{k2} & \cdots & a_{kn} & b_k
\end{matrix}
\right\|,
$$

called the *augmented matrix of the system* (3). The basic theorem on the compatibility of the system (3) reads as follows:

THEOREM 22. *The system* (3) *is compatible if and only if the rank of the augmented matrix of the system equals the rank of the coefficient matrix.*

Proof. Assume first that the system (3) is compatible. Then if $c_1, c_2, \ldots, c_n$ is a solution of the system, we have the equations

$$
\begin{aligned}
a_{11}c_1 + a_{12}c_2 + \cdots + a_{1n}c_n &= b_1, \\
a_{21}c_1 + a_{22}c_2 + \cdots + a_{2n}c_n &= b_2, \\
\cdots \quad \cdots \quad \cdots \quad \cdots \quad \cdot \\
a_{k1}c_1 + a_{k2}c_2 + \cdots + a_{kn}c_n &= b_k.
\end{aligned}
$$

These equations imply that the last column of A_1 is a linear combination of the other columns of A_1 (with coefficients $c_1, c_2, \ldots, c_n$). By Theorem 20, we can delete the last column of A_1 without changing its rank. But when the last column of A_1 is deleted, it becomes just A. Therefore, if the system (3) is compatible, the matrices A and A_1 have the same rank.

We now assume that the matrices A and A_1 have the same rank, and show that the system (3) is compatible.   Let r be the rank of the matrix A (and consequently also of the matrix A_1).   Consider r basis columns of A; they will also be basis columns of A_1.   By Theorem 4, the last column of A_1 can be written as a linear combination of the basis columns, and hence it can be written as a linear combination of *all* the columns of A.   If we denote the coefficients of this linear combination by $c_1, c_2, \ldots, c_n$, we find that the equations

$$a_{11}c_1 + a_{12}c_2 + \cdots + a_{1n}c_n = b_1,$$
$$a_{21}c_1 + a_{22}c_2 + \cdots + a_{2n}c_n = b_2,$$
$$\cdots \qquad \cdots \qquad \cdots \qquad \cdots \ .$$
$$a_{k1}c_1 + a_{k2}c_2 + \cdots + a_{kn}c_n = b_k$$

are satisfied.   Thus, the values

$$x_1 = c_1, x_2 = c_2, \ldots, x_n = c_n$$

satisfy the system (3), which is therefore compatible.   The theorem just proved is called the *Kronecker-Capelli theorem.*

22. The General Solution of a Linear System

The Kronecker-Capelli theorem, which gives the general condition for the compatibility of a linear system, does not give a method for solving the system.   In this section, we derive a formula which constitutes a general solution of a linear system.

Suppose we have a compatible linear system (3) with a coefficient matrix $A = \|a_{ij}\|$ of rank r.   We can assume that the basis minor M of the matrix A appears in its upper left-hand corner; if this is not the case, we can achieve this configuration by interchanging rows and columns of A, which corresponds to renumbering some of the equations and unknowns in the system (3). We take the first r equations of the system (3) and rewrite them in the following form:

$$a_{11}x_1 + a_{12}x_2 + \cdots + a_{1r}x_r = b_1 - a_{1,r+1}x_{r+1} - \cdots - a_{1n}x_n,$$
$$a_{21}x_1 + a_{22}x_2 + \cdots + a_{2r}x_r = b_2 - a_{2,r+1}x_{r+1} - \cdots - a_{2n}x_n, \qquad (4)$$
$$\cdots \quad \cdots \quad \cdots \quad \cdots \qquad \cdots \quad \cdots \quad \cdots$$
$$a_{r1}x_1 + a_{r2}x_2 + \cdots + a_{rr}x_r = b_r - a_{r,r+1}x_{r+1} - \cdots - a_{rn}x_n.$$

Next we assign the unknowns $x_{r+1}, \ldots, x_n$ the completely arbitrary values $c_{r+1}, \ldots, c_n$.   Then (4) becomes a system of r equations in the r unknowns $x_1, x_2, \ldots, x_r$, with a determinant M which is nonvanishing (a basis minor of the matrix A).   This system can be solved by using Cramer's rule (Sec. 7).   Therefore, there exist numbers $c_1, c_2, \ldots, c_n$ which, when substituted for the unknowns $x_1, x_2, \ldots, x_n$ of the system (4), reduce all the equations of the system to identities. We now show that these values $c_1, c_2, \ldots, c_n$ satisfy all the other equations of the system (3) as well.

The first r rows of the augmented matrix A_1 of the system (3) are basis rows of this matrix, since by the compatibility condition, the rank of the augmented matrix is r, while by construction, the nonvanishing minor M appears in the first r rows of A_1. By Theorem 4 (applied to rows), each of the last $n - r$ rows of A_1 is a linear combination of the first r rows. This means that every equation of the system (3) beginning with the $(r + 1)$th equation is a linear combination of the first r equations of the system. Therefore, if the values

$$x_1 = c_1, \ldots, x_n = c_n$$

satisfy the first r equations of the system (3), they also satisfy all the other equations.

To write a formula for the solution just obtained for the system (3), we denote by $M_j(\alpha_i)$ the determinant obtained from the basis minor

$$M = \det \|a_{ij}\| \qquad (i, j = 1, 2, \ldots, r)$$

by replacing its jth column by the column consisting of the quantities $\alpha_1, \alpha_2, \ldots, \alpha_r$. Then, using Cramer's rule to write the solution of the system (4), we obtain

$$
\begin{aligned}
c_j &= \frac{1}{M} M_j(b_i - a_{i,r+1}c_{r+1} - \cdots - a_{in}c_n) \\
&= \frac{1}{M} [M_j(b_i) - c_{r+1}M_j(a_{i,r+1}) - \cdots - c_n M_j(a_{in})] \qquad (j = 1, 2, \ldots, r).
\end{aligned}
\tag{5}
$$

These formulas express the values of the unknowns $x_j = c_j$ ($j = 1, 2, \ldots, r$) in terms of the coefficients of the system, the constant terms and the arbitrary quantities (parameters)

$$c_{r+1}, c_{r+2}, \ldots, c_n.$$

Finally, we show that (5) comprises any solution of the system (3). In fact, let $c_1^0, c_2^0, \ldots, c_r^0, c_{r+1}^0, \ldots, c_n^0$ be an arbitrary solution of the system (3). Obviously, it is also a solution of the system (4). But, using Cramer's rule to solve the system (4), we obtain unique expressions for the quantities $c_1^0, c_2^0, \ldots, c_r^0$ in terms of the quantities $c_{r+1}^0, \ldots, c_n^0$, namely the formulas (5). Thus, if we set

$$c_{r+1} = c_{r+1}^0, \ldots, c_n = c_n^0,$$

these formulas together with the formulas (5) give us just the solution $c_1^0, c_2^0, \ldots, c_r^0$, Q.E.D.

Problem 1. Solve the system of equations

$$x_1 + x_2 + x_3 + x_4 + x_5 = 7,$$
$$3x_1 + 2x_2 + x_3 + x_4 - 3x_5 = -2,$$
$$x_2 + 2x_3 + 2x_4 + 6x_5 = 23,$$
$$5x_1 + 4x_2 + 3x_3 + 3x_4 - x_5 = 12.$$

Ans. $c_1 = -16 + c_3 + c_4 + 5c_5,$ $c_2 = 23 - 2c_3 - 2c_4 - 6c_5.$

Problem 2. Study the solutions of the system

$$\lambda x + y + z = 1,$$
$$x + \lambda y + z = \lambda,$$
$$x + y + \lambda z = \lambda^2$$

as a function of λ.

Ans.   If $(\lambda - 1)(\lambda + 2) \neq 0$, then

$$x = -\frac{\lambda + 1}{\lambda + 2}, \quad y = \frac{1}{\lambda + 2}, \quad z = \frac{(\lambda + 1)^2}{\lambda + 2}.$$

If $\lambda = 1$, the system has solutions depending on two parameters. If $\lambda = -2$, the system is incompatible.

Problem 3. What is the condition for the three straight lines

$$a_1x + b_1y + c_1 = 0, \quad a_2x + b_2y + c_2 = 0, \quad a_3x + b_3y + c_3 = 0$$

to pass through one point?

Ans.   The matrices

$$\begin{Vmatrix} a_1 & b_1 \\ a_2 & b_2 \\ a_3 & b_3 \end{Vmatrix} \quad \text{and} \quad \begin{Vmatrix} a_1 & b_1 & c_1 \\ a_2 & b_2 & c_2 \\ a_3 & b_3 & c_3 \end{Vmatrix}$$

must have the same rank.

Problem 4. What is the condition for the n straight lines

$$a_1x + b_1y + c_1 = 0, \quad a_2x + b_2y + c_2 = 0, \ldots, \quad a_nx + b_ny + c_n = 0$$

to pass through one point?

Ans.   The matrices

$$\begin{Vmatrix} a_1 & b_1 \\ a_2 & b_2 \\ \cdot & \cdot \\ \cdot & \cdot \\ \cdot & \cdot \\ a_n & b_n \end{Vmatrix} \quad \text{and} \quad \begin{Vmatrix} a_1 & b_1 & c_1 \\ a_2 & b_2 & c_2 \\ \cdot & \cdot & \cdot \\ \cdot & \cdot & \cdot \\ \cdot & \cdot & \cdot \\ a_n & b_n & c_n \end{Vmatrix}$$

must have the same rank.

23. Geometric Properties of the Set of Solutions of a Linear System

23.1. Consider first the case of the homogeneous linear system (2). As we have already seen (Sec. 15), the set of all solutions of this system forms a linear space, which we denote by R. We now calculate the dimension of R and construct a basis for R.

For a homogeneous system, the equations (5) become

$$-Mc_j = c_{r+1}M_j(a_{i,r+1}) + \cdots + c_nM_j(a_{in}) \qquad (j = 1, 2, \ldots, r), \qquad (6)$$

since $M_j(b_i) = M_j(0) = 0$. With every solution $c_1, c_2, \ldots, c_r, c_{r+1}, \ldots, c_n$ of the system (2), we associate a vector $(c_{r+1}, \ldots, c_n)$ of the space T_{n-r} defined in Sec. 11. Since the numbers $c_{r+1}, \ldots, c_n$ can be chosen arbitrarily and since they uniquely define a solution of the system (2), the correspondence between the space of solutions of the system (2) and the space T_{n-r} is one-to-one. This correspondence is an isomorphism, since it preserves linear operations, as is easily verified. Thus, *given a homogeneous system of linear equations which has n unknowns and a coefficient matrix of rank r, the space R of solutions of the system is isomorphic to the space T_{n-r}.* In particular, the dimension of the space R is $n - r$.

By Theorem 12, any system of $n - r$ linearly independent solutions of a homogeneous linear system of equations forms a basis in the space of all solutions, and will be called a *fundamental system of solutions*. To construct a fundamental system of solutions, we can use any basis of the space T_{n-r}. Because of the isomorphism, the corresponding solutions of the system (2) will form a basis in the space of all solutions of the system. The simplest basis of the space T_{n-r} consists of the vectors

$$e_1 = (1, 0, \ldots, 0), \; e_2 = (0, 1, \ldots, 0), \ldots, \; e_{n-r} = (0, 0, \ldots, 1)$$

(see Sec. 13). For example, to obtain the solution of the system (2) corresponding to the vector e_1, we set $c_{r+1} = 1, c_{r+2} = \cdots = c_n = 0$ in the formulas (6) and determine the corresponding values

$$c_i = c_i^{(1)} \qquad (i = 1, 2, \ldots, n).$$

Similarly, we construct the solution corresponding to any other basis vector e_j $(j = 2, \ldots, n - r)$. The set of solutions of the system (2) constructed in this way is called a *normal fundamental system of solutions*. If we denote these solutions by $x^{(1)}, x^{(2)}, \ldots, x^{(n-r)}$, then by the definition of a basis, any solution x is given by the formula

$$x = C_1x^{(1)} + C_2x^{(2)} + \cdots + C_{n-r}x^{(n-r)}. \qquad (7)$$

Since any solution of the system (2) is a special case of (7), this formula gives the *general solution* of (2).

Problem. Write down the normal fundamental system of solutions for the system of equations

$$x_1 + x_2 + x_3 + x_4 + x_5 = 0,$$
$$3x_1 + 2x_2 + x_3 + x_4 - 3x_5 = 0,$$
$$x_2 + 2x_3 + 2x_4 + 6x_5 = 0,$$
$$5x_1 + 4x_2 + 3x_3 + 3x_4 - x_5 = 0.$$

Ans. $x^{(1)} = (1, -2, 1, 0, 0),$
$x^{(2)} = (1, -2, 0, 1, 0),$
$x^{(3)} = (5, -6, 0, 0, 1).$

23.2. Consider now the general case of an inhomogeneous system (3). As was shown in Sec. 17, the geometric object H, corresponding to the set of all solutions of an inhomogeneous system, is a hyperplane in the n-dimensional space T_n. This hyperplane is obtained by shifting the subspace R of all solutions of the corresponding homogeneous system (R has been shown to be isomorphic to the space T_{n-r}) by a vector x_0 which is an arbitrary particular solution of the inhomogeneous system. From this, we can conclude that the dimension of the hyperplane H is the same as the dimension of the subspace R. Moreover, if r is the rank of the coefficient matrix of the system (3), then any vector y of the subspace R can be represented as a sum

$$y = C_1 y^{(1)} + C_2 y^{(2)} + \cdots + C_{n-r} y^{(n-r)},$$

where $y^{(1)}, y^{(2)}, \ldots, y^{(n-r)}$ are basis vectors of the subspace R (a fundamental system of solutions). Consequently, any vector x of the hyperplane H can be represented as a sum

$$x = x_0 + y = x_0 + C_1 y^{(1)} + C_2 y^{(2)} + \cdots + C_{n-r} y^{(n-r)}.$$

In the language appropriate to solutions of the systems (2) and (3), this result can be stated as follows:

The general solution of the inhomogeneous system (3) equals the sum of any particular solution of this system and the general solution of the corresponding homogeneous system (2).

Problem. Write down the general solution of the system given in Problem 1 of Sec. 22, using the normal fundamental system of solutions of the corresponding homogeneous system, found in the problem of Sec. 23.1.

Ans.

$$x = \begin{Vmatrix} -16 \\ 23 \\ 0 \\ 0 \\ 0 \end{Vmatrix} + C_1 \begin{Vmatrix} 1 \\ -2 \\ 1 \\ 0 \\ 0 \end{Vmatrix} + C_2 \begin{Vmatrix} 1 \\ -2 \\ 0 \\ 1 \\ 0 \end{Vmatrix} + C_3 \begin{Vmatrix} 5 \\ -6 \\ 0 \\ 0 \\ 1 \end{Vmatrix},$$

for example. Here the first column contains the components of a vector x_0 which is a particular solution of the inhomogeneous system; the other columns contain the components of the vectors $y^{(1)}$, $y^{(2)}$, $y^{(3)}$ which form a normal fundamental system of solutions of the corresponding homogeneous system.

24. Methods for Calculating the Rank of a Matrix and for Finding the Basis Minor

To make practical use of the methods for solving systems of linear equations developed in the preceding sections, one must be able to calculate the rank of a matrix and find its basis minor. Obviously, the definition of the rank of a matrix given in Sec. 9 cannot serve *per se* as a reasonable practical means of calculating the rank. For example, a square matrix of order 5 contains 1 minor of order 5, 25 minors of order 4, 100 minors of order 3 and 100 minors of order 2; it is clear that it would be a very laborious task to find the rank of such a matrix by direct calculation of all its minors. In this section, we shall give a simple method for calculating the rank of a matrix and determining its basis minor. This method is based on a study of some operations on rows and columns of a matrix which do not change its rank; these operations will be called *elementary operations.* Since the rank of a matrix does not change when it is transposed, we shall define these operations only for the columns of a matrix. In keeping with this, we shall use in our proofs the geometric interpretation of a matrix with n rows and k columns as the matrix formed from the components of a system of k vectors $x_1, x_2, \ldots, x_k$ in an n-dimensional space R; we shall also make use of Theorem 15, according to which the rank of this matrix equals the dimension of the linear manifold spanned by the vectors $x_1, x_2, \ldots, x_k$.

We now study the following elementary operations:

1. *Permutation of columns.* Suppose the columns of the matrix A are permuted in any way. This operation does not change the rank of A, since the dimension of the linear manifold spanned by the vectors $x_1, x_2, \ldots, x_k$ does not depend on the order in which they are written.

2. *Dividing out a nonzero common factor of the elements of a column.* Assume that the common factor $\lambda \neq 0$ which we are dividing out belongs to the first column of the matrix A. This operation is equivalent to replacing the system of vectors $\lambda x_1, x_2, \ldots, x_k$ by the system $x_1, x_2, \ldots, x_k$; obviously, the linear manifolds spanned by these systems have the same dimension (since the linear manifolds themselves are the same). Therefore, the rank of the matrix A does not change as a result of this elementary operation.

3. *Adding an arbitrary multiple of one column to another column.* Suppose we multiply the mth column of the matrix A by the number λ and add

it to the jth column. This means that the system of vectors $x_1, \ldots,$ $x_j, \ldots, x_m, \ldots, x_k$ has been replaced by the system

$$x_1, \ldots, x_j + \lambda x_m, \ldots, x_m, \ldots, x_k.$$

We have to show that the linear manifolds L_1 and L_2 spanned by these systems are the same. In the first place, all the vectors of the second system lie in the linear manifold spanned by the vectors of the first system; therefore, by Lemma 1 of Sec. 16.2, we have $L_2 \subset L_1$. On the other hand, the equation

$$x_j = (x_j + \lambda x_m) - \lambda x_m$$

shows that the vector x_j lies in the linear manifold spanned by the vectors of the second system; since all the other vectors of the first system obviously belong to this linear manifold, we have $L_1 \subset L_2$. It follows that $L_1 = L_2$. Therefore, the rank of A does not change as a result of this elementary operation.

4. *Deletion of a column consisting entirely of zeros.* A column consisting entirely of zeros corresponds to the zero vector of the space R. Obviously, eliminating the zero vector from the system $x_1, x_2, \ldots, x_k$ does not change the linear manifold $L(x_1, x_2, \ldots, x_k)$ and therefore does not change the rank of the matrix A.

5. *Deletion of a column which is a linear combination of the other columns.* The legitimacy of this elementary operation was proved in Theorem 20 of Sec. 19.

We emphasize once again that for every elementary operation on the columns of a matrix, we can establish a corresponding elementary operation on the rows.

We now show how to calculate the rank and find a basis minor of a given matrix A with n rows and m columns, by using the elementary operations above. If the matrix A consists only of zeros, then its rank is obviously zero. Assume that A contains a nonzero element. Then, by suitably permuting the rows and columns, we can bring this element over to the upper left-hand corner of the matrix. Then, subtracting from every column the first column multiplied by a suitable coefficient, we can make all the other elements of the first row vanish. We shall make no further changes in the first row and first column (except for the rearrangements described below). If there are no nonzero elements among the remaining elements (i.e., the elements which do not belong to the first row and the first column), then the rank of the matrix A is obviously 1. If there is a nonzero element among the remaining elements, then by suitably rearranging rows and columns, we can bring this element over to the intersection of the second row and the second column and then make all the elements following it in the second row vanish (just as before). (We note that these operations do not affect the first row and the first column.) Continuing in this fashion, and assuming that the number of columns in A does not exceed the number of rows in A (this can always be achieved by transposition), we reduce A to one of the following two forms:

$$
A_1 = \begin{Vmatrix}
\alpha_1 & 0 & 0 & \cdots & 0 & 0 & \cdots & 0 \\
c_{21} & \alpha_2 & 0 & \cdots & 0 & 0 & \cdots & 0 \\
c_{31} & c_{32} & \alpha_3 & \cdots & 0 & 0 & \cdots & 0 \\
\cdot & \cdot & \cdot & \cdots & \cdot & \cdot & \cdots & \cdot \\
\cdot & \cdot & \cdot & \cdots & \cdot & \cdot & \cdots & \cdot \\
\cdot & \cdot & \cdot & \cdots & \cdot & \cdot & \cdots & \cdot \\
c_{k1} & c_{k2} & c_{k3} & \cdots & \alpha_k & 0 & \cdots & 0 \\
c_{k+1,1} & c_{k+1,2} & c_{k+1,3} & \cdots & c_{k+1,k} & 0 & \cdots & 0 \\
\cdot & \cdot & \cdot & \cdots & \cdot & \cdot & \cdots & \cdot \\
\cdot & \cdot & \cdot & \cdots & \cdot & \cdot & \cdots & \cdot \\
\cdot & \cdot & \cdot & \cdots & \cdot & \cdot & \cdots & \cdot \\
c_{n1} & c_{n2} & c_{n3} & \cdots & c_{nk} & 0 & \cdots & 0
\end{Vmatrix}
$$

or

$$
A_2 = \begin{Vmatrix}
\alpha_1 & 0 & 0 & \cdots & 0 \\
c_{21} & \alpha_2 & 0 & \cdots & 0 \\
c_{31} & c_{32} & \alpha_3 & \cdots & 0 \\
\cdot & \cdot & \cdot & \cdots & \cdot \\
\cdot & \cdot & \cdot & \cdots & \cdot \\
\cdot & \cdot & \cdot & \cdots & \cdot \\
c_{m1} & c_{m2} & c_{m3} & \cdots & \alpha_m \\
\cdot & \cdot & \cdot & \cdots & \cdot \\
\cdot & \cdot & \cdot & \cdots & \cdot \\
\cdot & \cdot & \cdot & \cdots & \cdot \\
c_{n1} & c_{n2} & c_{n3} & \cdots & c_{nm}
\end{Vmatrix}.
$$

Here the numbers α_1, α_2, etc. are nonzero. In the first case, the rank of A_1 equals k and its basis minor (in the transformed matrix) stands in the upper left-hand corner. In the second case, the rank of A_2 equals m (the number of columns) and its basis minor (in the transformed matrix) appears in the first m rows. This determines the rank of A. The location of the basis minor of A is easily found by following back in reverse order all the operations performed on A.

As an example, consider the following matrix with five columns and six rows:

$$
A = \begin{Vmatrix}
1 & 2 & 6 & -2 & -1 \\
-2 & -1 & 0 & -5 & -1 \\
3 & 1 & -1 & 8 & 1 \\
-1 & 0 & 2 & -4 & -1 \\
-1 & -2 & -7 & 3 & 2 \\
-2 & -2 & -5 & -1 & 1
\end{Vmatrix}.
$$

There is one zero in the second row of A; by using the general method described above, we can produce three more zeros in this row. However, for convenience, we first interchange the first and second rows. Then, interchanging the first and second columns (so that an element -1 with the smallest nonzero absolute value again appears in the upper left-hand corner), we obtain[2]

$$A \sim \begin{Vmatrix} -2 & -1 & 0 & -5 & -1 \\ 1 & 2 & 6 & -2 & -1 \\ 3 & 1 & -1 & 8 & 1 \\ -1 & 0 & 2 & -4 & -1 \\ -1 & -2 & -7 & 3 & 2 \\ -2 & -2 & -5 & -1 & 1 \end{Vmatrix} \sim \begin{Vmatrix} -1 & -2 & 0 & -5 & -1 \\ 2 & 1 & 6 & -2 & -1 \\ 1 & 3 & -1 & 8 & 1 \\ 0 & -1 & 2 & -4 & -1 \\ -2 & -1 & -7 & 3 & 2 \\ -2 & -2 & -5 & -1 & 1 \end{Vmatrix}$$

To obtain three more zeros in the first row, we multiply the first column by 2, 5, and 1, and subtract the results from the second, fourth, and fifth columns, respectively. This gives

$$A \sim \begin{Vmatrix} -1 & 0 & 0 & 0 & 0 \\ 2 & -3 & 6 & -12 & -3 \\ 1 & 1 & -1 & 3 & 0 \\ 0 & -1 & 2 & -4 & -1 \\ -2 & 3 & -7 & 13 & 4 \\ -2 & 2 & -5 & 9 & 3 \end{Vmatrix}.$$

The simplest thing to do next is to produce additional zeros in the third row. First we interchange this row with the second row. Then we multiply the second column by 1 and -3 and add the results to the third and fourth columns, respectively. Thus we have

$$A \sim \begin{Vmatrix} -1 & 0 & 0 & 0 & 0 \\ 1 & 1 & -1 & 3 & 0 \\ 2 & -3 & 6 & -12 & -3 \\ 0 & -1 & 2 & -4 & -1 \\ -2 & 3 & -7 & 13 & 4 \\ -2 & 2 & -5 & 9 & 3 \end{Vmatrix} \sim \begin{Vmatrix} -1 & 0 & 0 & 0 & 0 \\ 1 & 1 & 0 & 0 & 0 \\ 2 & -3 & 3 & -3 & -3 \\ 0 & -1 & 1 & -1 & -1 \\ -2 & 3 & -4 & 4 & 4 \\ -2 & 2 & -3 & 3 & 3 \end{Vmatrix} = A_1.$$

The fourth and fifth columns of the matrix A_1 are proportional to the third column and can be deleted. The matrix which is left obviously has rank 3, so that the original matrix A also has rank 3. Moreover, A_1 has a basis minor in its first three rows and first three columns. By reversing the successive transformations which led from A to A_1, we can easily verify that none of the transformations which were carried out has any effect on the absolute value of this minor. Therefore, the minor appearing in the first three rows and the first three columns of the original matrix is also a basis minor.

---

[2] Here the symbol $\sim$ written between two matrices means that they have the same rank.

Problem 1. Determine the rank and basis minor of the following matrices:

$$A_1 = \begin{Vmatrix} 1 & -2 & 3 & -1 & -1 & -2 \\ 2 & -1 & 1 & 0 & -2 & -2 \\ -2 & -5 & 8 & -4 & 3 & -1 \\ 6 & 0 & -1 & 2 & -7 & -5 \\ -1 & -1 & 1 & -1 & 2 & 1 \end{Vmatrix}, \quad A_2 = \begin{Vmatrix} 1 & 0 & 1 & 0 & 0 \\ 1 & 1 & 0 & 0 & 0 \\ 0 & 1 & 1 & 0 & 0 \\ 0 & 0 & 1 & 1 & 0 \\ 0 & 1 & 0 & 1 & 1 \end{Vmatrix}.$$

Ans. The rank of A_1 is 3, and there is a basis minor in the upper left-hand corner (for example). The rank of A_2 is 5, and the basis minor is the same as the determinant of the matrix.

Problem 2. Suppose the matrix A has a nonvanishing minor M of order r, while every minor of order $r + 1$ containing all the elements of M vanishes. Prove that A has rank r.

Hint. Move the minor M into the upper left-hand corner and then, by using the procedure described in the text, show that all the columns of A from the $(r + 1)$th column on can be made into zero columns.

4

LINEAR FUNCTIONS
OF A VECTOR ARGUMENT

In courses on elementary mathematical analysis, one studies functions of one or more real variables. Such functions can be regarded as functions of a *vector argument*. For example, a function of three variables can be regarded as a function whose argument is a vector of the space V_3. This suggests studying functions whose arguments are vectors from an *arbitrary* linear space. In making this study, we shall for the time being restrict ourselves to the simplest functions of this kind, namely *linear* functions. We shall first study linear *numerical* functions of a vector argument, i.e., functions whose values are numbers, and then linear *vector* functions of a vector argument, i.e., functions whose values are vectors (in fact, vectors from the same space as the space containing the argument). Linear vector functions, otherwise known as *linear operators*, are of great importance in linear algebra and its applications.

25. Linear Forms

25.1. A numerical (or scalar) function $f(x)$ of the vector argument x, defined on a linear space R, is called a *linear form* if it satisfies the following conditions:

1. $f(x + y) = f(x) + f(y)$ for any $x, y \in R$;

2. $f(\alpha x) = \alpha f(x)$ for any $x \in R$ and any real number α.

By using induction, we can easily deduce from conditions 1 and 2 that if $f(x)$ is a linear form, then

$$f(\alpha_1 x_1 + \alpha_2 x_2 + \cdots + \alpha_k x_k) = \alpha_1 f(x_1) + \alpha_2 f(x_2) + \cdots + \alpha_k f(x_k), \quad (1)$$

for any $x_1, x_2, \ldots, x_k \in R$ and any real numbers $\alpha_1, \alpha_2, \ldots, \alpha_k$.

Example 1. Suppose a basis is chosen in an n-dimensional space R, so that every vector $x \in R$ can be specified by its components $\xi_1, \xi_2, \ldots, \xi_n$. Then $f(x) = \xi_1$ (the first component) is obviously a linear form in x.

Example 2. A more general linear form in the same space is the expression

$$f(x) = \sum_{k=1}^{n} c_k \xi_k,$$

with arbitrary fixed coefficients $c_1, c_2, \ldots, c_n$.

Example 3. An example of a linear form defined on the space $C(a, b)$ (see Sec. 11) is the expression
$$f(x) = x(t_0),$$
where t_0 is a fixed point of the interval $a \leqslant t \leqslant b$.

Example 4. In the same space, we can study the linear form

$$f(x) = \int_a^b c(t)x(t)\, dt,$$

where $c(t)$ is a fixed continuous function.   Linear forms defined on infinite-dimensional spaces are usually called *linear functionals*.

Example 5. In the space V_3, the scalar product (x, x_0) of the vector x with a fixed vector $x_0 \in V_3$ is a linear form in x.

25.2. We now find the general representation of a linear form $f(x)$ defined on an n-dimensional space R.   Let $e_1, e_2, \ldots, e_n$ be an arbitrary basis of the space R, and denote the quantity $f(e_k)$ by c_k $(k = 1, 2, \ldots, n)$.   Then, by (1), for any

$$x = \sum_{k=1}^{n} \xi_k e_k,$$

we have

$$f(x) = f\left(\sum_{k=1}^{n} \xi_k e_k\right) = \sum_{k=1}^{n} \xi_k f(e_k) = \sum_{k=1}^{n} \xi_k c_k,$$

i.e., *the value of the linear form $f(x)$ is a linear combination of the components of the vector x, with the fixed coefficients $c_1, c_2, \ldots, c_n$.*   Thus, in Example 2, we have already encountered the general representation of a linear form in n-dimensional space.

Problem. After defining in the natural way addition of linear forms and multiplication of a linear form by a real number, construct a new linear space R^*, consisting of all the linear forms defined on some linear space R. If the dimension of the space R is n, what is the dimension of the space R^*?

Ans. Also n.

26. Linear Operators

By an operator A defined on a linear space R, we mean a function which associates with every vector $x \in R$ a vector $y = Ax$ of the same space. The operator A is called *linear* if the following conditions are met:

1. $A(x + y) = Ax + Ay$ for any $x, y \in R$;
2. $A(\alpha x) = \alpha Ax$ for any $x \in R$ and any real number α.

Just as for linear forms, from the conditions 1 and 2 we can easily deduce the more general formula

$$A(\alpha_1 x_1 + \alpha_2 x_2 + \cdots + \alpha_k x_k) = \alpha_1 Ax_1 + \alpha_2 Ax_2 + \cdots + \alpha_k Ax_k \quad (2)$$

for any $x_1, x_2, \ldots, x_k \in R$ and any real numbers $\alpha_1, \alpha_2, \ldots, \alpha_k$.

Example 1. The operator which associates the zero vector with every vector of the space R is obviously linear, and is called the *zero operator*.

Example 2. The operator E which carries every vector x into x itself is obviously linear, and is called the *unit operator* or the *identity operator*.

Example 3. The linear operator A which carries every vector x into λx, where λ is a fixed number, is called a *similarity operator*.

Example 4. Write the vectors of the Euclidean plane V_2 in polar coordinates, i.e., write $x = \{\varphi, \rho\}$. The operator carrying the vector $x = \{\varphi, \rho\}$ into $Ax = \{\varphi + \varphi_0, \rho\}$, where φ_0 is a fixed angle, is linear; this is easily verified by a geometric argument. This operator is called the *rotation operator through the angle* φ_0.

Example 5. Let $e_1, e_2, \ldots, e_n$ be a basis in an n-dimensional space R_n. Associate with the vector

$$x = \sum_{k=1}^{n} \xi_k e_k$$

the new vector

$$Px = \sum_{k=1}^{m} \xi_k e_k,$$

where $m < n$. The operator P is a linear operator called the *projection operator* onto the subspace R_m spanned by the vectors $e_1, e_2, \ldots, e_m$.

Example 6. Let $e_1, e_2, \ldots, e_n$ be a basis in an n-dimensional space R_n, and let $\lambda_1, \lambda_2, \ldots, \lambda_n$ be n fixed numbers. Define the operator A for the basis vectors by the relations

$$Ae_1 = \lambda_1 e_1, \; Ae_2 = \lambda_2 e_2, \ldots, Ae_n = \lambda_n e_n,$$

and for any other vector

$$x = \sum_{k=1}^{n} \xi_k e_k$$

by the relation

$$Ax = \sum_{k=1}^{n} \lambda_k \xi_k e_k$$

(which automatically makes A linear). This operator is said to be *diagonal*; the significance of this designation will appear later.

Example 7. In the space $C(a, b)$, multiplication of a given function by any fixed function $\varphi(t)$, in particular by $\varphi(t) = t$, is a linear operator.

Example 8. In the same space, one often studies the *Fredholm integral operator*, which carries a given function $x(t)$ into

$$y(t) = Ax(t) = \int_a^b K(t, s)x(s)\,ds,$$

where $K(t, s)$ is a fixed function of two variables (the "kernel" of the Fredholm operator).

Example 9. In infinite-dimensional spaces, it is often necessary to consider linear operators which are not defined on the whole space. An example of such an operator is the *differentiation operator* D, defined on the space $C(a, b)$ by the formula

$$Dx(t) = x'(t);$$

D is defined only for the differentiable functions of the space $C(a, b)$.

Problem 1. Which of the following vector functions defined on the space V_3 are linear operators?

a) $Ax = x + a$ (a is a fixed nonzero vector);
b) $Ax = a$;
c) $Ax = (a, x)a$;
d) $Ax = (a, x)x$;
e) $Ax = (\xi_1^2, \xi_2 + \xi_3, \xi_3^2)$, where $x = (\xi_1, \xi_2, \xi_3)$;
f) $Ax = (\sin \xi_1, \cos \xi_2, 0)$;
g) $Ax = (2\xi_1 - \xi_3, \xi_2 + \xi_3, \xi_1)$.

Ans. c) and g).

Problem 2. Consider the following operations on the space of all polynomials in t: (a) multiplication by t, (b) multiplication by t^2, (c) differentiation. Are these linear operators?

Ans. Yes.

27. General Form of a Linear Operator on an n-Dimensional Space

Let A be a linear operator, and let $e_1, e_2, \ldots, e_n$ be a basis in the space R_n. The vector e_1 is carried by the operator A into a vector Ae_1 of the space R_n, which, like every vector in this space, can be represented in terms of the basis vectors:

$$Ae_1 = a_1^{(1)}e_1 + a_2^{(1)}e_2 + \cdots + a_n^{(1)}e_n. \tag{3}$$

The operator A has a similar effect on the other basis vectors:

$$
\begin{aligned}
Ae_2 &= a_1^{(2)}e_1 + a_2^{(2)}e_2 + \cdots + a_n^{(2)}e_n, \\
&\ \cdots \quad\quad \cdots \quad\quad \cdots \quad\quad \cdots \\
Ae_n &= a_1^{(n)}e_1 + a_2^{(n)}e_2 + \cdots + a_n^{(n)}e_n.
\end{aligned}
\tag{4}
$$

Equations (3) and (4) can be written more briefly as

$$Ae_j = \sum_{i=1}^{n} a_i^{(j)}e_i \qquad (j = 1, 2, \ldots, n). \tag{5}$$

The coefficients $a_i^{(j)}$ $(i, j = 1, 2, \ldots, n)$ define a matrix

$$
A = A_{(e)} = \begin{Vmatrix}
a_1^{(1)} & a_1^{(2)} & \ldots & a_1^{(n)} \\
a_2^{(1)} & a_2^{(2)} & \ldots & a_2^{(n)} \\
\cdot & \cdot & \ldots & \cdot \\
a_n^{(1)} & a_n^{(2)} & \ldots & a_n^{(n)}
\end{Vmatrix},
$$

called the *matrix of the operator A relative to the basis* $\{e\} = \{e_1, e_2, \ldots, e_n\}$. The components of the vectors

$$f_1 = Ae_1, f_2 = Ae_2, \ldots, f_n = Ae_n$$

serve as the columns of this matrix.

Now let

$$x = \sum_{j=1}^{n} \xi_j e_j$$

be an arbitrary vector, and let

$$y = Ax = \sum_{i=1}^{n} \eta_i e_i.$$

We wish to express the components η_i of the vector y in terms of the components ξ_j of the vector x. We have

$$\sum_{i=1}^{n} \eta_i e_i = Ax = A\left(\sum_{j=1}^{n} \xi_j e_j\right) = \sum_{j=1}^{n} \xi_j Ae_j$$

$$= \sum_{j=1}^{n} \xi_j \sum_{i=1}^{n} a_i^{(j)}e_i = \sum_{i=1}^{n} \left(\sum_{j=1}^{n} a_i^{(j)}\xi_j\right)e_i.$$

Comparing coefficients of the vectors e_i, we find

$$\eta_i = \sum_{j=1}^{n} a_i^{(j)} \xi_j \qquad (i = 1, 2, \ldots, n), \tag{6}$$

or, in expanded form

$$\begin{aligned}
\eta_1 &= a_1^{(1)}\xi_1 + a_1^{(2)}\xi_2 + \cdots + a_1^{(n)}\xi_n, \\
\eta_2 &= a_2^{(1)}\xi_1 + a_2^{(2)}\xi_2 + \cdots + a_2^{(n)}\xi_n, \\
&\; \cdots \quad\quad \cdots \quad\quad \cdots \quad\quad \cdots \\
\eta_n &= a_n^{(1)}\xi_1 + a_n^{(2)}\xi_2 + \cdots + a_n^{(n)}\xi_n.
\end{aligned} \tag{7}$$

Therefore, if we know the matrix of the operator A relative to the basis $e_1, e_2, \ldots, e_n$, we can determine the result of applying A to any vector

$$x = \sum_{k=1}^{n} \xi_k e_k$$

of the space R_n, i.e., *the equations* (7) *give the components of the vector* $y = Ax$ *as linear combinations of the components of* x. *Moreover, the coefficient matrix of the equations* (7) *is just the matrix* $A_{(e)}$.

We now form the matrices corresponding to the operators introduced in Examples 1 to 6 of the preceding section.

1. Obviously, the matrix of the zero operator relative to any basis consists entirely of zeros.

2. The unit operator carries the vector e_1 into

$$1 \cdot e_1 + 0 \cdot e_2 + \cdots + 0 \cdot e_n = e_1,$$

the vector e_2 into

$$0 \cdot e_1 + 1 \cdot e_2 + \cdots + 0 \cdot e_n = e_2,$$

etc., and finally the vector e_n into

$$0 \cdot e_1 + 0 \cdot e_2 + \cdots + 1 \cdot e_n = e_n.$$

Therefore, the matrix of the unit operator has the form

$$\begin{Vmatrix}
1 & 0 & 0 & \ldots & 0 \\
0 & 1 & 0 & \ldots & 0 \\
0 & 0 & 1 & \ldots & 0 \\
\cdot & \cdot & \cdot & \ldots & \cdot \\
0 & 0 & 0 & \ldots & 1
\end{Vmatrix}.$$

3. Analogously, the matrix of the similarity operator has the form

$$\begin{Vmatrix} \lambda & 0 & \dots & 0 \\ 0 & \lambda & \dots & 0 \\ \cdot & \cdot & \dots & \cdot \\ \cdot & \cdot & \dots & \cdot \\ 0 & 0 & \dots & \lambda \end{Vmatrix}.$$

4. In V_2 choose a basis consisting of two orthogonal unit vectors e_1 and e_2. By drawing a picture, we can easily verify that after rotation through the angle φ_0 the vector e_1 goes into the vector $e_1 \cos \varphi_0 + e_2 \sin \varphi_0$ and the vector e_2 goes into $-e_1 \sin \varphi_0 + e_2 \cos \varphi_0$. Therefore, the matrix of the rotation operator has the form

$$\begin{Vmatrix} \cos \varphi_0 & -\sin \varphi_0 \\ \sin \varphi_0 & \cos \varphi_0 \end{Vmatrix}.$$

5. In Example 5, the vectors $e_1, e_2, \dots, e_m$ go into themselves, while the vectors $e_{m+1}, \dots, e_n$ go into the zero vector. Therefore, the matrix of the projection operator in the basis $e_1, e_2, \dots, e_n$ has the form

$$m\text{th row} \quad \begin{Vmatrix} 1 & 0 & \dots & 0 & 0 & \dots & 0 \\ 0 & 1 & \dots & 0 & 0 & \dots & 0 \\ \cdot & \cdot & \dots & \cdot & \cdot & \dots & \cdot \\ 0 & 0 & \dots & 1 & 0 & \dots & 0 \\ 0 & 0 & \dots & 0 & 0 & \dots & 0 \\ \cdot & \cdot & \dots & \cdot & \cdot & \dots & \cdot \\ 0 & 0 & \dots & 0 & 0 & \dots & 0 \end{Vmatrix}.$$

6. The matrix of the diagonal operator of Example 6 has the form

$$\begin{Vmatrix} \lambda_1 & 0 & \dots & 0 \\ 0 & \lambda_2 & \dots & 0 \\ \cdot & \cdot & \dots & \cdot \\ 0 & 0 & \dots & \lambda_n \end{Vmatrix}$$

in the basis $e_1, e_2, \dots, e_n$. In this matrix, there are nonzero elements only on the principal diagonal. Such a matrix is said to be *diagonal*; this explains why the corresponding operator is called a diagonal operator. We note that in another basis, the matrix of a diagonal operator will not be diagonal in general.

Problem. Suppose the operator A defined on V_3 carries the vectors

$$x_1 = (0, 0, 1), \, x_2 = (0, 1, 1), \, x_3 = (1, 1, 1)$$

into

$$y_1 = (2, 3, 5), \, y_2 = (1, 0, 0), \, y_3 = (0, 1, -1).$$

Form the matrix of the operator A in the following bases:

a) $e_1 = (1, 0, 0)$, $e_2 = (0, 1, 0)$, $e_3 = (0, 0, 1)$;

b) x_1, x_2, x_3.

$$Ans. \quad A_{(e)} = \begin{Vmatrix} -1 & -1 & 2 \\ 1 & -3 & 3 \\ -1 & -5 & 5 \end{Vmatrix}, \quad A_{(x)} = \begin{Vmatrix} 2 & 0 & -2 \\ 1 & -1 & 1 \\ 2 & 1 & 0 \end{Vmatrix}.$$

We have seen that (5) associates a matrix of order n with every linear operator A acting on an n-dimensional space. Now let $\|a_i^{(j)}\|$ be an arbitrary matrix of order n, where the superscript is the column index and the subscript is the row index. If

$$x = \sum_{j=1}^{n} \xi_j e_j,$$

we can construct the vector

$$y = \sum_{i=1}^{n} \eta_i e_i$$

by the formulas

$$\begin{aligned}
\eta_1 &= a_1^{(1)}\xi_1 + a_1^{(2)}\xi_2 + \cdots + a_1^{(n)}\xi_n, \\
\eta_2 &= a_2^{(1)}\xi_1 + a_2^{(2)}\xi_2 + \cdots + a_2^{(n)}\xi_n, \\
& \cdots \quad \cdots \quad \cdots \quad \cdots \\
\eta_n &= a_n^{(1)}\xi_1 + a_n^{(2)}\xi_2 + \cdots + a_n^{(n)}\xi_n.
\end{aligned} \tag{8}$$

It is easily verified that the operator A which carries the vector x into the vector y is a linear operator. We now construct the matrix of the operator A in the basis $e_1, e_2, \ldots, e_n$. Since the vector e_1 has components

$$\xi_1 = 1, \xi_2 = \cdots = \xi_n = 0,$$

according to (8), the components of the vector $f_1 = Ae_1$ will be the numbers $a_1^{(1)}, a_2^{(1)}, \ldots, a_n^{(1)}$, i.e.,

$$f_1 = Ae_1 = a_1^{(1)}e_1 + a_2^{(1)}e_2 + \cdots + a_n^{(1)}e_n.$$

Similarly, we have

$$f_j = Ae_j = a_1^{(j)}e_1 + a_2^{(j)}e_2 + \cdots + a_n^{(j)}e_n \quad (j = 1, 2, \ldots, n). \tag{9}$$

Therefore, the matrix of the operator A is the same as the original matrix $\|a_i^{(j)}\|$, and we have shown that *every matrix of order n is the matrix of a linear operator acting in an n-dimensional linear space.* Thus, (5) establishes a one-to-one correspondence between linear operators acting in an n-dimensional space and matrices of order n.

Finally, we note that (9) can be used to construct the operator A directly (and uniquely) from the matrix $A = \|a_i^{(j)}\|$. This formula shows that the

jth column of the matrix A consists of the components of the vector $f_j = Ae_j$. Since the jth column of the matrix A is quite arbitrary, we can draw the following conclusion: *Given any n vectors $f_1, f_2, \ldots, f_n$ in an n-dimensional space R, there exists a unique linear operator which acts in this space and carries the basis vectors $e_1, e_2, \ldots, e_n$ into $f_1, f_2, \ldots, f_n$, respectively.*

28. Sums and Products of Linear Operators

Consider the linear operators defined on a linear space R. In this section, we shall discuss how to add and multiply such operators. Before proceeding, we define equality of operators: *Two operators A and B are said to be equal if $Ax = Bx$ for every $x \in R$.*

1. *Addition of operators.* Given two linear operators A and B, the operator $C = A + B$ is defined by the formula

$$Cx = (A + B)x = Ax + Bx.$$

To verify that C is a linear operator, set $x = \alpha y + \beta z$. Then we have

$$C(\alpha y + \beta z) = A(\alpha y + \beta z) + B(\alpha y + \beta z) = \alpha Ay + \beta Az + \alpha By + \beta Bz$$
$$= \alpha(Ay + By) + \beta(Az + Bz) = \alpha Cy + \beta Cz.$$

Thus, both conditions 1 and 2 of Sec. 26 are clearly satisfied. The following relations are easily verified:

$$\begin{aligned}
A + B &= B + A, \\
(A + B) + C &= A + (B + C), \\
A + 0 &= A, \\
A + (-A) &= 0.
\end{aligned} \tag{10}$$

Here A, B, and C are arbitrary linear operators, 0 is the zero operator and $-A$ is the negative of the operator A, i.e., the operator which carries every vector $x \in R$ into $-Ax$.

2. *Multiplication of an operator by a number.* If A is a linear operator and λ is a real number, then the operator $B = \lambda A$ is defined by the formula

$$Bx = (\lambda A)x = \lambda(Ax).$$

Just as before, it is easily verified that this operator is linear. We have the relations

$$\begin{aligned}
\lambda_1(\lambda_2 A) &= (\lambda_1 \lambda_2)A, \\
1 \cdot A &= A, \\
(\lambda_1 + \lambda_2)A &= \lambda_1 A + \lambda_2 A, \\
\lambda(A + B) &= \lambda A + \lambda B.
\end{aligned} \tag{11}$$

The relations (10) and (11) show that *the set of all operators defined on a linear space R forms a new linear space.*

3. *Multiplication of operators.* If A and B are linear operators, then the operator $C = AB$ is defined by the condition

$$Cx = ABx = A(Bx),$$

i.e., first the operator B acts on the vector x and then the operator A acts on the result. C is a linear operator, since

$$C(\alpha x + \beta y) = AB(\alpha x + \beta y) = A(\alpha Bx + \beta By)$$
$$= \alpha ABx + \beta ABy = \alpha Cx + \beta Cy.$$

It is easy to verify the relations

$$\begin{array}{ll} \lambda(AB) = (\lambda A)B, & \\ A(BC) = (AB)C, & \text{(associative laws)} \end{array} \tag{12}$$
$$\begin{array}{ll} (A + B)C = AC + BC, & \\ A(B + C) = AB + AC. & \text{(distributive laws)} \end{array}$$

As an example, we verify the second associative law. In keeping with our definition of equality of operators, we have to prove the formula

$$[A(BC)]x = [(AB)C]x$$

for any $x \in R$. But by the very definition of the operator product, we have

$$[A(BC)]x = A[(BC)x] = A[B(Cx)],$$
$$[(AB)C]x = (AB)(Cx) = A[B(Cx)],$$

from which the required formula follows. All the other formulas are proved similarly.

The associative law allows us to define *powers of an operator A* as follows:

$$A^1 = A,$$
$$A^2 = AA,$$
$$A^3 = A^2A = (AA)A = A(AA) = AA^2,$$
$$\cdots\cdots$$
$$A^n = A^{n-1}A = AA^{n-1}.$$

Then we have the formula

$$A^{m+n} = A^m A^n \qquad (m, n = 1, 2, \ldots), \tag{13}$$

which can easily be proved by induction. We define

$$A^0 = E,$$

where E is the identity operator, and show that (13) remains valid in the case where one of the indices is zero. In fact, if B is any operator, we have

$$(BE)x = B(Ex) = Bx = E(Bx),$$

so that

$$BE = EB = B.$$

Setting $B = A^n$, we obtain

$$A^n E = EA^n = A^n,$$

as required. We emphasize that in general the operator product is not commutative, i.e., $AB \neq BA$; however, this does not exclude the possibility that *some* pairs of operators do commute, e.g., $AE = EA = A$ for any A.

Example. Noncommuting operators can be constructed even in the space V_2, i.e., even in the plane. Let the operator denote rotation through the angle $90°$ (see Example 4 of Secs. 27, 28) and let P denote the operator corresponding to projection on the e_1 axis (see Example 5). Consider the result of applying the operators AP and PA to the vector e_1. Then we have

$$APe_1 = A(Pe_1) = Ae_1 = e_2,$$
$$PAe_1 = P(Ae_1) = Pe_2 = 0,$$

so that $AP \neq PA$.

Problem 1. In three-dimensional space, let A denote the operator corresponding to rotation through $90°$ about the axis OX (taking OY into OZ), let B denote the operator corresponding to rotation through $90°$ about the axis OY (taking OZ into OX) and let C denote the operator corresponding to rotation through $90°$ about OZ (taking OX into OY). Show that

$$A^4 = B^4 = C^4 = E, \qquad AB \neq BA, \qquad A^2B^2 = B^2A^2.$$

Is the relation $ABAB = A^2B^2$ valid?

Problem 2. In the space of all polynomials in t, let A denote the differentiation operator and let B denote the operator corresponding to multiplication by the independent variable t, i.e.

$$AP(t) = P'(t), \qquad BP(t) = tP(t).$$

Is the relation $AB = BA$ valid? Find the operator $AB - BA$.

                   Ans. $AB - BA = E.$

Problem 3. Assuming that $AB = BA$, prove the formulas

$$(A + B)^2 = A^2 + 2AB + B^2,$$
$$(A + B)^3 = A^3 + 3A^2B + 3AB^2 + B^3.$$

How must these formulas be changed if $AB \neq BA$?

   Ans. $(A + B)^2 = A^2 + AB + BA + B^2,$
               $(A + B)^3 = A^3 + A^2B + ABA + AB^2 + BA^2 + BAB + B^2A + B^3.$

Problem 4. Assuming that $AB - BA = E$, prove the formula

$$A^m B - B A^m = m A^{m-1} \qquad (m = 1, 2, \ldots).$$

29. Corresponding Operations on Matrices

29.1. As we have seen, there is a one-to-one correspondence between linear operators on an n-dimensional space R and matrices of order n. We now study the matrix analogs of algebraic operations involving linear operators. First, we examine the definition of operator equality introduced in the beginning of Sec. 28. Suppose that the two operators A and B are equal, i.e., $Ax = Bx$ for all $x \in R$. Let $\|a_i^{(j)}\|$ be the matrix of the operator A relative to the basis

$$\{e\} = \{e_1, e_2, \ldots, e_n\}$$

and let $\|b_i^{(j)}\|$ be the matrix of the operator B relative to the same basis. Setting $x = e_j$ in the equality $Ax = Bx$, we obtain

$$\sum_{i=1}^{n} a_i^{(j)} e_i = A e_j = B e_j = \sum_{i=1}^{n} b_i^{(j)} e_i.$$

This implies that $a_i^{(j)} = b_i^{(j)}$ for all indices i and j, i.e.,

$$\|a_i^{(j)}\| = \|b_i^{(j)}\|.$$

Thus, *equal operators have equal matrices.*

29.2. We now consider *operator addition.* Suppose, as before, that we have two operators A and B on an n-dimensional space R, with matrices $\|a_i^{(j)}\|$ and $\|b_i^{(j)}\|$, respectively, relative to the basis

$$\{e\} = \{e_1, e_2, \ldots, e_n\}.$$

Then we have

$$A e_j = \sum_{i=1}^{n} a_i^{(j)} e_i, \quad B e_j = \sum_{i=1}^{n} b_i^{(j)} e_i \qquad (j = 1, 2, \ldots, n),$$

so that

$$(A + B) e_j = A e_j + B e_j = \sum_{i=1}^{n} (a_i^{(j)} + b_i^{(j)}) e_i,$$

whence it follows that the matrix $\|a_i^{(j)} + b_i^{(j)}\|$ corresponds to the operator $A + B$. This matrix is called the *sum* of the matrices $\|a_i^{(j)}\|$ and $\|b_i^{(j)}\|$.

Next we consider multiplication of an operator A by a number λ. If λ is a real number, then

$$(\lambda A) e_j = \lambda (A e_j) = \sum_{i=1}^{n} \lambda a_i^{(j)} e_i.$$

Therefore, the operator λA corresponds to the matrix $\|\lambda a_i^{(j)}\|$ obtained by multiplying all the elements of the matrix $\|a_i^{(j)}\|$ by the number λ. This matrix is called the *product of the matrix* $\|a_i^{(j)}\|$ *with the number* λ.

Since there is a one-to-one correspondence between matrices of order n and linear operators on an n-dimensional space (Sec. 27), there is a one-to-one correspondence between algebraic relations involving operators and those involving matrices. Since operators obey the rules (10) and (11) of the preceding section, we conclude that matrices obey the same rules; of course, this can easily be verified directly. Thus, we see that *the set of all matrices of order n is a linear space*, which, by its very construction, is isomorphic to the linear space of all linear operators defined on an n-dimensional space R.

Problem 1. Find the dimension of the linear space $\bar{R}$ of all linear operators on an n-dimensional space R, and construct a basis for $\bar{R}$.

Ans. The dimension of the space $\bar{R}$ is n^2. For basis operators we can take the operators corresponding to the matrices $A_{11}, A_{12}, \ldots, A_{nn}$, where A_{ij} has a single nonzero element at the intersection of the ith row and the jth column.

Problem 2. Show that for any operator A on an n-dimensional space R, we can find a polynomial

$$P(x) = a_0 + a_1 x + \cdots + a_m x^m$$

of degree not higher than n^2, for which

$$P(A) = a_0 E + a_1 A + \cdots + a_m A^m$$

is the zero operator.

Hint. Use the result of Problem 1.

Comment. Every polynomial $P(x)$ which has the property $P(A) = 0$ is called an *annihilating polynomial of the operator A*. One of the annihilating polynomials of degree n will be pointed out later (Problem 2 of Sec. 38).

Problem 3. Show that the relation $AB - BA = E$ is impossible for operators A and B in an n-dimensional space R.

Hint. Use the result of Problem 2, where the polynomial $P(x)$ has the smallest degree possible (for the operator A); then use the result of Problem 4 of Sec. 28.

Comment. The result of Problem 2 of Sec. 28 shows that here the assumption that the space R is finite-dimensional is essential.

29.3. Next we consider *operator multiplication*. With our previous notation we have

$$(AB)e_j = A(Be_j) = A \sum_{i=1}^{n} b_i^{(j)} e_i = \sum_{i=1}^{n} b_i^{(j)} A e_i$$

$$= \sum_{i=1}^{n} b_i^{(j)} \sum_{k=1}^{n} a_k^{(i)} e_k = \sum_{k=1}^{n} \left(\sum_{i=1}^{n} a_k^{(i)} b_i^{(j)} \right) e_k.$$

Therefore, the elements $c_k^{(j)}$ of the matrix C corresponding to the operator $C = AB$ have the form

$$c_k^{(j)} = \sum_{i=1}^{n} a_k^{(i)} b_i^{(j)} \qquad (j, k = 1, 2, \ldots, n), \tag{14}$$

which is the desired result. In other words, *the element of the matrix C belonging to the k'th row and j'th column is equal to the sum of the products of the elements of the k'th row of the matrix A with the corresponding elements of the j'th column of the matrix B.*

The matrix $\|c_k^{(j)}\|$ which is obtained from the matrices $\|a_k^{(j)}\|$ and $\|b_k^{(j)}\|$ by (14) is called the *product of* $\|a_k^{(j)}\|$ *and* $\|b_k^{(j)}\|$ (in that order). Since multiplication of linear operators is noncommutative, matrix multiplication is also noncommutative. Therefore, it is important to specify the order of the factors in a matrix product.

As we have seen, operator multiplication obeys the associative and distributive laws. Since there is a one-to-one correspondence between matrices and operators, and since matrix multiplication corresponds to operator multiplication, we can conclude that matrix multiplication also obeys the associative and distributive laws. Thus, in particular, we can define the operation of raising a matrix to a power, just as in Sec. 28. Obviously, (13) will also be valid for powers of a matrix.

Problem 1. Find the product of the matrices A and B, where

$$A = \begin{Vmatrix} 1 & 2 & 3 \\ 2 & 4 & 6 \\ 3 & 6 & 9 \end{Vmatrix}, \qquad B = \begin{Vmatrix} -1 & -2 & -4 \\ -1 & -2 & -4 \\ 1 & 2 & 4 \end{Vmatrix}.$$

Ans. $AB = \begin{Vmatrix} 0 & 0 & 0 \\ 0 & 0 & 0 \\ 0 & 0 & 0 \end{Vmatrix}.$

Problem 2. Raise the following matrices to the nth power:

$$A = \begin{Vmatrix} 1 & 1 \\ 0 & 1 \end{Vmatrix}, \qquad B = \begin{Vmatrix} \cos \varphi & -\sin \varphi \\ \sin \varphi & \cos \varphi \end{Vmatrix}.$$

Ans. $A^n = \begin{Vmatrix} 1 & n \\ 0 & 1 \end{Vmatrix}, \qquad B^n = \begin{Vmatrix} \cos n\varphi & -\sin n\varphi \\ \sin n\varphi & \cos n\varphi \end{Vmatrix}.$

Problem 3. Relative to some basis, the operators A and B have the matrices

$$A = \begin{Vmatrix} \sigma_1 & 0 & \ldots & 0 \\ 0 & \sigma_2 & \ldots & 0 \\ \cdot & \cdot & \ldots & \cdot \\ 0 & 0 & \ldots & \sigma_n \end{Vmatrix}, \qquad B = \begin{Vmatrix} \sigma & 1 & 0 & \ldots & 0 \\ 0 & \sigma & 1 & \ldots & 0 \\ \cdot & \cdot & \cdot & \ldots & \cdot \\ 0 & 0 & 0 & \ldots & \sigma \end{Vmatrix}.$$

Find the matrices of the operators A^k and B^k relative to this basis.

Ans.[1]
$$A^k = \begin{Vmatrix} \sigma_1^k & 0 & \dots & 0 \\ 0 & \sigma_2^k & \dots & 0 \\ . & . & . & . \\ 0 & 0 & \dots & \sigma_n^k \end{Vmatrix},$$

$$B^k = \begin{Vmatrix} \sigma^k & \binom{k}{1}\sigma^{k-1} & \binom{k}{2}\sigma^{k-2} & \dots & \binom{k}{n-1}\sigma^{k-(n-1)} \\ 0 & \sigma^k & \binom{k}{1}\sigma^{k-1} & \dots & \binom{k}{n-2}\sigma^{k-(n-2)} \\ . & . & . & \dots & . \\ 0 & 0 & 0 & \dots & \sigma^k \end{Vmatrix}.$$

Problem 4. Find all second-order matrices A which satisfy the condition

$$A^2 = \begin{Vmatrix} 0 & 0 \\ 0 & 0 \end{Vmatrix}.$$

Ans. $A = \begin{Vmatrix} a & b \\ c & -a \end{Vmatrix}$, where $bc = -a^2$.

Problem 5. Verify the associative and distributive laws for matrix multiplication directly from (14).

Problem 6. Calculate $AB - BA$ for the following matrices:
(a)

$$A = \begin{Vmatrix} 1 & 2 & 2 \\ 2 & 1 & 2 \\ 1 & 2 & 3 \end{Vmatrix}, \qquad B = \begin{Vmatrix} 4 & 1 & 1 \\ -4 & 2 & 0 \\ 1 & 2 & 1 \end{Vmatrix};$$

(b)

$$A = \begin{Vmatrix} 2 & 1 & 0 \\ 1 & 1 & 2 \\ -1 & 2 & 1 \end{Vmatrix}, \qquad B = \begin{Vmatrix} 3 & 1 & -2 \\ 3 & -2 & 4 \\ -3 & 5 & -1 \end{Vmatrix}.$$

Ans.
(a) $\begin{Vmatrix} -9 & -2 & -10 \\ 6 & 14 & 8 \\ -7 & 5 & -5 \end{Vmatrix}$, (b) $\begin{Vmatrix} 0 & 0 & 0 \\ 0 & 0 & 0 \\ 0 & 0 & 0 \end{Vmatrix}.$

We conclude this section by giving two more results concerning matrix multiplication; however, these results have no direct geometric meaning.

29.4. Multiplication of transposed matrices. Let A and B be two matrices of order n, and let $C = AB$. Let A', B', C' denote the transposes of the matrices A, B, C, respectively. We wish to show that

$$(AB)' = B'A'. \tag{15}$$

To prove this, denote the elements of the matrices A, B, C, A', B', C' by

---

[1] $\binom{n}{k} = \dfrac{n!}{k!(n-k)!}$, where $0! = 1$, and $\binom{n}{k} = 0$ if $k > n$.

a_{ij}, b_{ij}, c_{ij}, $a'_{ij} = a_{ji}$, $b'_{ij} = b_{ji}$, $c'_{ij} = c_{ji}$, respectively. Then (14), which gives the elements c_{ik}, can be rewritten as

$$c_{ik} = c'_{ki} = \sum_{j=1}^{n} a_{ij}b_{jk} = \sum_{j=1}^{n} a'_{ji}b'_{kj} = \sum_{j=1}^{n} b'_{kj}a'_{ji}.$$

The sum is over the index j, with the indices i and k fixed. Thus, to form the element c'_{ki} (appearing in the kth row and ith column of C'), the elements of the kth row of B' are multiplied by the elements of the ith column of A', and then added. By the definition of matrix multiplication, this means that C' is the product of B' and A', which proves (15).

29.5. The determinant of the product of two matrices. Let $A = \|a_{ij}\|$ and $B = \|b_{ij}\|$ be any two matrices, and let $C = \|c_{ij}\|$ be the product of the first of these matrices with the second, i.e., $C = AB$. We calculate the determinant of the matrix C, obtaining

$$\det C = \begin{Vmatrix} a_{11}b_{11} + \cdots + a_{1n}b_{n1} & a_{11}b_{12} + \cdots + a_{1n}b_{n2} & \cdots & a_{11}b_{1n} + \cdots + a_{1n}b_{nn} \\ a_{21}b_{11} + \cdots + a_{2n}b_{n1} & a_{21}b_{12} + \cdots + a_{2n}b_{n2} & \cdots & a_{21}b_{1n} + \cdots + a_{2n}b_{nn} \\ \cdots & \cdots & \cdots & \cdots \\ \cdots & \cdots & \cdots & \cdots \\ a_{n1}b_{11} + \cdots + a_{nn}b_{n1} & a_{n1}b_{12} + \cdots + a_{nn}b_{n2} & \cdots & a_{n1}b_{1n} + \cdots + a_{nn}b_{nn} \end{Vmatrix}.$$

To calculate this determinant, we use the linear property of determinants (Sec. 3.3). Each column of $\det C$ is a sum of n "elementary columns" with elements of the form $a_{ij}b_{jk}$, where j and k are fixed and i ranges from 1 to n as we go down the column. Therefore, the whole determinant is the sum of n^n "elementary determinants," consisting only of "elementary columns." Since, in each elementary column, the factor b_{jk} does not change as we go down the column, it can be factored out of the elementary determinant. After doing this, each of the elementary determinants takes the form

$$b_{i_1 1}b_{i_2 2} \cdots b_{i_n n} \begin{vmatrix} a_{1i_1} & a_{1i_2} & \cdots & a_{1i_n} \\ a_{2i_1} & a_{2i_2} & \cdots & a_{2i_n} \\ \cdot & \cdot & \cdots & \cdot \\ a_{ni_1} & a_{ni_2} & \cdots & a_{ni_n} \end{vmatrix}, \tag{16}$$

where the $i_1, i_2, \ldots, i_n$ are numbers from 1 to n. If some of these numbers are the same, then clearly the corresponding elementary determinant vanishes. Thus, we need only consider those elementary determinants for which the indices $i_1, i_2, \ldots, i_n$ are all different. In this case, the determinant

$$A(i_1, i_2, \ldots, i_n) = \begin{vmatrix} a_{1i_1} & a_{1i_2} & \cdots & a_{1i_n} \\ a_{2i_1} & a_{2i_2} & \cdots & a_{2i_n} \\ \cdot & \cdot & \cdots & \cdot \\ a_{ni_1} & a_{ni_2} & \cdots & a_{ni_n} \end{vmatrix}$$

is the same as the determinant of the matrix A, except possibly for sign. Next we find the sign which must be assigned to the determinant $A(i_1, i_2, \ldots, i_n)$. Successively interchanging adjacent columns of the determinant $A(i_1, i_2, \ldots, i_n)$, we finally arrive at the normal arrangement of the columns, i.e., the arrangement they have in the matrix A itself. At each interchange of two adjacent columns, the determinant $A(i_1, i_2, \ldots, i_n)$ changes sign, and the number of inversions in the permutation $i_1, i_2, \ldots, i_n$ changes by unity. Since in the final arrangement of the columns, the second indices are in natural order, i.e., without inversions, *the number of successive changes of sign is equal to the number of inversions in the permutation* $i_1, i_2, \ldots, i_n$. (It is assumed that the change in the indices $i_1, i_2, \ldots, i_n$ produced by every column interchange causes a smaller index to appear before a larger index, with the result that the total number of inversions changes by just unity.) We denote the number of sign changes by N. Then the expression (16) takes the form

$$(-1)^N b_{i_1 1} b_{i_2 2} \ldots b_{i_n n} \det A. \tag{17}$$

To obtain $\det C$, we add up all the expressions of the form (17) and divide through by the common factor $\det A$. The result is $\det A$ multiplied by a sum of terms of the form

$$(-1)^N b_{i_1 1} b_{i_2 2} \ldots b_{i_n n}.$$

Each of these terms is clearly a term in the determinant of the matrix B, with the appropriate sign (Sec. 2), so that the sum of all these terms is equal to just $\det B$. As a result, we obtain the following important formula:

$$\det AB = \det A \det B, \tag{18}$$

i.e., *the determinant of the product of two matrices equals the product of the determinants of the separate matrices.*

Problem 1. Let

$$x_j = \sum_{i=1}^{n} \xi_i^{(j)} e_i \qquad (j = 1, 2, \ldots, m)$$

be m linearly independent vectors in an n-dimensional space, and let the operator A carry x_j into

$$y_j = Ax_j = \sum_{k=1}^{m} a_k^{(j)} x_k \qquad (j = 1, 2, \ldots, m).$$

Show that every minor of order k in the matrix formed from the components of y_j (with respect to the basis $e_1, e_2, \ldots, e_n$) equals the product of $\det \|a_k^{(j)}\|$ with the corresponding minor of the matrix formed from the components of the vectors x_j.

Problem 2. Show that if the basis minor of a matrix of rank r appears in the upper left-hand corner, then the ratio of any minor M of order r to the minor appearing in the same columns but in the first r rows, depends only on the column indices of the minor M.

Hint. Express the elements of the matrix M in terms of the elements appearing in the first r rows, and then use the theorem of Sec. 29.5.

Problem 3. Show that if A is a matrix of rank r, then any second-order determinant of the form

$$\begin{vmatrix} M_{i_1, i_2, \ldots, i_r}^{i_1, i_2, \ldots, i_r} & M_{k_1, k_2, \ldots, k_r}^{i_1, i_2, \ldots, i_r} \\ M_{i_1, i_2, \ldots, i_r}^{k_1, k_2, \ldots, k_r} & M_{k_1, k_2, \ldots, k_r}^{k_1, k_2, \ldots, k_r} \end{vmatrix},$$

consisting of minors of order r of the matrix A, vanishes.

Problem 4. Show that every minor of order k of the matrix $C = AB$ equals the sum of the products of certain minors of A with certain minors of B, and write the corresponding formula.

Hint. Apply the method used to prove (18).

Ans. $M_{j_1, j_2, \ldots, j_k}^{i_1, i_2, \ldots, i_k}(AB) = \displaystyle\sum_{\alpha_1, \alpha_2, \ldots, \alpha_k} M_{\alpha_1, \alpha_2, \ldots, \alpha_k}^{i_1, i_2, \ldots, i_k}(A) \cdot M_{j_1, j_2, \ldots, j_k}^{\alpha_1, \alpha_2, \ldots, \alpha_k}(B).$

Problem 5. Show that every minor of order k of the matrix ABC equals a sum of products of certain minors of order k of the matrices A, B, and C.

30. Inverse Operators and Inverse Matrices

An operator B in a linear space R is called the *inverse* (more precisely, the *left inverse*) of the operator A in the same space, if the relation

$$BA = E \tag{19}$$

holds, where E is the identity operator. It is far from being true that every linear operator has an inverse. For example, if the operator A carries a nonzero vector $x \in R$ into zero, then for *any* linear operator B we have

$$(BA)x = B(Ax) = B \cdot 0 = 0,$$

and therefore, the relation $BA = E$ is impossible for any linear operator B. Thus, the operator A has no inverse. If R is a finite-dimensional space, and A, B are the matrices of the operators A, B relative to a fixed basis, then according to Sec. 29.3, (19) is equivalent to the equation

$$BA = E, \tag{20}$$

where now A, B, and E are matrices. It is natural to call the matrix B satisfying (20) the *inverse* of the matrix A. According to the multiplication theorem for determinants (Sec. 29.5), (20) implies

$$\det B \det A = \det E = 1. \tag{21}$$

This shows that *a necessary condition for the inverse matrix to exist is that the original matrix have a nonvanishing determinant.* Obviously, if $\det A = 0$, (21) cannot be satisfied for any matrix B. Moreover, it follows from (21) that *the inverse matrix, if it exists, always has a nonvanishing determinant.*

Henceforth, we shall call a matrix with a nonvanishing determinant a *nonsingular (nondegenerate)* matrix and a matrix with a vanishing determinant a *singular (degenerate)* matrix. We now show that *every nonsingular matrix has a unique inverse.* To show this, it is convenient to go over to geometric language. The operator A, corresponding to the matrix A, carries the basis vectors $e_1, e_2, \ldots, e_n$ into certain vectors $f_1, f_2, \ldots, f_n$, whose components are just the columns of the matrix A (Sec. 27). Since $\det A \neq 0$, its columns are linearly independent, and therefore the vectors $f_1, f_2, \ldots, f_n$ form a new basis for the n-dimensional space R. Now assume that the desired inverse operator B exists. Then the conditions

$$BAe_k = e_k \qquad (k = 1, 2, \ldots, n) \tag{22}$$

or

$$Bf_k = e_k \qquad (k = 1, 2, \ldots, n) \tag{23}$$

have to be satisfied. Thus, we see that the desired operator B must carry the vectors f_k into e_k. But the vectors f_k form a basis in the space R. Therefore, (23) uniquely specifies the operator B. Since (22) and (23) are equivalent, the operator BA has the unit matrix relative to the basis $\{e\}$, and is therefore the unit operator. Thus, the operator constructed by using the conditions (23) is actually the inverse of the operator A and is uniquely determined by the construction. We note that here *the operator A is the inverse of the operator B*; in fact, for the basis $\{f\}$ we have

$$ABf_k = Ae_k = f_k, \tag{24}$$

which shows that the operator AB has the unit matrix relative to the basis f_k and is therefore the unit operator.

We now construct the matrix of the operator B relative to the basis $\{e\}$. Denote the unknown elements of this matrix by $b_i^{(j)}$ $(i, j = 1, 2, \ldots, n)$, so that

$$Be_j = \sum_{i=1}^{n} b_i^{(j)} e_i. \tag{25}$$

Applying the operator A to the equations (25) and using the relation $AB = E$, we find

$$e_j = \sum_{i=1}^{n} b_i^{(j)} Ae_i = \sum_{i=1}^{n} b_i^{(j)} f_i. \tag{26}$$

Thus, the coefficients $b_i^{(j)}$ are the coefficients of the expansions of the vectors e_j with respect to the system $\{f_i\}$. But then these coefficients can be obtained by solving the following system of equations with known coefficients:

$$f_j = \sum_{i=1}^{n} a_i^{(j)} e_i.$$

Writing $D = \det \|a_i^{(j)}\|$ and solving this system by Cramer's rule (Sec. 7), we obtain

$$e_j = \frac{1}{D} \begin{vmatrix} a_1^{(1)} & \cdots & a_{j-1}^{(1)} & f_1 & a_{j+1}^{(1)} & \cdots & a_n^{(1)} \\ a_1^{(2)} & \cdots & a_{j-1}^{(2)} & f_2 & a_{j+1}^{(2)} & \cdots & a_n^{(2)} \\ \cdot & \cdots & \cdot & \cdot & \cdot & \cdots & \cdot \\ \cdot & \cdots & \cdot & \cdot & \cdot & \cdots & \cdot \\ \cdot & \cdots & \cdot & \cdot & \cdot & \cdots & \cdot \\ a_1^{(n)} & \cdots & a_{j-1}^{(n)} & f_n & a_{j+1}^{(n)} & \cdots & a_n^{(n)} \end{vmatrix}$$

$$= \frac{1}{D} (A_{1j} f_1 + A_{2j} f_2 + \cdots + A_{nj} f_j),$$

where A_{ij} denotes the cofactor of the element $a_j^{(i)}$ in the matrix A. Comparing this result with (26), we find

$$b_i^{(j)} = \frac{A_{ij}}{D},$$

i.e., *the element $b_i^{(j)}$ of the inverse matrix equals the ratio of the cofactor of the element $a_j^{(i)}$ of the original matrix A to the determinant of A.*

The inverse of the matrix A is denoted by A^{-1}. Moreover, $(A^{-1})^k$ is written as A^{-k}. It is easy to prove by induction that (13) can also be extended to the case of negative powers. Similar notation is used to denote the inverse operator and its powers. The extension of (13) to negative powers of operators in a finite-dimensional space is an immediate consequence of the validity of such an extension in the case of matrices. In the general (infinite-dimensional) case, this extension can be made only if it is assumed that the operator A commutes with its inverse.

Problem 1. Find the inverses of the following matrices:

$$A = \begin{Vmatrix} 1 & 2 \\ 2 & 5 \end{Vmatrix}, \qquad B = \begin{Vmatrix} 1 & 2 & -3 \\ 0 & 1 & 2 \\ 0 & 0 & 1 \end{Vmatrix}, \qquad C = \begin{Vmatrix} \frac{1}{2} & \frac{1}{2} & \frac{1}{2} & \frac{1}{2} \\ \frac{1}{2} & \frac{1}{2} & -\frac{1}{2} & -\frac{1}{2} \\ \frac{1}{2} & -\frac{1}{2} & \frac{1}{2} & -\frac{1}{2} \\ \frac{1}{2} & -\frac{1}{2} & -\frac{1}{2} & \frac{1}{2} \end{Vmatrix}.$$

Ans. $A^{-1} = \begin{Vmatrix} 5 & -2 \\ -2 & 1 \end{Vmatrix}, \qquad B^{-1} = \begin{Vmatrix} 1 & -2 & 7 \\ 0 & 1 & -2 \\ 0 & 0 & 1 \end{Vmatrix}, \qquad C^{-1} = C.$

Problem 2. Prove the relation

$$(A')^{-1} = (A^{-1})'$$

for any nonsingular matrix A.

Problem 3. Investigate the equation $XA = 0$, where A is a given second-order matrix, X is an unknown second-order matrix and 0 is the zero matrix (the matrix all of whose elements vanish).

Ans. If A is the zero matrix, then X is arbitrary. If $\det A \neq 0$, then X is the zero matrix. If $\det A = 0$ and A is not the zero matrix, then its rows are proportional. Let α/β be the ratio of the corresponding elements of the first and second rows of the matrix A. Then we have

$$X = \left\|\begin{array}{cc} -\beta p & \alpha p \\ -\beta q & \alpha q \end{array}\right\|$$

for any p and q.

Problem 4. Let $A = \|a_i^{(j)}\|$ be any square matrix of order n and let $A_i^{(j)}$ be the cofactor of the element $a_i^{(j)}$ in the determinant of A. The matrix $\tilde{A} = \|A_i^{(j)}\|$ is called the *adjugate* of the matrix A. Show that

$$\tilde{A}A = A\tilde{A} = (\det A) E.$$

Finally, we emphasize another important fact. We have already seen that in an n-dimensional space, an operator A which has an inverse has a nonsingular matrix relative to *any* basis. On the other hand, we have shown that if the matrix of an operator A is nonsingular in *any* basis, then the operator A has an inverse, and therefore has a nonsingular matrix relative to *any other* basis. Thus, whether or not a matrix is singular does not depend on the choice of a basis, but is determined solely by the operator itself. An operator which has an inverse has a nonsingular matrix in any basis; an operator which does not have an inverse has a singular matrix in any basis.

In infinite-dimensional spaces, the problem of the existence of inverse operators and their properties cannot be solved so easily. In particular, we generally cannot infer from the condition $BA = E$ that $AB = E$ as well. Moreover, the operator B which is the inverse of an operator A may itself have no inverse.

Problem 1. Let R be the linear space of *all* polynomials in the variable t. Consider the operators A and B defined by the relations

$$A[a_0 + a_1 t + \cdots + a_n t^n] = a_1 + a_2 t + \cdots + a_n t^{n-1},$$
$$B[a_0 + a_1 t + \cdots + a_n t^n] = a_0 t + a_1 t^2 + \cdots + a_n t^{n+1}.$$

Show that A and B are linear operators and that

$$AB = E,$$
$$BA \neq E.$$

Does the operator A have an inverse?

Ans. No.

Problem 2. Prove that if the operator A has an inverse B and is itself the inverse of an operator C, then $B = C$, and B is the unique inverse of A.

31. Simple Geometric Characteristics of Linear Operators

31.1. Given a linear operator A in the space R, the set of vectors $y = Ax$ for all $x \in R$ is called the *range of the operator* A and is denoted by $T(A)$. We now show that $T(A)$ is a *subspace* of the space R. In fact, if $y_1 = Ax_1$, $y_2 = Ax_2$, then

$$y_1 + y_2 = Ax_1 + Ax_2 = A(x_1 + x_2).$$

Moreover, for any real λ, we obviously have

$$\lambda y_1 = \lambda Ax_1 = A(\lambda x_1).$$

Thus, if y_1 and y_2 are in the range of the operator A, so are the vectors $y_1 + y_2$ and λy_1, as asserted. If an operator A in an n-dimensional space R is given by specifying its matrix relative to a basis

$$\{e\} = \{e_1, e_2, \ldots, e_n\},$$

then the dimension of its range can easily be calculated as follows: Setting

$$x = \sum_{k=1}^{n} \xi_k e_k,$$

we obtain

$$y = Ax = \sum_{k=1}^{n} \xi_k Ae_k.$$

Therefore, the range of the operator A is just the linear manifold spanned by the vectors $Ae_1, Ae_2, \ldots, Ae_n$. According to Sec. 16.3, the dimension of the linear manifold $L(Ae_1, Ae_2, \ldots, Ae_n)$ equals the maximum number of linearly independent vectors in the system $Ae_1, Ae_2, \ldots, Ae_n$. We know that the matrix of the operator A has for its columns the components of the vectors Ae_i with respect to the basis $\{e\}$. Thus, the problem of finding the maximum number of linearly independent vectors in the system $Ae_1, Ae_2, \ldots, Ae_n$ reduces at once to the problem of finding the maximum number of linearly independent columns in the matrix of the operator A. But by Theorem 18 (Sec. 19), this latter number is just the rank of the operator A. Thus, *the dimension of the range of a linear operator in an n-dimensional space R equals the rank of the matrix of the operator A relative to some basis of R.*

We note that the choice of this basis is immaterial. Consequently, *the rank of an operator A does not depend on the choice of a basis but only on the operator A itself.* Hereafter, we shall call the rank of the matrix of the operator A (in any basis) simply the *rank of the operator A.*

31.2. We turn next to a second geometric characteristic of the operator A. Consider the set L of all the vectors $x \in R$ which the operator A carries into zero. This set is called the *null space* (or *null manifold*) *of the operator A* and is denoted by $N(A)$. The null space is a subspace since it is obvious that $Ax_1 = 0$, $Ax_2 = 0$ imply that $A(x_1 + x_2) = Ax_1 + Ax_2 = 0$ and that $A(\lambda x_1) = \lambda A x_1 = 0$. We now find the dimension of the null space of an operator A defined on an n-dimensional space R, by using the matrix of A relative to some basis

$$\{e\} = \{e_1, e_2, \ldots, e_n\}.$$

Let

$$x = \sum_{i=1}^{n} \xi_i\, e_i \in N(A).$$

Then the system (7) of Sec. 27 takes the form

$$
\begin{aligned}
a_1^{(1)}\xi_1 + a_1^{(2)}\xi_2 + \cdots + a_1^{(n)}\xi_n &= 0, \\
a_2^{(1)}\xi_1 + a_2^{(2)}\xi_2 + \cdots + a_2^{(n)}\xi_n &= 0, \\
\cdots \qquad \cdots \qquad \cdots \qquad \cdots \qquad & \\
a_n^{(1)}\xi_1 + a_n^{(2)}\xi_2 + \cdots + a_n^{(n)}\xi_n &= 0.
\end{aligned}
\tag{27}
$$

Conversely, it is clear that every vector $x \in R$ whose components satisfy the system (27) belongs to the null space of the operator A. Thus, finding the dimension of the null space of the operator A is equivalent to finding the dimension of the subspace of solutions of the system (27). In Sec. 23, it was shown that the dimension of the latter subspace is equal to $n - r$, where r is the rank of the coefficient matrix of the system, i.e., the rank of the operator A. Thus, *the dimension of the null space of a linear operator A in an n-dimensional space R equals the dimension of R minus the rank of the operator A.*

Problem. Verify the following facts:

a) The relation $N(A) \supset T(A)$ is necessary and sufficient for the equality $A^2 = 0$ to hold;

b) For any operator $N(A) \subset N(A^2) \subset N(A^3) \subset \ldots$;

c) For any operator $T(A) \supset T(A^2) \supset T(A^3) \supset \ldots$;

d) If $T(A^k) \subset N(A^m)$, then $T(A) \subset N(A^{m+k-1})$, $T(A^{m+k-1}) \subset N(A)$.

Hint. The relation $T(A^k) \subset N(A^m)$ is necessary and sufficient for the equality $A^{k+m} = 0$ to hold.

31.3. Remark. The following theorem is a consequence of the results of Secs. 31.1 and 31.2:

Given any two subspaces N and T of an n-dimensional space R such that the sum of the dimensions of N and T equals the dimension of R, there exists a linear operator A for which N(A) = N and T(A) = T.

Proof. Denote the dimensions of the subspaces N and T by k and $m = n - k$, respectively. Choose m linearly independent vectors $f_1, f_2, \ldots, f_m$ in the subspace T. Then choose any basis $e_1, e_2, \ldots, e_n$ in the space R such that the first k basis vectors lie in the subspace N. The operator A is defined by the conditions

$$
\begin{aligned}
Ae_i &= 0 & (i = 1, 2, \ldots, k), \\
Ae_{i+k} &= f_i & (i = 1, 2, \ldots, m).
\end{aligned} \tag{28}
$$

We now show that this operator A satisfies the requirements of the theorem.

First of all, it is obvious that $T(A)$ is the linear manifold spanned by the vectors f_i $(i = 1, 2, \ldots, m)$ and therefore coincides with the subspace T. Then, by construction, every vector of the subspace N belongs to $N(A)$. It remains only to show that any vector of the subspace $N(A)$ belongs to N. Assume that $Ax = 0$ for some

$$
x = \sum_{i=1}^{n} \xi_i e_i.
$$

Using the conditions (28), we obtain

$$
0 = Ax = A(\xi_1 e_1 + \cdots + \xi_n e_n) = \xi_{k+1} f_1 + \cdots + \xi_n f_m.
$$

Since the vectors f_i $(i = 1, 2, \ldots, m)$ are linearly independent, we have

$$
\xi_{k+1} = \cdots = \xi_n = 0.
$$

But then

$$
x = \xi_1 e_1 + \cdots + \xi_k e_k \in N,
$$

as required.

31.4. The following theorems on the rank of the product of two matrices are consequences of the geometric notions just introduced.

THEOREM 23. *The rank of the product of two matrices does not exceed the rank of each of the factors.*

Proof. Let A and B be the linear operators corresponding to the matrices being multiplied. Then, by its very definition, the range of the operator AB is contained in the range of the operator A. Since, as we have just seen, the dimension of the range of any operator equals the rank of the corresponding matrix, we find that *the rank of the product of two matrices does not exceed the rank of the first factor.* To prove that it also does not exceed the rank of the second factor, we take transposed matrices. Using equation (15) of Sec. 29, we find that the rank of AB = the rank of $(AB)'$ = the rank of $B'A' \leqslant$ the rank of B' = the rank of B, Q.E.D.

The rank of the product of two matrices can actually be less than the rank of each factor. For example, the matrices

$$A = \left\| \begin{matrix} 0 & 1 \\ 0 & 0 \end{matrix} \right\|, \qquad B = \left\| \begin{matrix} 1 & 0 \\ 0 & 0 \end{matrix} \right\|$$

both have rank one, but their product

$$AB = \left\| \begin{matrix} 0 & 0 \\ 0 & 0 \end{matrix} \right\|$$

has rank zero. Therefore, the following theorem, which gives a lower bound rather than an upper bound for the rank of the product of two matrices, is of interest:

THEOREM 24. *The rank of the product of two matrices of order n is not less than* $(r_1 + r_2) - n$, *where* r_1 *and* r_2 *are the ranks of the factors.*

Proof. We first show that an operator A of rank r carries every k-dimensional subspace $R' \subset R$ into a subspace $A(R')$ of dimension not less than $r - (n - k)$. Choose a basis $e_1, e_2, \ldots, e_n$ in the space R such that the first k basis vectors lie in the subspace R' (see Sec. 15.2). The components of the vectors $Ae_1, Ae_2, \ldots, Ae_k$, which span the subspace $A(R')$, occupy the first k columns of the matrix of the operator A. By hypothesis, there are r linearly independent columns in the matrix of A. We divide these columns into two groups, the first consisting of columns whose numbers lie in the range 1 to k, the second consisting of columns whose numbers lie in the range $k + 1$ to n. The second group contains no more than $n - k$ columns, and therefore the first group contains no more than $r - (n - k)$ columns. Thus, the subspace $A(R')$ has no more than $r - (n - k)$ linearly independent vectors, as asserted.

Now consider the product of two matrices A and B of ranks r_1 and r_2, respectively. Let the corresponding operators be denoted by the same symbols, so that A is an operator of rank r_1 and B an operator of rank r_2. As was shown earlier in this section, an estimate of the rank of the matrix AB is equivalent to an estimate of the dimension of the range of the corresponding operator AB. The operator B carries the whole space R into a subspace $T(B)$ of dimension r_2. By what has just been proved, the operator A carries the subspace $T(B)$ into a subspace whose dimension is not less than

$$r_1 - (n - r_2) = r_1 + r_2 - n.$$

Thus, the dimension of the range of the operator AB, and therefore the rank of the matrix AB, is not less than $r_1 + r_2 - n$, Q.E.D.

COROLLARY 1. *The rank of the product of two matrices of order n, one of which is nonsingular (i.e., has rank n), equals the rank of the other matrix.* In fact, in this case the upper and lower estimates of the rank of a product, given by Theorems 23 and 24, yield the same result, namely the rank of the other matrix.

COROLLARY 2. *A product of two nonsingular matrices is nonsingular.* This result is also a consequence of the theorem on the determinant of the product of two matrices, proved in Sec. 29.

*32. The Algebra of Linear Operators on an n-Dimensional Space and its Ideals

We have already seen that the set of all linear operators defined on an n-dimensional space R is itself a new linear space, which we denote here by $\mathscr{R}$. For the elements of the space $\mathscr{R}$, we have defined not only the usual linear operations but also the operation of multiplication (of one element by another), which satisfies the associative and distributive laws. A linear space with a multiplication operation obeying the associative and distributive laws is called an *algebra*, and our space $\mathscr{R}$ is an example of such an algebra.

Suppose that we have an algebra $\mathscr{R}$, whose elements we shall denote by the letters A, B, ... A subspace $\mathscr{R}' \subset \mathscr{R}$ is called a *left ideal* of the algebra $\mathscr{R}$ if $B \in \mathscr{R}'$ implies $AB \in \mathscr{R}'$ for any element $A \in \mathscr{R}$. Similarly, a subspace $\mathscr{R}'' \subset \mathscr{R}$ is called a *right ideal* of the algebra $\mathscr{R}$, if $B \in \mathscr{R}''$ implies $BA \in \mathscr{R}''$ for any element $A \in \mathscr{R}$.

We now show some examples of left and right ideals of the algebra of all linear operators defined on a linear space R. First of all, the whole algebra itself and its zero subspace (i.e., the subspace consisting of just the zero operator) are obviously both left and right ideals. Next, we fix a subspace $R' \subset R$ and consider all the operators $B \in \mathscr{R}$ which carry every vector of the subspace R' into zero. We denote this set of operators by $\mathscr{R}'$. It is easily verified that the set $\mathscr{R}'$ is a subspace of the algebra $\mathscr{R}$; moreover, $\mathscr{R}'$ is a left ideal of $\mathscr{R}$. In fact, if $Bx = 0$ for any $x \in R'$, then $ABx = 0$ for any $x \in R'$, where $A \in \mathscr{R}$ is an arbitrary operator. Finally, we fix a subspace $R'' \subset R$ and consider the operators $B \in \mathscr{R}$ whose ranges are contained in the subspace R''. We denote this set of operators by $\mathscr{R}''$. The set $\mathscr{R}''$ is obviously a subspace of the algebra $\mathscr{R}$; moreover, $\mathscr{R}''$ is a right ideal of $\mathscr{R}$. In fact, if $Bx \in R''$ for any $x \in R$, then $BAx \in R''$ for any $x \in R$, where $A \in \mathscr{R}$ is an arbitrary operator.

It turns out that in the case where the space R is finite-dimensional, the examples just given are the most general examples of ideals of the algebra $\mathscr{R}$. In fact, we have the following theorem:

THEOREM 25. *Every left ideal $\mathscr{R}'$ of the algebra $\mathscr{R}$ of all linear operators defined on a finite-dimensional space R is a set consisting of all the operators $B \in \mathscr{R}$ which carry every vector of some subspace $R' \subset R$ into zero. Every right ideal $\mathscr{R}''$ of the algebra $\mathscr{R}$ is a set consisting of all the operators $B \in \mathscr{R}$ whose ranges belong to some subspace $R'' \subset R$.*

To prove this theorem, we need some lemmas concerning null spaces and ranges of linear operators in an n-dimensional space. We denote the null spaces of the operators A, B, C by $N(A)$, $N(B)$, $N(C)$, and the dimensions of these null spaces by α, β, γ, respectively. Similarly, we denote the ranges of the operators A, B, C by $T(A)$, $T(B)$, $T(C)$, and the dimensions of these ranges by a, b, c, respectively. According to Sec. 31, we have

$$\alpha + a = \beta + b = \gamma + c = n.$$

LEMMA 1. *Let $e_1, e_2, \ldots, e_n$ be a basis of the space R, and let the vectors $e_1, e_2, \ldots, e_\alpha$ belong to $N(A)$. Then the vectors $Ae_{\alpha+1}, \ldots, Ae_n$ are linearly independent.*

Proof. We wish to show that the linear relation

$$\lambda_{\alpha+1}Ae_{\alpha+1} + \cdots + \lambda_n Ae_n = 0$$

implies that $\lambda_{\alpha+1} = \cdots = \lambda_n = 0$. This relation is equivalent to

$$A(\lambda_{\alpha+1}e_{\alpha+1} + \cdots + \lambda_n e_n) = 0,$$

i.e.,

$$x = \lambda_{\alpha+1}e_{\alpha+1} + \cdots + \lambda_n e_n \in N(A).$$

But then x is a linear combination of $e_1, e_2, \ldots, e_\alpha$:

$$x = \lambda_{\alpha+1}e_{\alpha+1} + \cdots + \lambda_n e_n = \lambda_1 e_1 + \lambda_2 e_2 + \cdots + \lambda_\alpha e_\alpha.$$

Since the vectors $e_1, e_2, \ldots, e_n$ are linearly independent,

$$\lambda_{\alpha+1} = \cdots = \lambda_n = 0,$$

as required.

LEMMA 2. *Choose a linearly independent vectors $f_1, f_2, \ldots, f_a$ in $T(A)$, where $f_i = Ae_i$ $(i = 1, 2, \ldots, a)$. Moreover, choose $n - a$ linearly independent vectors $e_{a+1}, \ldots, e_n$ in $N(A)$. Then, the vectors $e_1, e_2, \ldots, e_n$ are linearly independent and therefore form a basis of the space R.*

Proof. We wish to show that the linear relation

$$\lambda_1 e_1 + \lambda_2 e_2 + \cdots + \lambda_n e_n = 0$$

implies that $\lambda_1 = \lambda_2 = \cdots = \lambda_n = 0$. If we apply the operator A to this relation, then since the vectors $e_{a+1}, \ldots, e_n$ belong to $N(A)$ and $Ae_i = f_i$ for $i = 1, 2, \ldots, a$, we obtain

$$\lambda_1 f_1 + \lambda_2 f_2 + \cdots + \lambda_a f_a = 0.$$

Because of the linear independence of the vectors f_i, we have

$$\lambda_1 = \lambda_2 = \cdots = \lambda_a = 0.$$

But then $\lambda_{a+1} = \cdots = \lambda_n = 0$ also, since the vectors $e_{a+1}, \ldots, e_n$ are linearly independent, by hypothesis. Thus, $\lambda_i = 0$, for $i = 1, 2, \ldots, n$, which implies the linear independence of the vectors $e_1, e_2, \ldots, e_n$, Q.E.D.

LEMMA 3. *If $N(B) \subset N(C)$, there exists an operator A satisfying the condition $C = AB$.*

Proof. Construct a basis $e_1, e_2, \ldots, e_n$ such that the vectors $e_1, e_2, \ldots, e_\beta$ lie in $N(B)$. Then, by Lemma 1, the vectors

$$f_{\beta+1} = Be_{\beta+1}, \ldots, f_n = Be_n$$

are linearly independent. Now construct vectors $f_1, f_2, \ldots, f_\beta$ such that the system $f_1, f_2, \ldots, f_n$ is a basis for the space R. Define the operator A by the relations

$$Af_i = Ce_i \qquad (i = 1, 2, \ldots, n).$$

It is asserted that this is the desired operator. In fact, the relation

$$ABe_i = Ce_i$$

is satisfied for $i \leqslant \beta$, since both the left- and right-hand sides are zero; for $i > \beta$, the relation is also satisfied, since $Be_i = f_i$ for these i. Since the vectors $e_1, e_2, \ldots, e_n$ form a basis, we have $AB = C$, as required.

LEMMA 4. *If $T(B) \supset T(C)$, there exists an operator A satisfying the condition $C = BA$.*

Proof. Choose c linearly independent vectors $f_1, f_2, \ldots, f_c$ in $T(C)$. Then define the vectors $e_1, e_2, \ldots, e_c$ by the conditions

$$f_i = Be_i \qquad (i = 1, 2, \ldots, c);$$

this can be done, since $T(B) \supset T(C)$. Using Lemma 2, as applied to the vectors $f_1, f_2, \ldots, f_c$ and the operator C, construct a basis $g_1, g_2, \ldots, g_n$ such that

$$Cg_i = f_i \qquad (i = 1, 2, \ldots, c),$$
$$Cg_i = 0 \qquad (i > c).$$

Now define the operator A by the conditions

$$Ag_i = e_i \quad \text{for } i \leqslant c,$$
$$Ag_i = 0 \quad \text{for } i > c.$$

It is asserted that this is the desired operator. In fact, the relation

$$BAg_i = Cg_i$$

is satisfied for $i \leqslant c$, since for these i we have

$$BAg_i = Be_i = f_i = Cg_i;$$

for $i > c$, the relation is also satisfied, since then both the left- and right-hand sides are zero. Since $g_1, g_2, \ldots, g_n$ is a basis for the space R, we have $BA = C$, as required.

LEMMA 5. *Let $N(A)$ be the intersection of a finite number of subspaces $P, Q, \ldots$. Then there exist operators $B, C, \ldots$ such that $N(B) \supset P$, $N(C) \supset Q, \ldots$ and $B + C + \cdots = A$.*

Proof. It is sufficient to consider the case of two subspaces P and Q. Choose α linearly independent vectors $e_1, e_2, \ldots, e_\alpha$ in $N(A)$. Supplement these vectors with the vectors $p_1, p_2, \ldots, p_r$ to make a basis for the subspace P, and supplement them with the vectors $q_1, q_2, \ldots, q_s$ to make a basis for the subspace Q. By Sec. 15.3, the vectors $e_1, \ldots, e_\alpha$, $p_1, \ldots, p_r, q_1, \ldots, q_s$ are linearly independent. Supplement these vectors with the vectors $f_1, f_2, \ldots, f_m$ to make a basis for the whole space R. Then define the operators B and C by the conditions

$$Be_i = 0, \quad Bp_i = 0, \quad Bq_i = Aq_i, \quad Bf_i = \tfrac{1}{2}Af_i,$$
$$Ce_i = 0, \quad Cp_i = Ap_i, \quad Cq_i = 0, \quad Cf_i = \tfrac{1}{2}Af_i.$$

It is clear that $B + C = A$. Moreover, $N(B) \supset P$, $N(C) \supset Q$, as required.

LEMMA 6. *Let $T(A)$ be the sum of a finite number of subspaces $P, Q, \ldots$. Then there exist operators $B, C, \ldots$, such that $T(B) \subset P$, $T(C) \subset Q, \ldots$ and $B + C + \cdots = A$.*

Proof. It is sufficient to consider the case of two subspaces P and Q. Let S denote the intersection of P and Q. Construct a basis $e_1, \ldots, e_k$, $p_1, \ldots, p_r, q_1, \ldots, q_s$ in the subspace $T(A)$ such that the vectors $e_1, \ldots, e_k$ form a basis for the subspace S, the vectors $e_1, \ldots, e_k, p_1, \ldots, p_r$ form a basis for the subspace P and the vectors $e_1, \ldots, e_k, q_1, \ldots, q_s$ form a basis for the subspace Q. Moreover, let the vectors e_i^*, p_i^*, q_i^* be defined by the relations

$$e_i = Ae_i^*, \quad p_i = Ap_i^*, \quad q_i = Aq_i^*.$$

According to Lemma 2, the vectors e_i^*, p_i^*, q_i^* together with α linearly independent vectors f_i^* taken from $N(A)$ form a basis for the space R. We define the operators B and C by the formulas

$$Be_i^* = \tfrac{1}{2} e_i, \quad Bp_i^* = p_i, \quad Bq_i^* = 0, \quad Bf_i^* = 0,$$
$$Ce_i^* = \tfrac{1}{2} e_i, \quad Cp_i^* = 0, \quad Cq_i^* = q_i, \quad Cf_i^* = 0.$$

It is clear that $B + C = A$. Moreover, $T(B)$ is spanned by the vectors e_i and p_i and hence is contained in the subspace P; similarly, we have $T(C) \subset Q$. This completes the proof of Lemma 6.

We now turn to the proof of Theorem 25. First suppose that $\mathscr{R}'$ is a left ideal of the algebra $\mathscr{R}$. Consider the set $N(\mathscr{R}')$ of all vectors $x \in R$ for which $Bx = 0$ for any $B \in \mathscr{R}'$. It is clear that $N(\mathscr{R}')$ is a subspace. We shall show that every operator A for which $N(\mathscr{R}') \subset N(A)$ is an element of the ideal $\mathscr{R}'$; this will prove the first assertion in Theorem 25. First we assume that $N(A) = N(\mathscr{R}')$; the existence of such an operator A follows from Sec. 31.3, for example.

The subspace $N(\mathscr{R}')$ can be constructed by means of the following finite process: Choose any operator $B_1 \in \mathscr{R}'$ and form $N(B_1)$. If for every $C_1 \in \mathscr{R}', N(C_1) \supset N(B_1)$, then $N(\mathscr{R}') = N(B_1)$ and the construction is finished. If we can find a $C_1 \in \mathscr{R}'$ such that $N(C_1)$ does not contain $N(B_1)$, we then form the intersection $N(B_1, C_1)$ of the subspaces $N(B_1)$ and $N(C_1)$. If for every $D_1 \in \mathscr{R}', N(D_1) \supset N(B_1, C_1)$, then

$$N(\mathscr{R}') = N(B_1, C_1)$$

and the construction is finished. If not, we can continue along the lines indicated by using an appropriate operator. After a finite number of steps, the process must come to an end, since at each step the dimension of the candidate for $N(\mathscr{R}')$ is decreased.

Thus, we can assume that the subspace $N(\mathscr{R}')$ is the intersection of a finite number of subspaces $P = N(B_1)$, $Q = N(C_1), \ldots$, where $B_1, C_1, \ldots$ are elements of the ideal $\mathscr{R}'$. By Lemma 5, we can construct operators $B, C, \ldots$, such that

$$N(B) \supset P = N(B_1),$$
$$N(C) \supset Q = N(C_1), \ldots, B + C + \cdots = A.$$

By Lemma 3, the operators $B, C, \ldots$ are obtained from the operators $B_1, C_1, \ldots$ by multiplying them on the *left* by certain other operators. Therefore, the operators $B, C, \ldots$, as well as the operators $B_1, C_1, \ldots$, are elements of the ideal $\mathscr{R}'$, and hence their sum $B + C + \cdots = A$ is also an element of $\mathscr{R}'$. Thus, every operator A for which $N(A) = N(\mathscr{R}')$ is an element of the ideal $\mathscr{R}'$. Applying Lemma 3 again, we find that any operator A_1 for which

$$N(A_1) \supset N(A) = N(\mathscr{R}')$$

also belongs to the ideal $\mathscr{R}'$. This proves the first assertion in Theorem 25.

Now let $\mathscr{R}''$ be a right ideal of the algebra $\mathscr{R}$. Consider the linear manifold $T(\mathscr{R}'')$ of all the vectors Bx where $x \in R$, $B \in \mathscr{R}''$. *The subspace $T(\mathscr{R}'')$ can be obtained as a finite sum of subspaces $T(B_1)$, $T(C_1)$, ...* To show this, we use a method similar to that just used for intersections in the proof of the first part of the theorem. To prove the second assertion in Theorem 25, we have to show that every operator A for which $T(A) \subset T(\mathscr{R}'')$ is an element of the ideal $\mathscr{R}''$. First suppose that $T(A) = T(\mathscr{R}'')$. Then by Lemma 6, there exists operators $B, C, \ldots$, such that

$$
\begin{aligned}
T(B) &\subset T(B_1), \\
T(C) &\subset T(C_1), \ldots, \\
B + C + \cdots &= A.
\end{aligned}
$$

By Lemma 4, the operators $B, C, \ldots$ are obtained from the operators $B_1, C_1, \ldots$ by multiplying them on the *right* by certain other operators. Therefore, the operators $B, C, \ldots$ as well as the operators $B_1, C_1, \ldots$ are elements of the ideal $\mathscr{R}''$ and hence their sum $B + C + \cdots = A$ is also an element of $\mathscr{R}''$. Thus, every operator A for which $T(A) = T(\mathscr{R}'')$ is an element of the ideal $\mathscr{R}''$. Applying Lemma 4 again, we find that any operator A_1 for which

$$
T(A_1) \subset T(A) = T(\mathscr{R}'')
$$

also belongs to the ideal $\mathscr{R}''$. This completes the proof of Theorem 25.

Remark. The reader has undoubtedly noticed a certain parallelism between the proofs of the first and second assertions in Theorem 25, and in fact, the second assertion can be deduced from the first assertion. However, this derivation requires the introduction of new concepts, with which we shall first become acquainted in Chapter 7 (see Problems 8 and 9 of Sec. 35).

In concluding this section, we consider an ideal $\mathscr{R}' \subset \mathscr{R}$ which is simultaneously a right and a left ideal (a *two-sided ideal*). Since $\mathscr{R}'$ is a left ideal, according to Theorem 25 it coincides with the set of all operators which carry all the vectors of some subspace $R' \subset R$ into zero. Since $\mathscr{R}'$ is also a right ideal, it coincides with the set of all operators whose ranges belong to some subspace $R'' \subset R$. But in this case, every operator which carries the subspace R' into zero takes all its values in the subspace R'', and conversely. Such a situation occurs, for example, in the case $R' = R$, $R'' = 0$. In this case, $\mathscr{R}'$ is the so-called *null ideal*, which as we have already seen, is two-sided. If $R' \neq R$, then according to Sec. 31.3, the subspace R'' contains any preassigned vector. It follows that $R'' = R$ and the ideal $\mathscr{R}'$ consists of all linear operators, i.e., coincides with the whole algebra $\mathscr{R}$.

Thus, we have shown that the algebra $\mathscr{R}$ of linear operators defined on an n-dimensional space has no two-sided ideals except the trivial ones, i.e., the null ideal and the whole algebra $\mathscr{R}$.

*33. General Linear Operators

So far, we have considered linear functions of a vector argument such that when the argument varies over a linear space R, the values of the functions are either numbers or vectors in the *same* space R. However, we can also adopt a more general point of view and consider functions of a vector argument whose values are vectors in *any* linear space R'. If such a function satisfies the conditions

$$A(x + y) = Ax + Ay,$$
$$A(\alpha x) = \alpha Ax,$$

(where the operations on the right-hand side are performed in the space R'), it is called a *general linear operator*. In particular, if R' coincides with R, we obtain an ordinary linear operator; if R' is one-dimensional, the general linear operator reduces to an ordinary linear form.

The operations of addition and multiplication by real numbers are defined in the natural way for general linear operators (defined on a space R with values in a space R'). The set of all such operators constitutes a new linear space. The operation of *multiplication* can be defined in a meaningful way only for two linear operators A, B, such that the first operator is defined on the space containing the range of the second operator. Then, the product $C = AB$ is defined by the formula

$$Cx = (AB)x = A(Bx).$$

Problem 1. Construct the general form of a linear operator which is defined on an n-dimensional space R and takes its values in a k-dimensional space R'.

Hint. Associate with each such operator a matrix with k rows and n columns.

Problem 2. What operation with rectangular matrices corresponds to multiplication of general linear operators?

Ans. Multiplication of $k \times n$ matrices by $n \times m$ matrices by a rule analogous to the ordinary rule of matrix multiplication.

5

COORDINATE
TRANSFORMATIONS

As is well known, in solving geometric problems by the methods of analytic geometry, a very important role is played by the proper choice of a coordinate system. Proper choice of a coordinate system also plays a very important role in a much wider class of problems connected with the geometry of n-dimensional linear spaces. This chapter is devoted to a study of the rules governing coordinate transformations in n-dimensional spaces. In particular, the results obtained here are fundamental for the classification of quadratic forms which will be made in the next chapter.

34. Transformation to a New Basis

Let

$$\{e\} = \{e_1, e_2, \ldots, e_n\}$$

be a basis in an n-dimensional space R, and let

$$\{f\} = \{f_1, f_2, \ldots, f_n\}$$

be another basis in the same space. The vectors of the system $\{f\}$ are uniquely determined by their expansions in terms of the vectors of the original basis:

$$
\begin{aligned}
f_1 &= a_1^{(1)}e_1 + a_2^{(1)}e_2 + \cdots + a_n^{(1)}e_n, \\
f_2 &= a_1^{(2)}e_1 + a_2^{(2)}e_2 + \cdots + a_n^{(2)}e_n, \\
&\quad \cdots \quad \cdots \quad \cdots \quad \cdots \\
f_n &= a_1^{(n)}e_1 + a_2^{(n)}e_2 + \cdots + a_n^{(n)}e_n,
\end{aligned}
\tag{1}
$$

which can be abbreviated to

$$f_j = \sum_{i=1}^{n} a_i^{(j)} e_i \qquad (j = 1, 2, \ldots, n). \qquad (2)$$

The coefficients $a_i^{(j)}$ $(i, j = 1, 2, \ldots, n)$ in (1) and (2) define a matrix

$$A = \|a_i^{(j)}\| = \begin{Vmatrix} a_1^{(1)} & a_1^{(2)} & \ldots & a_1^{(n)} \\ a_2^{(1)} & a_2^{(2)} & \ldots & a_2^{(n)} \\ \cdot & \cdot & \ldots & \\ a_n^{(1)} & a_n^{(2)} & \ldots & a_n^{(n)} \end{Vmatrix},$$

called the *transformation matrix from the basis* $\{e\}$ *to the basis* $\{f\}$. As was done previously in similar cases (Sec. 26 et seq.), we write the components of the vectors f_j (with respect to the basis $\{e\}$) as the columns of the matrix A.

The determinant of the matrix A is nonvanishing; in fact, if this were not the case, the columns of A and hence the vectors $f_1, f_2, \ldots, f_n$ would be linearly dependent (Sec. 19). A matrix with a nonvanishing determinant was called *nonsingular* in Sec. 30. Thus, *the transformation from one basis of an n-dimensional space R to another basis is always accomplished by using a nonsingular matrix.*

Conversely, let $\{e\} = \{e_1, e_2, \ldots, e_n\}$ be a given basis of the *n*-dimensional space R and let $A = \|a_i^{(j)}\|$ be a nonsingular matrix of order n. Using the equations (1), construct the system of vectors $f_1, f_2, \ldots, f_n$. It is clear that these vectors are linearly independent, since the columns of every nonsingular matrix are linearly independent (Sec. 19). Consequently, the vectors $f_1, f_2, \ldots, f_n$ form a new basis for the space R, i.e., *every nonsingular matrix* $A = \|a_i^{(j)}\|$ *determines via* (1) *a transformation from one basis of an n-dimensional space R to another basis.*

Since the matrix A is nonsingular, the equations (1) can be solved for the vectors $e_1, e_2, \ldots, e_n$; the resulting system of equations

$$\begin{aligned} e_1 &= b_1^{(1)} f_1 + b_2^{(1)} f_2 + \cdots + b_n^{(1)} f_n, \\ e_2 &= b_1^{(2)} f_1 + b_2^{(2)} f_2 + \ldots + b_n^{(2)} f_n, \\ \cdot &\quad \ldots \quad \ldots \quad \ldots \quad \ldots \\ e_n &= b_1^{(n)} f_1 + b_2^{(n)} f_2 + \cdots + b_n^{(n)} f_n \end{aligned} \qquad (3)$$

obviously defines a transformation from the basis $\{f\}$ to the basis $\{e\}$.

The equations (1) together with the matrix A also determine a corresponding linear operator A, defined by the relations

$$f_i = A e_i \qquad (i = 1, 2, \ldots, n).$$

This operator is called the *transformation operator from the basis* $\{e\}$ *to the basis* $\{f\}$. The equations (3) define the inverse operator A^{-1}.

Finally, we note a particular case of a transformation to a new basis, i.e., the case where every vector f_k is just the corresponding vector e_k multiplied by a number $\lambda_k \neq 0$ $(k = 1, 2, \ldots, n)$. Then the equations (1) take the form

$$\begin{aligned}
f_1 &= \lambda_1 e_1, \\
f_2 &= \lambda_2 e_2, \\
&\ \cdot \quad \cdots \\
f_n &= \lambda_n e_n,
\end{aligned}$$

and the matrix A has the diagonal form

$$A = \begin{Vmatrix} \lambda_1 & 0 & \ldots & 0 \\ 0 & \lambda_2 & \ldots & 0 \\ \cdot & \cdot & \cdots & \cdot \\ 0 & 0 & \ldots & \lambda_n \end{Vmatrix}.$$

In particular, for $\lambda_1 = \lambda_2 = \cdots = \lambda_n = 1$, we obtain the matrix of the *identity transformation* or the *unit* matrix

$$E = \begin{Vmatrix} 1 & 0 & \ldots & 0 \\ 0 & 1 & \ldots & 0 \\ \cdot & \cdot & \cdots & \cdot \\ 0 & 0 & \ldots & 1 \end{Vmatrix}. \tag{4}$$

The original basis is not changed by the identity transformation.

35. Transformation of the Components of a Vector when the Basis is Changed

Let $\{e\} = \{e_1, e_2, \ldots, e_n\}$ and $\{f\} = \{f_1, f_2, \ldots, f_n\}$ be two bases in an n-dimensional linear space R. Any vector $x \in R$ has the expansions

$$x = \xi_1 e_1 + \xi_2 e_2 + \cdots + \xi_n e_n = \eta_1 f_1 + \eta_2 f_2 + \cdots + \eta_n f_n, \tag{5}$$

where $\xi_1, \xi_2, \ldots, \xi_n$ are the components of the vector x with respect to the basis $\{e\}$ and $\eta_1, \eta_2, \ldots, \eta_n$ are its components with respect to the basis $\{f\}$. We now show how to calculate the components of the vector x with respect to the basis $\{f\}$ in terms of its components with respect to the basis $\{e\}$.

Suppose we have the transformation matrix $A = \|a_i^{(j)}\|$ from the basis $\{e\}$ to the basis $\{f\}$. Then, according to (3), the vectors $\{e\}$ are linear combinations of the vectors $\{f\}$:

$$\begin{aligned}
e_1 &= b_1^{(1)} f_1 + b_2^{(1)} f_2 + \cdots + b_n^{(1)} f_n, \\
e_2 &= b_1^{(2)} f_1 + b_2^{(2)} f_2 + \cdots + b_n^{(2)} f_n, \\
&\ \cdot \quad\quad \cdots \quad\quad\quad \cdots \quad\quad \cdots \\
e_n &= b_1^{(n)} f_1 + b_2^{(n)} f_2 + \cdots + b_n^{(n)} f_n,
\end{aligned}$$

or more briefly

$$e_j = \sum_{k=1}^{n} b_k^{(j)} f_k \qquad (j = 1, 2, \ldots, n), \tag{6}$$

where $B = \|b_k^{(j)}\|$ is the inverse of the matrix A (Sec. 30). Substituting the equations (6) into the expression (5), we obtain

$$x = \sum_{j=1}^{n} \xi_j e_j = \sum_{k=1}^{n} \eta_k f_k = \sum_{j=1}^{n} \xi_j \left(\sum_{k=1}^{n} b_k^{(j)} f_k \right) = \sum_{k=1}^{n} \left(\sum_{j=1}^{n} b_k^{(j)} \xi_j \right) f_k,$$

from which it follows by the uniqueness of the expansion of the vector x with respect to the basis $\{f\}$ that

$$\eta_k = \sum_{j=1}^{n} b_k^{(j)} \xi_j \qquad (k = 1, 2, \ldots, n); \tag{7}$$

in expanded form, this is equivalent to the system of equations

$$
\begin{aligned}
\eta_1 &= b_1^{(1)}\xi_1 + b_1^{(2)}\xi_2 + \cdots + b_1^{(n)}\xi_n, \\
\eta_2 &= b_2^{(1)}\xi_1 + b_2^{(2)}\xi_2 + \cdots + b_2^{(n)}\xi_n, \\
&\;\cdots \qquad \cdots \qquad \cdots \qquad \cdots \\
\eta_n &= b_n^{(1)}\xi_1 + b_n^{(2)}\xi_2 + \cdots + b_n^{(n)}\xi_n.
\end{aligned}
\tag{7a}
$$

Thus, the components of the vector x with respect to the basis $\{f\}$ are linear combinations of the components of x with respect to the basis $\{e\}$. Moreover, the coefficients of these linear combinations define a matrix C which is the transpose of the transformation matrix from the basis $\{f\}$ to the basis $\{e\}$, i.e., the transpose of the inverse of the matrix A.[1]

Denoting the inverse of a matrix A by A^{-1} and the transpose by A', we can write the matrix defined by (7a) as $(A^{-1})'$.

The converse theorem is also valid:

Let $\xi_1, \xi_2, \ldots, \xi_n$ be the components of an arbitrary vector x with respect to the basis

$$\{e\} = \{e_1, e_2, \ldots, e_n\}$$

of the n-dimensional space R, and let the quantities $\eta_1, \eta_2, \ldots, \eta_n$ be defined by the formulas

$$
\begin{aligned}
\eta_1 &= c_1^{(1)}\xi_1 + c_2^{(1)}\xi_2 + \cdots + c_n^{(1)}\xi_n, \\
\eta_2 &= c_1^{(2)}\xi_1 + c_2^{(2)}\xi_2 + \cdots + c_n^{(2)}\xi_n, \\
&\;\cdots \qquad \cdots \qquad \cdots \qquad \cdots \\
\eta_n &= c_1^{(n)}\xi_1 + c_2^{(n)}\xi_2 + \cdots + c_n^{(n)}\xi_n,
\end{aligned}
$$

where $\det \|c_i^{(j)}\| \neq 0$. *Then we can find a new basis $\{f\} = \{f_1, f_2, \ldots, f_n\}$ in the space R such that the quantities $\eta_1, \eta_2, \ldots, \eta_n$ are the components of the vector x with respect to the basis $\{f\}$.*

Proof. Introduce the matrices $C = \|c_i^{(j)}\|$ and $A = (C')^{-1}$, and then construct a new basis, using the matrix A and the formulas (1); it is asserted that this is the desired basis. In fact, consider the trans-

---

[1] By the matrix " defined " by the linear combinations (7a) is meant, as on p. 98, the matrix whose jth column consists of the coefficients in the jth row of (7a).

formation formulas (7), which give the components of the vector x with respect to the new basis. As we have seen, these formulas can be written in terms of the matrix $(A^{-1})'$. But in the present case, $(A^{-1})'$ coincides with C, since

$$(A^{-1})' = ([(C')^{-1}]^{-1})' = (C')' = C.$$

Therefore, for any vector x, the quantities $\eta_1, \eta_2, \ldots, \eta_n$ are just the components of the vector x with respect to the basis $\{f\}$, Q.E.D.

Problem 1. The vector $x \in R$ has components $\xi_1, \xi_2, \ldots, \xi_n$ with respect to the basis $e_1, e_2, \ldots, e_n$. How does one construct a new basis in R such that the components of x with respect to this basis equal $1, 0, 0, \ldots, 0$?

Problem 2. A basis $e_1, e_2, \ldots, e_n$ is chosen in the n-dimensional space R. Show that every subspace $R' \subset R$ can be specified as the set of all vectors $x \in R$ whose components (with respect to the basis $e_1, e_2, \ldots, e_n$) satisfy a system of equations of the form

$$\sum_{j=1}^{n} a_{ij}\xi_j = 0 \qquad (i = 1, 2, \ldots, k).$$

Hint. Choose a new basis $f_1, f_2, \ldots, f_n$ such that the first $n - k$ vectors form a basis for the subspace R'. Write the condition $x \in R'$ in the form of a system of equations involving the components of x in the new basis. Use the transformation formulas to construct the corresponding system of equations involving the components of x in the original basis.

Problem 3. (Continuation) Show that every hyperplane $H \subset R$ can be specified as the set of all vectors $x \in R$ whose components (with respect to the basis $e_1, e_2, \ldots, e_n$) satisfy a system of equations of the form

$$\sum_{j=1}^{n} a_{ij}\xi_j = b_i \qquad (i = 1, 2, \ldots, k).$$

36. Consecutive Transformations

36.1. Let $A = \|a_i^{(j)}\|$ be the transformation matrix from the basis

$$\{e\} = \{e_1, e_2, \ldots, e_n\}$$

to the basis

$$\{f\} = \{f_1, f_2, \ldots, f_n\},$$

and let $B = \|b_j^{(k)}\|$ be the transformation matrix from the basis $\{f\}$ to the basis

$$\{g\} = \{g_1, g_2, \ldots, g_n\}.$$

We now determine the transformation matrix which goes directly from the basis $\{e\}$ to the basis $\{g\}$. The transformation formula from the basis $\{e\}$ to the basis $\{f\}$ has the form (2):

$$f_j = \sum_{i=1}^{n} a_i^{(j)} e_i \qquad (j = 1, 2, \ldots, n). \tag{8}$$

Similarly, the transformation from the basis $\{f\}$ to the basis $\{g\}$ is

$$g_k = \sum_{j=1}^{n} b_j^{(k)} f_j \qquad (k = 1, 2, \ldots, n). \tag{9}$$

Substituting (8) into (9), we obtain

$$g_k = \sum_{j=1}^{n} b_j^{(k)} \sum_{i=1}^{n} a_i^{(j)} e_i = \sum_{i=1}^{n} \left(\sum_{j=1}^{n} a_i^{(j)} b_j^{(k)} \right) e_i \qquad (k = 1, 2, \ldots, n). \tag{10}$$

On the other hand, if $C = \|c_i^{(k)}\|$ denotes the desired transformation matrix from the basis $\{e\}$ to the basis $\{g\}$, we can write

$$g_k = \sum_{i=1}^{n} c_i^{(k)} e_i \qquad (k = 1, 2, \ldots, n). \tag{11}$$

Comparing (10) and (11), we obtain

$$c_i^{(k)} = \sum_{j=1}^{n} a_i^{(j)} b_j^{(k)} \qquad (i, k = 1, 2, \ldots, n). \tag{12}$$

Equation (12) differs from equation (14) of Sec. 29.3 only by the choice of indices, but the role of the indices is the same. Therefore, the *desired matrix C is the product of the matrices A and B.*

36.2. Similarly, we can construct the matrix corresponding to consecutive transformations of the components of a vector. Let $\xi_1, \xi_2, \ldots, \xi_n$ be the components of the vector x with respect to the basis $\{e\}$ and let the quantities $\eta_1, \eta_2, \ldots, \eta_n$ and $\tau_1, \tau_2, \ldots, \tau_n$ be defined by the equations

$$\eta_j = \sum_{i=1}^{n} a_i^{(j)} \xi_i \qquad (j = 1, 2, \ldots, n),$$

$$\tau_k = \sum_{j=1}^{n} b_j^{(k)} \eta_j \qquad (k = 1, 2, \ldots, n),$$

with corresponding matrices $A = \|a_i^{(j)}\|$ and $B = \|b_j^{(k)}\|$. Then, just as before, we can express the quantities $\tau_1, \tau_2, \ldots, \tau_n$ directly in terms of the quantities $\xi_1, \xi_2, \ldots, \xi_n$ by the formulas

$$\tau_k = \sum_{i=1}^{n} \left(\sum_{j=1}^{n} a_i^{(j)} b_j^{(k)} \right) \xi_i = \sum_{i=1}^{n} c_i^{(k)} \xi_i \qquad (k = 1, 2, \ldots, n),$$

where the quantities $c_i^{(k)}$ $(i, k = 1, 2, \ldots, n)$ form a matrix C equal to the product of the matrices A and B. To prove this, it suffices to replace the vectors e_i, f_j, g_k everywhere in the proof given in the first part of this section by the numbers ξ_i, η_j, τ_k, respectively.

Problem. Choose three bases in the plane, and let the components of the vector x with respect to these three bases be ξ_1 and ξ_2, η_1 and η_2, τ_1 and τ_2, respectively. Suppose that

$$\eta_1 = a_{11}\xi_1 + a_{12}\xi_2, \qquad \eta_2 = a_{21}\xi_1 + a_{22}\xi_2;$$
$$\tau_1 = b_{11}\xi_1 + b_{12}\xi_2, \qquad \tau_2 = b_{21}\xi_1 + b_{22}\xi_2,$$

and $A = \|a_{ij}\|$, $B = \|b_{ij}\|$. Express the components τ_1, τ_2 in terms of the components η_1, η_2.

Ans. The matrix of the desired transformation is $C = BA^{-1}$.

37. Transformation of the Coefficients of a Linear Form

Let $f(x)$ be a linear form defined on an n-dimensional space R. As we saw in Sec. 25, if a basis

$$\{e\} = \{e_1, e_2, \ldots, e_n\}$$

is chosen in R, then the values of the form $f(x)$ can be calculated from the formula

$$f(x) = \sum_{k=1}^{n} c_k \xi_k,$$

where the ξ_k $(k = 1, 2, \ldots, n)$ are the components of the vector x with respect to the basis $\{e\}$, and the coefficients c_k are given by

$$c_k = f(e_k) \qquad (k = 1, 2, \ldots, n).$$

The coefficients c_k obviously depend on the choice of the basis $\{e\}$. We now derive the rule which governs the transformation of the coefficients of a linear form when we go over to a new basis. Suppose the formulas

$$g_j = \sum_{i=1}^{n} a_i^{(j)} e_i \qquad (j = 1, 2, \ldots, n) \tag{13}$$

define the transformation from the basis $\{e\}$ to the new basis $\{g\}$. We wish to find the coefficients of the linear form $f(x)$ in the basis $\{g\}$. These coefficients are the numbers $d_j = f(g_j)$; using (13) to calculate these numbers, we find

$$d_j = f(g_j) = \sum_{i=1}^{n} a_i^{(j)} f(e_i) = \sum_{i=1}^{n} a_i^{(j)} c_i.$$

Thus, *the coefficients of a linear form transform in the same way as the basis vectors themselves.*

Problem.   Given a linear form $f(x) \not\equiv 0$ in an n-dimensional space, choose a basis $g_1, g_2, \ldots, g_n$ such that the relation

$$f(x) = \eta_1$$

holds for every vector

$$x = \sum_{k=1}^{n} \eta_k g_k.$$

38. Transformation of the Matrix of a Linear Operator

38.1.   Given a linear operator A in an n-dimensional space, let $A_{(e)} = \|a_i^{(j)}\|$ be its matrix relative to the basis

$$\{e\} = \{e_1, e_2, \ldots, e_n\}$$

and let $A_{(f)} = \|\alpha_q^{(p)}\|$ be its matrix relative to the basis

$$\{f\} = \{f_1, f_2, \ldots, f_n\}.$$

Suppose that the transformation formulas from the basis $\{e\}$ to the basis $\{f\}$ have the form

$$f_k = \sum_{j=1}^{n} c_j^{(k)} e_j \qquad (k = 1, 2, \ldots, n), \tag{14}$$

and let C denote the matrix $\|c_j^{(k)}\|$.   We now find the relation between the matrices $A_{(e)}, A_{(f)},$ and C.   The matrix $A_{(e)}$ is defined by the system of equations

$$A e_j = \sum_{i=1}^{n} a_i^{(j)} e_i \qquad (j = 1, 2, \ldots, n), \tag{15}$$

and the matrix $A_{(f)}$ by the system of equations

$$A f_p = \sum_{q=1}^{n} \alpha_q^{(p)} f_q \qquad (p = 1, 2, \ldots, n).$$

In the last equation, we replace the vectors f_q by their expressions in terms of the vectors $\{e\}$ by using (14).   The result is

$$A f_p = \sum_{q=1}^{n} \alpha_q^{(p)} \sum_{i=1}^{n} c_i^{(q)} e_i = \sum_{i=1}^{n} \left(\sum_{q=1}^{n} c_i^{(q)} \alpha_q^{(p)} \right) e_i.$$

Using (14) and (15), we can also write Af_p as

$$Af_p = A \sum_{j=1}^{n} c_j^{(p)} e_j = \sum_{j=1}^{n} c_j^{(p)} A e_j$$

$$= \sum_{j=1}^{n} c_j^{(p)} \sum_{i=1}^{n} a_i^{(j)} e_i = \sum_{i=1}^{n} \left(\sum_{j=1}^{n} a_i^{(j)} c_j^{(p)} \right) e_i.$$

Comparing coefficients of e_i in the last two expansions, we obtain

$$\sum_{q=1}^{n} c_i^{(q)} \alpha_q^{(p)} = \sum_{j=1}^{n} a_i^{(j)} c_j^{(p)},$$

or in matrix form

$$CA_{(f)} = A_{(e)}C. \tag{16}$$

This is the desired relation between the matrices $A_{(e)}$, $A_{(f)}$ and C. Multiplying on the left by the matrix C^{-1}, we obtain an expression for the matrix $A_{(f)}$:

$$A_{(f)} = C^{-1}A_{(e)}C.$$

38.2. Using the theorem on the determinant of a product of two matrices (Sec. 29.5), we obtain from (16) the following relation

$$\det C \det A_{(f)} = \det A_{(e)} \det C,$$

or, since $\det C \neq 0$,

$$\det A_{(e)} = \det A_{(f)}.$$

Thus, *the determinant of the matrix of an operator does not depend on the choice of a basis in the space.* Hence, we can talk about the determinant of an operator, meaning thereby the determinant of the matrix of the operator relative to any basis.

38.3. In addition to the determinant, there exist other functions of the matrix elements of an operator which remain unchanged under transformations to a new basis. To construct these functions, consider the operator $A - \lambda E$, where λ is a parameter. This operator obviously has the matrices $A_{(e)} - \lambda E$ and $A_{(f)} - \lambda E$ in the bases $\{e\}$ and $\{f\}$. By what was just proved, we have

$$\det(A_{(e)} - \lambda E) = \det(A_{(f)} - \lambda E)$$

for any λ. Both sides of this equation are polynomials of degree n in λ. Since these polynomials are identically equal, they have the same coefficients for any power of λ. Therefore, these coefficients are functions of the matrix elements of the operator which are invariant under changes of basis. We now examine the nature of these functions. The determinant of the matrix $A_{(e)} - \lambda E$ has the form

$$\begin{vmatrix} a_1^{(1)} - \lambda & a_1^{(2)} & \cdots & a_1^{(n)} \\ a_2^{(1)} & a_2^{(2)} - \lambda & \cdots & a_2^{(n)} \\ \cdot & \cdot & \cdots & \cdot \\ a_n^{(1)} & a_n^{(2)} & \cdots & a_n^{(n)} - \lambda \end{vmatrix} = (-1)^n \lambda^n + \Delta_1 \lambda^{n-1} + \cdots + \Delta_{n-1}\lambda + \Delta_n.$$

It is an easy consequence of the definition of a determinant that the coefficient Δ_1 of λ^{n-1} equals the sum[2]

$$a_1^{(1)} + a_2^{(2)} + \cdots + a_n^{(n)}$$

of the diagonal elements, taken with the sign $(-1)^{n-1}$. The coefficient Δ_2 of λ^{n-2} is the sum of all the *principal minors*[3] of order 2, taken with the sign $(-1)^{n-2}$. Similarly, the coefficient Δ_k of λ^{n-k} is the sum of all the principal minors of order k, taken with the sign $(-1)^{n-k}$. Finally, the coefficient Δ_n of λ^0, i.e., the constant term, is obviously equal to just the determinant of the operator. The polynomial $\det(A_{(e)} - \lambda E)$, which, as we have just seen is independent of the choice of basis, is called the *characteristic polynomial of the operator A*.

Problem 1. Denote by $B(\lambda)$ the adjugate of the matrix $A - \lambda E$ (Problem 4 of Sec. 30). The matrix $B(\lambda)$ can be written in the form

$$B(\lambda) = B^{(0)} + \lambda B^{(1)} + \cdots + \lambda^{n-1} B^{(n-1)},$$

where $B^{(0)}, B^{(1)}, \ldots, B^{(n-1)}$ are numerical matrices (not containing λ). Show that these matrices satisfy the equations

$$-B^{(n-1)} = (-1)^n E,$$
$$B^{(n-1)}A - B^{(n-2)} = \Delta_1 E,$$
$$\cdots \qquad \qquad \cdots$$
$$B^{(1)}A - B^{(0)} = \Delta_{n-1}E,$$
$$B^{(0)}A = \Delta_n E,$$

where $(-1)^n$, $\Delta_1, \Delta_2, \ldots, \Delta_n$ are the coefficients of the characteristic polynomial of the operator A.

Hint. Use the result of Problem 4 of Sec. 30.

Problem 2. Show that the operator A satisfies the equation

$$(-1)^n A^n + \Delta_1 A^{n-1} + \cdots + \Delta_{n-1}A + \Delta_n E = 0.$$

Hint. Multiply the first of the matrix equations in Problem 1 by A^n, the second by A^{n-1}, and so on, and then add.

---

[2] The sum $a_1^{(1)} + a_2^{(2)} + \cdots + a_n^{(n)}$ is called the *trace* of the operator A.

[3] The minor $M_{i_1, i_2, \ldots, i_k}^{j_1, j_2, \ldots, j_k}$ is said to be a *principal minor* if $i_1 = j_1, i_2 = j_2, \ldots, i_k = j_k$.

Comment. Problem 2 of Sec. 29.2 dealt with the existence of annihilating polynomials for the operator A. In this problem, we have given a simple construction of one of these annihilating polynomials, where the degree of the polynomial is n; for large n, this is much smaller than the bound n^2 given in the problem of Sec. 29.

The problem arises of the extent to which knowledge of the characteristic polynomial determines the geometric properties of the operator itself. As an example, consider the diagonal operator A (Example 6 of Sec. 26), where it is assumed that all the numbers $\lambda_1, \lambda_2, \ldots, \lambda_n$ are different. The characteristic polynomial of the operator A, calculated from the matrix of the operator relative to the basis $e_1, e_2, \ldots, e_n$ is equal to

$$\begin{vmatrix} \lambda_1 - \lambda & 0 & \ldots & 0 \\ 0 & \lambda_2 - \lambda & \ldots & 0 \\ \cdot & \cdot & \ldots & \cdot \\ 0 & 0 & \ldots & \lambda_n - \lambda \end{vmatrix} = (\lambda_1 - \lambda)(\lambda_2 - \lambda) \ldots (\lambda_n - \lambda).$$

This is a polynomial of degree n, which has the numbers $\lambda_1, \lambda_2, \ldots, \lambda_n$ as its roots. If we transform to a new basis, then in general the operator A will no longer be diagonal, but, by what has been proved, the characteristic polynomial of the operator A remains unchanged. If the characteristic polynomial is calculated from the new matrix, then it will still be a polynomial with the roots $\lambda_1, \lambda_2, \ldots, \lambda_n$. Later (Chapter 9) we shall show that an operator A whose characteristic polynomial has n distinct real roots has a diagonal matrix in some basis. Thus, in the case where the characteristic polynomial of the operator A has n distinct real roots, the geometric nature of the operator A is completely determined, i.e., A is a diagonal operator. In the general case, however, as we shall see in the examples given in Chapter 9, there can exist operators which are different in their geometric properties but have identical characteristic polynomials.

*39. Tensors

The components of a vector, the coefficients of a linear form, the elements of the matrix of a linear operator, these are all examples of a general class of geometric objects called *tensors*. Before giving the definition of a tensor, we shall first revise our notation somewhat. The basis vectors of an n-dimensional space R will be denoted, as before, by the symbols $e_1, e_2, \ldots, e_n$ (with subscripts). The components of vectors, e.g., x and y, will be denoted by $\xi^1, \xi^2, \ldots, \xi^n$ and $\eta^1, \eta^2, \ldots, \eta^n$ (with superscripts). The coefficients of a linear form will be denoted by $c_1, c_2, \ldots, c_n$ (with subscripts). The matrix elements of a linear operator will be denoted by a_i^j, where the superscript designates the *row number* and the subscript designates the *column number* (in contradistinction to the notation adopted in Sec. 27).   The convenience of this arrangement of indices is contained in the following summation convention: If we have a sum of terms such that the summation index i occurs twice in the general term, once as a superscript and once as a subscript, *then we shall omit the summation sign.*   For example, with our convention, the expansion of the vector x with respect to the basis $\{e_1, e_2, \ldots, e_n\}$ takes the form

$$x = \xi^i e_i$$

(although the summation sign is omitted, summation over i is implied). The expression for a linear form $f(x)$ in terms of the components of the vector x and the coefficients of the form becomes

$$f(x) = c_i \xi^i$$

(summation over i is implied).   The result of applying the operator A to a basis vector takes the form

$$Ae_i = a_i^j e_j$$

(summation over j is implied).   The components η^j of the vector Ax are expressed in terms of the components of the vector x as follows:

$$\eta^j = a_i^j \xi^i$$

(summation over i is implied).

We shall denote quantities pertaining to a new coordinate system by the same symbols as used in the case of the old coordinate system but with primes on the indices.   Thus, we denote new basis vectors by $e_{1'}, e_{2'}, \ldots, e_{n'}$, new components of a vector x by $\xi^{1'}, \xi^{2'}, \ldots, \xi^{n'}$, etc.   The elements of the transformation matrix from the basis e_i to the basis $e_{i'}$ will be denoted by $p_{i'}^i$, so that

$$e_{i'} = p_{i'}^i e_i \tag{17}$$

(summation over i is implied).   The elements of the matrix of the inverse transformation will be denoted by $q_i^{i'}$, i.e.,

$$e_i = q_i^{i'} e_{i'} \tag{18}$$

(summation over i' is implied).   The matrix $q_i^{i'}$ is the inverse of the matrix $p_{i'}^i$; this fact can be expressed by the equation

$$p_{i'}^i q_j^{i'} = \begin{cases} 0 \text{ for } i \neq j \\ 1 \text{ for } i = j, \end{cases} \tag{19}$$

or the equation

$$p_{i'}^i q_i^{j'} = \begin{cases} 0 \text{ for } i' \neq j' \\ 1 \text{ for } i' = j'. \end{cases} \tag{20}$$

To shorten the notation, we denote by δ_j^i the quantity which depends on the indices i and j in such a way that it equals 0 when the indices are different and 1 when the indices are the same. Then, (19) can be written in the form

$$p_{i'}^i q_j^{i'} = \delta_j^i, \tag{21}$$

and (20) can be written in the form

$$p_{i'}^i q_i^{j'} = \delta_{i'}^{j'}. \tag{22}$$

To show the advantages of using our new notation, we derive once more the formulas by which the components of a vector, the coefficients of a linear form and the matrix elements of an operator transform when we go over to a new basis. Thus, suppose we have a vector

$$x = \xi^i e_i = \xi^{i'} e_{i'}.$$

Using (18) to replace e_i by $q_i^{i'} e_{i'}$, we obtain

$$x = \xi^i q_i^{i'} e_{i'} = \xi^{i'} e_{i'}.$$

Since the $e_{i'}$ constitute a basis, we have

$$\xi^{i'} = q_i^{i'} \xi^i. \tag{23}$$

This is just the transformation formula for the components of a vector. Next, suppose we have a linear form $f(x)$. The numbers $c_{i'}$ are defined as usual by the relations $c_{i'} = f(e_{i'})$. Substituting (17) for $e_{i'}$ we obtain

$$c_{i'} = f(p_{i'}^i e_i) = p_{i'}^i f(e_i) = p_{i'}^i c_i,$$

so that

$$c_{i'} = p_{i'}^i c_i, \tag{24}$$

which is the formula we want. Finally, suppose we have an operator A. The elements of its matrix in the new basis are defined by the relations

$$A e_{i'} = a_{i'}^{j'} e_{j'}.$$

Using (17) to substitute for the quantities $e_{i'}$ and $e_{j'}$, we obtain

$$p_{i'}^i A e_i = a_{i'}^{j'} p_{j'}^j e_j.$$

But $A e_i = a_i^j e_j$, so that the result is

$$p_{i'}^i a_i^j e_j = a_{i'}^{j'} p_{j'}^j e_j.$$

Since the e_j are basis vectors, we have

$$p_{i'}^i a_i^j = a_{i'}^{j'} p_{j'}^j.$$

To get $a_{i'}^{k'}$ on the left we multiply both sides by $q_j^{k'}$ and sum over the index j. Using the relation (22), we obtain

$$p_{i'}^i a_i^j q_j^{k'} = a_{i'}^{j'} p_{j'}^j q_j^{k'} = a_{i'}^{j'} \delta_{j'}^{k'}.$$

By definition of the quantity $\delta^{k'}_{j'}$, the sum over j' reduces to only one term corresponding to the value $j' = k'$. Thus, we obtain

$$a^{k'}_{i'} = p^i_{i'} q^{k'}_j a^j_i,\qquad (25)$$

which is the desired formula.

It is not hard to verify that all three transformation formulas which we have just derived are the same as those which we derived earlier in the ordinary way (Secs. 35, 37, 38). Equations (23), (24), and (25) have much in common. In the first place, these formulas are linear in the transformed quantities. Secondly, the coefficients in these formulas are elements of the matrix transforming the old basis into the new basis or elements of the matrix of the inverse transformation or, finally, elements of both matrices.

We are now in a position to give the definition of a tensor. Tensors are divided into three classes, *covariant*, *contravariant* and *mixed*. Moreover, every tensor has a definite *order*. We begin by defining a covariant tensor, which, to be explicit, we take to have *order* three. Suppose there is a rule which in every coordinate system of an n-dimensional space R allows us to construct n^3 numbers (components) T_{ijk}, each of which is specified by giving the indices i, j, k definite values from 1 to n. By definition, these numbers T_{ijk} form a *covariant tensor of order three* if in going to a new basis, the quantities T_{ijk} transform according to the formula

$$T_{i'j'k'} = p^i_{i'} p^j_{j'} p^k_{k'} T_{ijk}.$$

A covariant tensor of any other order is defined similarly; a tensor of order m has n^m components instead of n^3 components, and in the transformation formula there appear m factors of the form $p^i_{i'}$ instead of three factors. In particular, the coefficients of a linear form, which transform by (24), constitute a covariant tensor of order one.

We now define a contravariant tensor of order three. Suppose we have a rule which in every coordinate system allows us to construct n^3 numbers T^{ijk}, each of which is specified by giving the indices i, j, k definite values from 1 to n. By definition, these numbers T^{ijk} form a *contravariant tensor of order three* if in going to a new basis, the quantities T^{ijk} transform according to the formula

$$T^{i'j'k'} = q^{i'}_i q^{j'}_j q^{k'}_k T^{ijk}.$$

A contravariant tensor of any other order is defined similarly. In particular, the components of a vector form a contravariant tensor of order one.

The terms "covariant" and "contravariant," which have just been introduced, are very simply explained. "Covariant" means "transforming in the same way" as the basis vectors, i.e., by using the coefficients $p^i_{i'}$. "Contravariant" means "transforming in the opposite direction," i.e., by using the coefficients $q^{i'}_i$.

There is still the case of mixed tensors to consider. For example, n^3 numbers T_{ij}^k, specified in every coordinate system, form a *mixed tensor of order three, with two covariant indices and one contravariant index*, if in going to a new basis, the quantities T_{ij}^k transform according to the formula

$$T_{i'j'}^{k'} = p_{i'}^i p_{j'}^j q_k^{k'} T_{ij}^k.$$

A mixed tensor with l covariant indices and m contravariant indices is defined similarly. In particular, the elements of the matrix of a linear operator form a mixed tensor of order two, with one covariant index and one contravariant index. We note that the arrangement of indices which we have chosen indicates at a glance the character of any tensor. Later, we shall encounter concrete examples of various kinds of tensors.

Operations on tensors. We can define the operation of *addition* for two tensors of the same structure, e.g., for two tensors T_{ij}^k and S_{ij}^k (with two covariant indices and one contravariant index). In this case, the sum is a tensor Q_{ij}^k of the same structure, defined as follows: In every coordinate system, the component of Q_{ij}^k with fixed indices i, j, k is the sum of the corresponding components of T_{ij}^k and S_{ij}^k. The fact that the quantities Q_{ij}^k actually form a tensor, and indeed one of the same structure as T_{ij}^k and S_{ij}^k, is implied by the following equality:

$$Q_{i'j'}^{k'} = T_{i'j'}^{k'} + S_{i'j'}^{k'} = p_{i'}^i p_{j'}^j q_k^{k'} T_{ij}^k + p_{i'}^i p_{j'}^j q_k^{k'} S_{ij}^k$$
$$= p_{i'}^i p_{j'}^j q_k^{k'} (T_{ij}^k + S_{ij}^k) = p_{i'}^i p_{j'}^j q_k^{k'} Q_{ij}^k.$$

The operation of *multiplication* is applicable to tensors of any structure. For example, let us multiply a tensor T_{ij} by a tensor S_k^l. The result is a tensor Q_{ijk}^l of order four. In any coordinate system its component with fixed indices i, j, k, l is defined to be equal to the product of the corresponding components of the factors T_{ij} and S_k^l. The tensor character of Q_{ijk}^l can be verified as follows:

$$Q_{i'j'k'}^{l'} = T_{i'j'} S_{k'}^{l'} = p_{i'}^i p_{j'}^j T_{ij} p_k^k q_l^{l'} S_k^l = p_{i'}^i p_{j'}^j p_k^k q_l^{l'} T_{ij} S_k^l = p_{i'}^i p_{j'}^j p_k^k q_l^{l'} Q_{ijk}^l.$$

Next, we consider still another operation called *contraction*. This operation can be applied to tensors which have at least one covariant index and one contravariant index. For example, suppose we have a tensor T_{ij}^k. To contract T_{ij}^k with respect to the superscript and the first subscript means to form the quantity

$$T_{ij}^i$$

in every coordinate system. Here summation over the index i is implied; as a result, the quantity $T_j = T_{ij}^i$ depends only on the index j. *Contraction of a tensor yields another tensor, whose order is two less than the order of the original tensor.* We verify this for the present example. We have

$$T_{j'} = T_{i'j'}^{i'} = p_{i'}^i p_{j'}^j q_k^{i'} T_{ij}^k = (p_{i'}^i q_k^{i'}) p_{j'}^j T_{ij}^k = \delta_k^i p_{j'}^j T_{ij}^k.$$

Here, the summation over k reduces to only one term, corresponding to the value $k = i$. Thus, we obtain

$$T_{j'} = p_{j'}^j T_{ij}^i = p_{j'}^j T_j,$$

as required.

What is the result of contracting a mixed tensor T_i^j of order two with respect to its two indices? The quantity $T = T_i^i$ no longer has even a single index, i.e., in every coordinate system it consists of just one number. This number is the same in every coordinate system, since

$$T' = T_{i'}^{i'} = p_{i'}^i q_j^{i'} T_i^j = \delta_j^i T_i^j = T_i^i = T.$$

Such a *scalar quantity*, which does not depend on the coordinate system, is called an *invariant*. Thus, by contracting tensors, we can obtain invariants of the tensors. For example, if we contract the tensor a_i^j corresponding to the linear operator A, the invariant a_i^i so obtained is the sum of the diagonal terms of the matrix of the operator A. We have already proved by another method that this quantity is invariant (Sec. 38).

Problem 1. Show that the quantities δ_i^j form a tensor of the second order, with one covariant index and one contravariant index.

Problem 2. A set of quantities S_{ij} is defined in every coordinate system by solving the system of equations

$$T^{ik} S_{ij} = \delta_j^k,$$

where T^{ik} is a tensor with two contravariant indices, with $\det \|T^{ik}\| \neq 0$. Show that S_{ij} is a tensor with two covariant indices.

Problem 3. If c_i and ξ^i have the same meaning as in the text, what is the geometric meaning of the contracted tensor $c_i \xi^i$?
Hint. See Sec. 25.2.

6

BILINEAR AND
QUADRATIC FORMS

In this chapter, we shall study linear numerical functions of two vector arguments. (The case of linear vector functions of two vector arguments will not be discussed.) Unlike the theory of numerical functions of *one* vector argument, the theory of numerical functions of *two* vector arguments has rich geometric content.

40. Bilinear Forms

40.1. A numerical function $A(x, y)$ of two vector arguments x, y in a linear space R is called a *bilinear form* or a *bilinear function* if it is a linear function of x for every fixed value of y and a linear function of y for every fixed value of x. In other words, $A(x, y)$ is a bilinear form in x and y if the following relations hold for any x, y, and z:

$$\left.\begin{array}{l} A(x + z, y) = A(x, y) + A(z, y) \\ \quad A(\alpha x, y) = \alpha A(x, y) \end{array}\right\} \quad \text{linearity in the first argument,}$$

$$\tag{1}$$

$$\left.\begin{array}{l} A(x, y + z) = A(x, y) + A(x, z) \\ \quad A(x, \alpha y) = \alpha A(x, y) \end{array}\right\} \quad \text{linearity in the second argument.}$$

Example 1. If $f_1(x)$ and $f_2(x)$ are linear forms, then $A(x, y) = f_1(x) f_2(y)$ is obviously a bilinear form in x and y.

Example 2. An example of a bilinear form in an n-dimensional linear space with a fixed basis $e_1, e_2, \ldots, e_n$ is the function

$$A(x, y) = \sum_{i=1}^{n} \sum_{k=1}^{n} a_{ik} \xi_i \eta_k,$$

where

$$x = \sum_{i=1}^{n} \xi_i e_i, \qquad y = \sum_{i=1}^{n} \eta_k e_k$$

are arbitrary vectors and a_{ik} ($i, k = 1, 2, \ldots, n$) are fixed numbers.

Example 3. In the space $C(a, b)$, the function

$$A(x, y) = \int_a^b \int_a^b K(s, t) x(t) y(s) \, ds \, dt,$$

where $K(s, t)$ is a given continuous function of s and t, is obviously a bilinear form in the vectors $x(t)$ and $y(t)$.

Example 4. In the space V_3, the scalar product of two vectors x and y is a bilinear form in x and y.

By using the relations (1) which define a bilinear form, we can easily obtain the general formula

$$A\left(\sum_{i=1}^{k} \alpha_i x_i, \ \sum_{j=1}^{m} \beta_j y_j \right) = \sum_{i=1}^{k} \sum_{j=1}^{m} \alpha_i \beta_j A(x_i, y_j), \tag{2}$$

where $x_1, x_2, \ldots, x_k$; $y_1, y_2, \ldots, y_m$ are arbitrary vectors of the space R, and $\alpha_1, \alpha_2, \ldots, \alpha_k$; $\beta_1, \beta_2, \ldots, \beta_m$ are any real numbers.

Bilinear forms defined on infinite-dimensional spaces are usually called *bilinear functionals*.

40.2. The general representation of a bilinear form defined on an n-dimensional linear space. Suppose we have a bilinear form $A(x, y)$ defined on an n-dimensional linear space R. Choose an arbitrary basis $e_1, e_2, \ldots, e_n$ in R, and write

$$A(e_i, e_k) = a_{ik} \qquad (i, k = 1, 2, \ldots, n).$$

Then for any two vectors

$$x = \sum_{i=1}^{n} \xi_i e_i, \qquad y = \sum_{k=1}^{n} \eta_k e_k,$$

according to (2) we have

$$A(x, y) = A\left(\sum_{i=1}^{n} \xi_i e_i, \ \sum_{k=1}^{n} \eta_k e_k \right) = \sum_{i=1}^{n} \sum_{k=1}^{n} \xi_i \eta_k A(e_i, e_k)$$

$$= \sum_{i=1}^{n} \sum_{k=1}^{n} a_{ik} \xi_i \eta_k. \tag{3}$$

Thus, in Example 2, we have already encountered the most general representation of a bilinear form defined on an n-dimensional linear space.

The coefficients a_{ik} form a square matrix

$$A = A_{(e)} = \begin{Vmatrix} a_{11} & a_{12} & \dots & a_{1n} \\ a_{21} & a_{22} & \dots & a_{2n} \\ \cdot & \cdot & \dots & \cdot \\ a_{n1} & a_{n2} & \dots & a_{nn} \end{Vmatrix} = \|a_{ik}\|,$$

which we shall call the *matrix of the bilinear form* $A(x, y)$ relative to the basis

$$\{e\} = \{e_1, e_2, \dots, e_n\}.$$

40.3. Symmetric bilinear forms. A bilinear form is called *symmetric* if

$$A(x, y) = A(y, x)$$

for any vectors x and y. If the bilinear form $A(x, y)$ is symmetric, then

$$a_{ik} = A(e_i, e_k) = A(e_k, e_i) = a_{ki},$$

so that the matrix $A_{(e)}$ of a symmetric bilinear form, relative to any basis $e_1, e_2, \dots, e_n$ of the space R, equals its own transpose $A'_{(e)}$. It is easily verified that the converse is also true, i.e., if $A'_{(e)} = A_{(e)}$ relative to any basis $e_1, e_2, \dots, e_n$, then the form $A(x, y)$ is symmetric. In fact, we have

$$A(y, x) = \sum_{i,k=1}^{n} a_{ik}\eta_i\xi_k = \sum_{i,k=1}^{n} a_{ki}\eta_i\xi_k = \sum_{i,k=1}^{n} a_{ik}\xi_i\eta_k = A(x, y),$$

as required. In particular, we have the following result: If the matrix of the bilinear form $A(x, y)$ calculated in any basis equals its own transpose, then the matrix of the form calculated in any other basis also equals its own transpose. A matrix which equals its own transpose will henceforth be called *symmetric*.

40.4. Transformation of the matrix of a bilinear form when the basis is changed. Of course, if we transform to a new basis, the matrix of a bilinear form changes according to a certain transformation law; we now find this law. Let $A_{(e)} = \|a_{ik}\|$ be the matrix of the bilinear form $A(x, y)$ relative to the basis

$$\{e\} = \{e_1, e_2, \dots, e_n\}$$

and let $A_{(f)} = \|b_{ik}\|$ be the matrix of the same form relative to the basis

$$\{f\} = \{f_1, f_2, \dots, f_n\}.$$

Assuming that the transformation from one basis to the other is described by the formula

$$f_i = \sum_{j=1}^{n} c_j^{(i)} e_j \qquad (i = 1, 2, \dots, n)$$

with the transformation matrix $C = \|c_j^{(i)}\|$, we have

$$b_{ik} = A(f_i, f_k) = A\left(\sum_{j=1}^{n} c_j^{(i)} e_j, \sum_{l=1}^{n} c_l^{(k)} e_l \right)$$

$$= \sum_{j,l=1}^{n} c_j^{(i)} c_l^{(k)} A(e_j, e_l) = \sum_{j,l=1}^{n} c_j^{(i)} c_l^{(k)} a_{jl}.$$

This formula can be written in the form

$$b_{ik} = \sum_{j=1}^{n} \sum_{l=1}^{n} c_i^{(j)'} a_{jl} c_l^{(k)}, \tag{4}$$

where $c_i^{(j)'} = c_j^{(i)}$ is an element of the matrix C' which is the transpose of the matrix C. Equation (4) corresponds to the following matrix relation (Sec. 29):

$$A_{(f)} = C' A_{(e)} C. \tag{5}$$

Since the matrices C and C' are nonsingular, then by Corollary 1 of Theorem 24 (Sec. 31), the rank of the matrix $A_{(f)}$ equals the rank of the matrix $A_{(e)}$, i.e., *the rank of the matrix of a bilinear form is independent of the choice of a basis.* By the theorem on the determinant of a product of matrices, we obtain the relation

$$\det A_{(f)} = \det A_{(e)} (\det C)^2. \tag{6}$$

Problem. Do the elements of the matrix of a bilinear form constitute a tensor (Sec. 39), and if so, of what type?

Ans. A tensor of order two, with two covariant indices.

41. Quadratic Forms

One of the basic problems of plane analytic geometry is to reduce the general equation of a second-degree curve to canonical form by transforming to a new coordinate system. The equation of a second-degree curve with center at the origin $x = 0$, $y = 0$, has the familiar form

$$Ax^2 + 2Bxy + Cy^2 = D. \tag{7}$$

A coordinate transformation is described by the formulas

$$x = a_{11}x' + a_{12}y',$$
$$y = a_{21}x' + a_{22}y',$$

where a_{11}, a_{12}, a_{21}, a_{22} are certain numbers (usually sines and cosines of the angle through which the axes are rotated). As a result of this coordinate transformation, (7) takes the simpler form

$$A'x'^2 + B'y'^2 = D.$$

An analogous problem can be stated for a space with any number of dimensions. The solution of this and related problems is the fundamental aim of the theory of quadratic forms, which we now present.

We begin with the following definition:

A *quadratic form* defined on a linear space R is a function $A(x, x)$ of one vector argument $x \in R$, obtained by changing y to x in any bilinear form $A(x, y)$.

According to (2), in an n-dimensional linear space with a fixed basis

$$\{e\} = \{e_1, e_2, \ldots, e_n\},$$

every quadratic form can be written as

$$A(x, x) = \sum_{i=1}^{n} \sum_{k=1}^{n} a_{ik} \xi_i \xi_k, \tag{8}$$

where $\xi_1, \xi_2, \ldots, \xi_n$ are the components of the vector x with respect to the basis $\{e\}$. Conversely, *a function $A(x, x)$ of the vector x, defined relative to the basis $\{e\}$ by* (8), *represents a quadratic form in x.* In fact, introduce the bilinear form

$$B(x, y) = \sum_{i=1}^{n} \sum_{k=1}^{n} a_{ik} \xi_i \eta_k,$$

where $\eta_1, \eta_2, \ldots, \eta_n$ are the components of the vector y with respect to the basis $\{e\}$. Then, obviously, the function $A(x, x)$ is just the quadratic form $B(x, x)$.

We can write the double sum (8) somewhat differently by combining similar terms. Let $b_{ii} = a_{ii}$ and $b_{ik} = a_{ik} + a_{ki}$ $(i \neq k)$. Then, since

$$a_{ik} \xi_i \xi_k + a_{ki} \xi_k \xi_i = (a_{ik} + a_{ki}) \xi_i \xi_k = b_{ik} \xi_i \xi_k,$$

the double sum (8) can be written as

$$A(x, x) = \sum_{k=1}^{n} \sum_{i \leqslant k} b_{ik} \xi_i \xi_k,$$

and has fewer terms. It follows that two different bilinear forms

$$A(x, y) = \sum_{i, k=1}^{n} a_{ik} \xi_i \eta_k, \quad C(x, y) = \sum_{i, k=1}^{n} c_{ik} \xi_i \eta_k$$

can reduce to the same quadratic form after y is replaced by x; all that is necessary is that $a_{ik} + a_{ki} = c_{ik} + c_{ki}$ for any i and k. Thus, in general, we cannot reconstruct uniquely the bilinear form generating a given quadratic form. However, in the case where it is known that the original bilinear form is *symmetric*, it *can* be reconstructed. In fact, if $a_{ik} = a_{ki}$, then the coefficients a_{ik} of the bilinear form are uniquely determined by the equation

$$a_{ik} + a_{ki} = b_{ik} \qquad (i \neq k),$$

i.e.,

$$a_{ik} = a_{ki} = \tfrac{1}{2} b_{ik} \qquad (i \neq k), \tag{9}$$
$$a_{ii} = b_{ii},$$

so that the bilinear form itself is uniquely determined. This assertion can be proved without recourse to bases and components. In fact, we have

$$A(x + y, x + y) = A(x, x) + A(x, y) + A(y, x) + A(y, y)$$

by the definition of a bilinear form, and

$$A(x, y) = \tfrac{1}{2}[A(x, y) + A(y, x)] = \tfrac{1}{2}[A(x + y, x + y) - A(x, x) - A(y, y)]$$

by the assumption that $A(x, y)$ is symmetric. Therefore, the value of the bilinear form $A(x, y)$ for any pair of vectors x, y is uniquely determined by the values of the corresponding quadratic form for the vectors x, y, and $x + y$. On the other hand, to obtain all possible quadratic forms, we need only use symmetric bilinear forms. In fact, if $A(x, y)$ is an arbitrary bilinear form, then

$$A_1(x, y) = \tfrac{1}{2}[A(x, y) + A(y, x)]$$

is a symmetric bilinear form, and

$$A_1(x, x) = \tfrac{1}{2}[A(x, x) + A(x, x)] = A(x, x),$$

i.e., the quadratic forms $A_1(x, x)$ and $A(x, x)$ coincide. These considerations show that in using bilinear forms to study the properties of quadratic forms, we need only consider the case of *symmetric* bilinear forms.

42. Reduction of Quadratic Forms to Canonical Form

Suppose we are given an arbitrary quadratic form $A(x, x)$ defined on an n-dimensional linear space R_n. We now show that *there exists a basis* $\{f\} = f_1, f_2, \ldots, f_n$ *in R_n such that, for any vector*

$$x = \sum_{k=1}^{n} \eta_k f_k,$$

the quadratic form $A(x, x)$ is given by

$$A(x, x) = \lambda_1 \eta_1^2 + \lambda_2 \eta_2^2 + \cdots + \lambda_n \eta_n^2, \tag{10}$$

where $\lambda_1, \lambda_2, \ldots, \lambda_n$ are certain fixed numbers. Every basis with this property will be called a *canonical basis* of the form $A(x, x)$, and the expression

(10) will be called the *canonical form* of $A(x, x)$; in particular, the numbers $\lambda_1, \lambda_2, \ldots, \lambda_n$ will be called the *canonical coefficients* of $A(x, x)$.

Let $\{e_1, e_2, \ldots, e_n\}$ be an arbitrary basis of the space R_n; if

$$x = \sum_{k=1}^{n} \xi_k e_k,$$

then as we have already seen, $A(x, x)$ can be written in the form

$$A(x, x) = \sum_{k=1}^{n} \sum_{i \leqslant k} b_{ik} \xi_i \xi_k. \tag{11}$$

According to Sec. 35, our assertion will be proved if we can write a system of equations

$$
\begin{aligned}
\eta_1 &= c_{11}\xi_1 + c_{12}\xi_2 + \cdots + c_{1n}\xi_n, \\
\eta_2 &= c_{21}\xi_1 + c_{22}\xi_2 + \cdots + c_{2n}\xi_n, \\
& \quad \cdots \quad \cdots \quad \cdots \quad \cdots \\
\eta_n &= c_{n1}\xi_1 + c_{n2}\xi_2 + \cdots + c_{nn}\xi_n,
\end{aligned}
\tag{12}
$$

with a nonsingular matrix $C = \|c_{ik}\|$, such that if we express the variables $\xi_1, \xi_2, \ldots, \xi_n$ appearing in (11) in terms of $\eta_1, \eta_2, \ldots, \eta_n$, we thereby transform (11) to the form (10). We shall carry out the proof by induction on the number of variables ξ_i actually appearing in (11), i.e., those which have nonzero coefficients.

If (11) actually contains only one variable ξ_1 (say), i.e., if (11) has the form

$$A(x, x) = b_{11}\xi_1^2,$$

then the basis $\{e_1, e_2, \ldots, e_n\}$ is already canonical, with

$$\lambda_1 = b_{11}, \lambda_2 = \lambda_3 = \cdots = \lambda_n = 0.$$

Suppose now that every form actually containing $m - 1$ variables $\xi_1, \xi_2 \ldots, \xi_{m-1}$, say, can be reduced to canonical form by using a transformation with a nonsingular matrix. Consider a form (11) which actually contains the m variables $\xi_1, \xi_2, \ldots, \xi_m$. First we assume that one of the numbers $b_{11}, b_{22}, \ldots, b_{mm}$, say b_{mm}, is nonzero. We group together all the terms in (11) which contain the variable ξ_m; these terms can be written in the form

$$b_{1m}\xi_1\xi_m + b_{2m}\xi_2\xi_m + \cdots + b_{m-1,m}\xi_{m-1}\xi_m + b_{mm}\xi_m^2$$

$$= b_{mm}\left(\frac{b_{1m}}{2b_{mm}}\xi_1 + \frac{b_{2m}}{2b_{mm}}\xi_2 + \cdots + \frac{b_{m-1,m}}{2b_{mm}}\xi_{m-1} + \xi_m\right)^2 + A_1(x, x),$$

where $A_1(x, x)$ denotes a quadratic form which depends only on the variables $\xi_1, \xi_2, \ldots, \xi_{m-1}$. Now consider the following coordinate transformation

$$\tau_1 = \xi_1,$$
$$\tau_2 = \xi_2,$$
$$\cdots\cdots$$
$$\tau_{m-1} = \xi_{m-1},$$
$$\tau_m = \frac{b_{1m}}{2b_{mm}}\,\xi_1 + \frac{b_{2m}}{2b_{mm}}\,\xi_2 + \cdots + \frac{b_{m-1,m}}{2b_{mm}}\,\xi_{m-1} + \xi_m,$$

The matrix of this transformation is nonsingular (its determinant is actually 1). In the new coordinate system, $A(x, x)$ clearly has the form

$$A(x, x) = B(x, x) + b_{mm}\tau_m^2,$$

where the quadratic form $B(x, x)$ depends only on the variables $\tau_1, \tau_2, \ldots, \tau_{m-1}$. By the induction hypothesis, there exists a new transformation

$$\begin{aligned}
\eta_1 &= c_{11}\tau_1 + c_{12}\tau_2 + \cdots + c_{1,m-1}\tau_{m-1}, \\
\eta_2 &= c_{21}\tau_1 + c_{22}\tau_2 + \cdots + c_{2,m-1}\tau_{m-1}, \\
&\cdots \quad\quad \cdots \quad\quad \cdots \quad\quad \cdots \\
\eta_{m-1} &= c_{m-1,1}\tau_1 + c_{m-1,2}\tau_2 + \cdots + c_{m-1,m-1}\tau_{m-1},
\end{aligned} \tag{13}$$

with a nonsingular matrix $C = \|c_{ik}\|$, which carries $B(x, x)$ into the canonical form

$$B(x, x) = \lambda_1\eta_1^2 + \lambda_2\eta_2^2 + \cdots + \lambda_{m-1}\eta_{m-1}^2.$$

If we supplement the system of equations (13) with the additional equation $\eta_m = \tau_m$, we obtain a nonsingular transformation of the variables $\tau_1, \tau_2, \ldots, \tau_m$ into the variables $\eta_1, \eta_2, \ldots, \eta_m$, which carries $A(x, x)$ into the canonical form

$$A(x, x) = B(x, x) + b_{mm}\tau_m^2 = \lambda_1\eta_1^2 + \lambda_2\eta_2^2 + \cdots + \lambda_{m-1}\eta_{m-1}^2 + b_{mm}\eta_m^2.$$

According to Sec. 36, the direct transformation from the variables ξ_i to the variables η_i is accomplished by using the matrix equal to the product of the transformation matrix from ξ_i to τ_i and the transformation matrix from τ_i to η_i. Since both of these matrices are nonsingular, the product of the matrices is also nonsingular.

It remains only to consider the case of a quadratic form $A(x, x)$ in m variables $\xi_1, \xi_2, \ldots, \xi_m$, which has all the numbers $b_{11}, b_{22}, \ldots, b_{mm}$ equal to zero. Consider one of the terms $b_{ik}\xi_i\xi_k$ which has a nonzero coefficient b_{ik}; for example, let $b_{12} \neq 0$. Then carry out the following coordinate transformation:

$$\begin{aligned}
\xi_1 &= \xi_1' + \xi_2', \\
\xi_2 &= \xi_1' - \xi_2', \\
\xi_3 &= \xi_3', \\
&\cdots\cdots \\
\xi_m &= \xi_m'.
\end{aligned} \tag{14}$$

The determinant of the matrix of the transformation (14) equals -2, and therefore, this transformation is again nonsingular; the term $b_{12}\xi_1\xi_2$ is transformed into

$$b_{12}\xi_1\xi_2 = b_{12}\xi_1'^2 - b_{12}\xi_2'^2.$$

Thus, two squared terms with nonzero coefficients are produced simultaneously in the new form. (Clearly, these terms cannot cancel any of the other terms, since all the other terms contain a variable ξ_i' with $i > 2$.) We can now apply our inductive method to the quadratic form (11) written in the new variables ξ_i'. Thus, we have actually proved our theorem for any integer m, but the case $m = n$ suffices to prove the theorem for an arbitrary quadratic form defined on an n-dimensional space.

The idea of our proof, i.e., consecutive isolation of complete squares, can be used as a practical method for reducing a given quadratic form to canonical form. However, in Sec. 45 we shall describe another method, which permits us to obtain directly both the canonical form and the vectors of the canonical basis.

Example. To reduce the form

$$A(x, x) = \xi_1^2 + 6\xi_1\xi_2 + 5\xi_2^2 - 4\xi_1\xi_3 - 12\xi_2\xi_3 + 4\xi_3^2 - 4\xi_2\xi_4 - 8\xi_3\xi_4 - \xi_4^2$$

to canonical form, we first complete the square in the group of terms containing ξ_1, writing

$$\eta_1 = \xi_1 + 3\xi_2 - 2\xi_3.$$

Then the form is transformed into

$$A(x, x) = \eta_1^2 - 4\xi_2^2 - 4\xi_2\xi_4 - 8\xi_3\xi_4 - \xi_4^2.$$

Next we complete the square in the group of terms containing ξ_2, writing

$$\eta_2 = 2\xi_2 + \xi_4.$$

This reduces the form to

$$A(x, x) = \eta_1^2 - \eta_2^2 - 8\xi_3\xi_4.$$

There are no squares of the variables ξ_3 and ξ_4. Therefore, we write

$$\xi_3 = \eta_3 - \eta_4,$$
$$\xi_4 = \eta_3 + \eta_4,$$

so that $\xi_3\xi_4 = \eta_3^2 - \eta_4^2$. Thus, the form $A(x, x)$ is reduced to the canonical form

$$A(x, x) = \eta_1^2 - \eta_2^2 - 8\eta_3^2 + 8\eta_4^2$$

by the transformation

$$\eta_1 = \xi_1 + 3\xi_2 - 2\xi_3, \quad \eta_2 = 2\xi_2 + \xi_4,$$
$$\eta_3 = \tfrac{1}{2}\xi_3 + \tfrac{1}{2}\xi_4, \quad \eta_4 = -\tfrac{1}{2}\xi_3 + \tfrac{1}{2}\xi_4.$$

Problem. Reduce the quadratic form

$$\xi_1\xi_2 + \xi_2\xi_3 + \xi_3\xi_1$$

to canonical form.

Ans. For example

$$\eta_1^2 - \eta_2^2 - \eta_3^2,$$

where

$$\eta_1 = \tfrac{1}{2}\xi_1 + \tfrac{1}{2}\xi_2 + \xi_3, \quad \eta_2 = \tfrac{1}{2}\xi_1 - \tfrac{1}{2}\xi_2, \quad \eta_3 = \xi_3.$$

43. Uniqueness Questions

Neither the canonical basis nor the canonical form of a quadratic form is uniquely determined. For example, any permutation of the vectors of a canonical basis gives another canonical basis. In Sec. 45, it will be shown, among other things, that with a few rare exceptions, a canonical basis for a given quadratic form can be constructed by choosing an *arbitrary* vector of the space as the first vector of the basis. Moreover, if $A(x, x)$ is written in the canonical form

$$A(x, x) = \lambda_1\eta_1^2 + \lambda_2\eta_2^2 + \cdots + \lambda_n\eta_n^2,$$

where $\eta_1, \eta_2, \ldots, \eta_n$ are the components of the vector x, then the transformation

$$\eta_1 = \alpha_1\tau_1,$$
$$\eta_2 = \alpha_2\tau_2,$$
$$\cdots\cdots$$
$$\eta_n = \alpha_n\tau_n$$

(where $\alpha_1, \alpha_2, \ldots, \alpha_n$ are fixed numbers all different from zero, and $\tau_1, \tau_2, \ldots, \tau_n$ are new components) carries $A(x, x)$ into the new form

$$A(x, x) = (\lambda_1\alpha_1^2)\tau_1^2 + (\lambda_2\alpha_2^2)\tau_2^2 + \cdots + (\lambda_n\alpha_n^2)\tau_n^2,$$

which is also canonical but has different coefficients. We note that in this example, the new coefficients of the form $A(x, x)$ have the same signs as the old coefficients. Thus, the total number of positive coefficients and the total number of negative coefficients remain unchanged. It turns out that this property holds in the general case as well, i.e, the following theorem, called the *law of inertia for quadratic forms*, is valid:

THEOREM 26. (*Law of inertia for quadratic forms.*) *The number of positive coefficients and the number of negative coefficients in the canonical form of a quadratic form $A(x, x)$ are invariants of the form, i.e., do not depend on the choice of a canonical basis.*

Proof. Consider the quadratic form $A(x, x)$, which in the basis

$$\{e\} = \{e_1, e_2, \ldots, e_n\}$$

has the form

$$A(x, x) = \sum_{i,k=1}^{n} a_{ik}\xi_i\xi_k,$$

where $\xi_1, \xi_2, \ldots, \xi_n$ are the components of the vector x with respect to the basis $\{e\}$. Assume that $A(x, x)$ has two canonical bases

$$\{f\} = \{f_1, f_2, \ldots, f_n\}$$

and

$$\{g\} = \{g_1, g_2, \ldots, g_n\}.$$

Let $\eta_1, \eta_2, \ldots, \eta_n$ denote the components of the vector x with respect to the basis $\{f\}$, and let $\tau_1, \tau_2, \ldots, \tau_n$ denote the components of x with respect to the basis $\{g\}$. Let the corresponding transformation formulas be

$$\begin{aligned}
\eta_1 &= b_{11}\xi_1 + b_{12}\xi_2 + \cdots + b_{1n}\xi_n, \\
\eta_2 &= b_{21}\xi_1 + b_{22}\xi_2 + \cdots + b_{2n}\xi_n, \\
&\quad \cdots \quad\quad \cdots \quad\quad \cdots \quad\quad \cdots \\
\eta_n &= b_{n1}\xi_1 + b_{n2}\xi_2 + \cdots + b_{nn}\xi_n,
\end{aligned} \tag{15a}$$

and

$$\begin{aligned}
\tau_1 &= c_{11}\xi_1 + c_{12}\xi_2 + \cdots + c_{1n}\xi_n, \\
\tau_2 &= c_{21}\xi_1 + c_{22}\xi_2 + \cdots + c_{2n}\xi_n, \\
&\quad \cdots \quad\quad \cdots \quad\quad \cdots \quad\quad \cdots \\
\tau_n &= c_{n1}\xi_1 + c_{n2}\xi_2 + \cdots + c_{nn}\xi_n,
\end{aligned} \tag{15b}$$

where the matrices $\|b_{ik}\|$ and $\|c_{ik}\|$ are nonsingular. In the basis $\{f\}$, $A(x, x)$ has the form

$$A(x, x) = \alpha_1\eta_1^2 + \cdots + \alpha_k\eta_k^2 - \alpha_{k+1}\eta_{k+1}^2 - \cdots - \alpha_m\eta_m^2, \tag{16}$$

and in the basis $\{g\}$, it has the form

$$A(x, x) = \beta_1\tau_1^2 + \cdots + \beta_p\tau_p^2 - \beta_{p+1}\tau_{p+1}^2 - \cdots - \beta_q\tau_q^2, \tag{17}$$

where the numbers $\alpha_1, \ldots, \alpha_m, \beta_1, \ldots, \beta_q$ are assumed to be *positive*. We wish to show that $k = p$, $m = q$. Equating the right-hand sides of (16) and (17) and transposing negative terms to the opposite sides of the equation, we obtain

$$\begin{aligned}
\alpha_1\eta_1^2 &+ \cdots + \alpha_k\eta_k^2 + \beta_{p+1}\tau_{p+1}^2 + \cdots + \beta_q\tau_q^2 \\
&= \alpha_{k+1}\eta_{k+1}^2 + \cdots + \alpha_m\eta_m^2 + \beta_1\tau_1^2 + \cdots + \beta_p\tau_p^2.
\end{aligned} \tag{18}$$

Now assume that $k < p$, and consider the vectors x which satisfy the conditions

$$\begin{aligned}
\eta_1 &= 0, \eta_2 = 0, \ldots, \eta_k = 0, \\
\tau_{p+1} &= 0, \ldots, \tau_q = 0, \tau_{q+1} = 0, \ldots, \tau_n = 0.
\end{aligned} \tag{19}$$

There are clearly less than n of these conditions, since $k < p$. Using the formulas (15) to express $\eta_1, \ldots, \eta_k, \tau_{p+1}, \ldots, \tau_n$ in terms of the variables $\xi_1, \xi_2, \ldots, \xi_n$, we obtain a homogeneous system of linear equations in the unknowns $\xi_1, \xi_2, \ldots, \xi_n$. The number of equations is less than the number of unknowns, and therefore this homogeneous system has a nontrivial solution $x = (\xi_1, \xi_2, \ldots, \xi_n)$. On the other hand, because of (18), every vector x satisfying the conditions (19) also satisfies the conditions

$$\tau_1 = \tau_2 = \cdots = \tau_p = 0.$$

However, since $\det \|c_{ik}\| \neq 0$, any vector x for which

$$\tau_1 = \tau_2 = \cdots = \tau_p = \tau_{p+1} = \cdots = \tau_n = 0$$

must be the zero vector, with all its components $\xi_1, \xi_2, \ldots, \xi_n$ equal to zero. Thus, the assumption that $k < p$ leads to a contradiction. Because of the complete symmetry of the role played by the numbers k and p in this problem, the assertion $p < k$ also leads to a contradiction. It follows that $k = p$. Moreover, if we examine the conditions

$$\tau_1 = 0, \tau_2 = 0, \ldots, \tau_p = 0,$$
$$\eta_{k+1} = 0, \ldots, \eta_m = 0,$$
$$\tau_{q+1} = 0, \ldots, \tau_n = 0,$$

we can refute the assumption that $m < q$ or that $q < m$ by using the same argument as before and the symmetry property. Thus, we finally find that $k = p$, $m = q$, Q.E.D.

The total number of terms appearing in the canonical form of the quadratic form $A(x, x)$ is called the *rank* of the form or its *index of inertia*. The number of positive terms is called the *positive index of inertia*, and the number of negative terms is called the *negative index of inertia*. In the next section, we shall show how to calculate the rank of a quadratic form without actually having to reduce it to canonical form. Determination of the positive and negative indices of inertia is more complicated; we shall return to this problem in Secs. 45 and 69.

Problem 1. Let p be the positive index of inertia of the quadratic form $A(x, x)$ and let q be its negative index of inertia. Let there be given p positive numbers $\lambda_1, \lambda_2, \ldots, \lambda_p$ and q negative numbers $\mu_1, \mu_2, \ldots, \mu_q$. Show that there exists a basis in which $A(x, x)$ takes the form

$$A(x, x) = \lambda_1 \tau_1^2 + \cdots + \lambda_p \tau_p^2 + \mu_1 \tau_{p+1}^2 + \cdots + \mu_q \tau_{p+q}^2.$$

Comment. This result shows that the numbers p and q are the only invariants of a quadratic form.

Problem 2. Show that the matrix of a quadratic form of rank r always has at least one nonvanishing principal minor of order r.

Hint. Use the result of Problem 3 of Sec. 29.5.

A quadratic form whose rank equals the number of dimensions of the space is called *nonsingular*. If in addition, the positive index of inertia equals the rank, then the form is called *positive definite*. In other words, a quadratic form is positive definite if all n of its canonical coefficients are positive. It follows that *a positive definite form takes a positive value at every point of the space except the origin of coordinates*. Conversely, if a quadratic form defined on an n-dimensional space takes positive values everywhere except at the origin, then its rank is n and its positive index of inertia is also n, i.e., the form is positive definite. In fact, for a form of rank less than n or with less than n positive canonical coefficients, it is easy to find points in the space other than the origin, where the form takes either the value 0 or negative values. For example, the quadratic form

$$A(x, x) = \xi_1^2 + \xi_3^2$$

of rank two, defined on a three-dimensional space, takes the value zero for any nonzero vector with components $\xi_1 = 0, \xi_2 \neq 0, \xi_3 = 0$. For these vectors, the form

$$A(x, x) = \xi_1^2 - \xi_2^2 + \xi_3^2$$

of rank three, defined on a three-dimensional space, takes negative values. Clearly, these examples illustrate the full generality of the situation.

44. The Canonical Basis of a Bilinear Form

44.1. The vector x_1 is said to be *conjugate* to the vector y_1 with respect to the bilinear form $A(x, y)$ if

$$A(x_1, y_1) = 0.$$

If the vectors $x_1, x_2, \ldots, x_k$ are all conjugate to the vector y_1, then any vector of the linear manifold $L(x_1, x_2, \ldots, x_k)$ spanned by $x_1, x_2, \ldots, x_k$ is also conjugate to y_1. In fact, by the properties of a bilinear form, we have

$$A(\alpha_1 x_1 + \alpha_2 x_2 + \cdots + \alpha_k x_k, y_1)$$
$$= \alpha_1 A(x_1, y_1) + \alpha_2 A(x_2, y_1) + \cdots + \alpha_k A(x_k, y_1) = 0.$$

In general, a vector y_1 which is conjugate to every vector of a subspace $R' \subset R$ is said to be *conjugate to the subspace R'*. A basis $e_1, e_2, \ldots, e_n$ is called a *canonical basis for the bilinear form $A(x, y)$* if the basis vectors are conjugate to each other, i.e., if

$$A(e_i, e_k) = 0 \quad \text{for } i \neq k.$$

Example. In the space V_3, let the bilinear form $A(x, y)$ be the scalar product of the vectors x and y. Two vectors being conjugate with respect to this bilinear form obviously means that they are orthogonal. In this case, any orthogonal basis of the space V_3 serves as a canonical basis.

44.2. The matrix of a bilinear form relative to a canonical basis is diagonal, since

$$a_{ik} = A(e_i, e_k) = 0 \quad \text{for } i \neq k.$$

Since a diagonal matrix coincides with its own transpose, a bilinear form which has a canonical basis must be symmetric. (We recall that whether or not the matrix of a bilinear form is symmetric does not depend on the choice of a basis (Sec. 40.3).) We now prove the converse, i.e.,

Every symmetric bilinear form $A(x, y)$ has a canonical basis.

Proof. Consider the quadratic form $A(x, x)$ corresponding to the given bilinear form $A(x, y)$. We know that there exists a basis $e_1, e_2, \ldots, e_n$ for the space R, relative to which $A(x, x)$ can be written in the canonical form

$$A(x, x) = \sum_{i=1}^{n} \lambda_i \xi_i^2.$$

According to (9), the corresponding symmetric bilinear form $A(x, y)$ has the canonical form

$$A(x, y) = \sum_{i=1}^{n} \lambda_i \xi_i \eta_i, \tag{20}$$

where

$$y = \sum_{i=1}^{n} \eta_i e_i;$$

therefore, its matrix is diagonal. But this just means that the basis $e_1, e_2, \ldots, e_n$ is canonical for the form $A(x, y)$.

Problem. Reduce the bilinear form

$$A(x, y) = \xi_1 \eta_1 + \xi_1 \eta_2 + \xi_2 \eta_1 + 2\xi_2 \eta_2 + 2\xi_2 \eta_3 + 2\xi_3 \eta_2 + 5\xi_3 \eta_3$$

to canonical form.

Ans. $A(x, y) = \sigma_1 \tau_1 + \sigma_2 \tau_2 + \sigma_3 \tau_3$, where σ_i and τ_i $(i = 1, 2, 3)$ are the new components of the vectors x and y; the transformation formulas to the new basis are

$$\sigma_1 = \xi_1 + \xi_2, \quad \sigma_2 = \xi_2 + 2\xi_3, \quad \sigma_3 = \xi_3.$$

44.3. The law of inertia which we proved for quadratic forms generalizes immediately to the case of symmetric bilinear forms, i.e., *the number of* *positive coefficients and the number of negative coefficients in the canonical form of a symmetric bilinear form $A(x, y)$ is independent of the choice of a canonical basis.* Thus, the *rank* of a symmetric bilinear form and its *positive and negative indices of inertia* are well-defined concepts. The rank of a symmetric bilinear form, i.e., the total number of terms in its canonical form, is clearly just the rank of the matrix of the bilinear form in a canonical basis. Consequently, since the rank of the matrix of a bilinear form does not depend on the choice of a basis (Sec. 40.4), the rank of a bilinear form can be found, without having to reduce it to canonical form, by finding the rank of the matrix of the form in any basis.

44.4. In analytic geometry, it is shown that the midpoints of the chords of a second-degree curve which are parallel to a given vector **r** lie on a straight line. The vector **s** which defines the direction of this line is said to be *conjugate* to the vector **r**. We now show that this definition coincides with our definition of conjugate vectors, at least in the case of central curves.

As is well known, after shifting the origin of coordinates to the center of the curve, the equation of a second-degree curve takes the form

$$a\xi_1^2 + 2b\xi_1\xi_2 + c\xi_2^2 = d$$

or

$$A(x, x) = d,$$

where $A(x, x)$ denotes the symmetric quadratic form in the left-hand side of the equation. Let the vector **z** be the position vector of the midpoint of a chord parallel to the vector **r**. This means that the equations

$$A(\mathbf{z} + t\mathbf{r}, \mathbf{z} + t\mathbf{r}) = d, \qquad A(\mathbf{z} - t\mathbf{r}, \mathbf{z} - t\mathbf{r}) = d$$

hold for some t. Writing these equations in expanded form, we have

$$A(\mathbf{z}, \mathbf{z}) + 2tA(\mathbf{z}, \mathbf{r}) + t^2 A(\mathbf{r}, \mathbf{r}) = d,$$
$$A(\mathbf{z}, \mathbf{z}) - 2tA(\mathbf{z}, \mathbf{r}) + t^2 A(\mathbf{r}, \mathbf{r}) = d.$$

Subtracting the second equation from the first, we obtain

$$A(\mathbf{z}, \mathbf{r}) = 0,$$

i.e., the vectors **z** and **r** are conjugate in the sense defined earlier in this section. The locus of all vectors **z** conjugate to a given vector **r** is defined by the linear homogeneous equation $A(\mathbf{z}, \mathbf{r}) = 0$, and therefore represents a straight line going through the origin.

45. Construction of a Canonical Basis by Jacobi's Method

The construction of a canonical basis given in Sec. 42 has the shortcoming that it does not permit us to determine beforehand the coefficients

λ_i and the components of the vectors of a canonical basis, when we are given the elements of the matrix $A_{(f)}$ of the symmetric bilinear form $A(x, y)$ relative to a given basis

$$\{f\} = \{f_1, f_2, \ldots, f_n\}.$$

Jacobi's method, which will now be presented, does allow us to find these coefficients and the components of the vectors of the desired canonical basis. However, we must now impose the following supplementary condition on the matrix $A_{(f)}$: The *descending principal minors* of $A_{(f)}$ of the form

$$\delta_1 = a_{11}, \quad \delta_2 = \begin{vmatrix} a_{11} & a_{12} \\ a_{21} & a_{22} \end{vmatrix}, \ldots,$$

$$\delta_{n-1} = \begin{vmatrix} a_{11} & a_{12} & \cdots & a_{1,n-1} \\ a_{21} & a_{22} & \cdots & a_{2,n-1} \\ \cdot & \cdot & \cdots & \cdot \\ a_{n-1,1} & a_{n-1,2} & \cdots & a_{n-1,n-1} \end{vmatrix},$$

up to and including order $n - 1$, must all be *nonvanishing*.

The vectors $e_1, e_2, \ldots, e_n$ are constructed by the formulas

$$\begin{aligned} e_1 &= f_1, \\ e_2 &= \alpha_1^{(1)} f_1 + f_2, \\ e_3 &= \alpha_1^{(2)} f_1 + \alpha_2^{(2)} f_2 + f_3, \\ &\ldots\ldots\ldots \\ e_{k+1} &= \alpha_1^{(k)} f_1 + \alpha_2^{(k)} f_2 + \alpha_3^{(k)} f_3 + \cdots + \alpha_k^{(k)} f_k + f_{k+1}, \\ &\ldots\ldots\ldots\ldots \\ e_n &= \alpha_1^{(n-1)} f_1 + \alpha_2^{(n-1)} f_2 + \alpha_3^{(n-1)} f_3 + \cdots + \alpha_{n-1}^{(n-1)} f_{n-1} + f_n, \end{aligned} \tag{21}$$

where the coefficients $\alpha_i^{(k)}$ ($i = 1, 2, \ldots, k$; $k = 1, 2, \ldots, n - 1$) are still to be determined. First of all, we note that the transformation from the vectors $f_1, f_2, \ldots, f_k$ to the vectors $e_1, e_2, \ldots, e_k$ is accomplished by using the matrix

$$\begin{Vmatrix} 1 & 0 & 0 & \ldots & 0 & 0 \\ \alpha_1^{(1)} & 1 & 0 & \ldots & 0 & 0 \\ \cdot & \cdot & \cdot & \cdots & \cdot & \cdot \\ \cdot & \cdot & \cdot & \cdots & \cdot & \cdot \\ \alpha_1^{(k-1)} & \alpha_2^{(k-1)} & \alpha_3^{(k-1)} & \ldots & \alpha_{k-1}^{(k-1)} & 1 \end{Vmatrix},$$

whose determinant is unity. Therefore, for $k = 1, 2, \ldots, n$, the vectors $f_1, f_2, \ldots, f_k$ can be expressed as linear combinations of the $e_1, e_2, \ldots, e_k$, so that the linear manifolds $L(f_1, f_2, \ldots, f_k)$ and $L(e_1, e_2, \ldots, e_k)$ coincide.

We now impose on the coefficients $\alpha_i^{(k)}$ ($i = 1, 2, \ldots, k$) the condition that the vector e_{k+1} be conjugate to the subspace $L(e_1, e_2, \ldots, e_k)$; a necessary and sufficient condition for this is that the relations

$$A(e_{k+1}, f_1) = 0, \quad A(e_{k+1}, f_2) = 0, \quad \ldots, \quad A(e_{k+1}, f_k) = 0 \tag{22}$$

be satisfied. In fact, it follows from the conditions (22) that the vector e_{k+1} is conjugate to the linear manifold spanned by the vectors $f_1, f_2, \ldots, f_k$, which, as we have just proved, coincides with the linear manifold spanned by the vectors $e_1, e_2, \ldots, e_k$. Conversely, if the vector e_{k+1} is conjugate to the subspace $L(e_1, e_2, \ldots, e_k)$, it is conjugate to every vector in the subspace, in particular, to the vectors $f_1, f_2, \ldots, f_k$, so that the relations (22) are satisfied.

Substituting the expression (21) for e_{k+1} into the formulas (22) and using the definition of a bilinear form, we obtain the following system of equations in the quantities $\alpha_i^{(k)}$ $(i = 1, 2, \ldots, k)$:

$$A(e_{k+1}, f_1) = \alpha_1^{(k)} A(f_1, f_1) + \alpha_2^{(k)} A(f_2, f_1) + \cdots + \alpha_k^{(k)} A(f_k, f_1) + A(f_{k+1}, f_1) = 0,$$

$$A(e_{k+1}, f_2) = \alpha_1^{(k)} A(f_1, f_2) + \alpha_2^{(k)} A(f_2, f_2) + \cdots + \alpha_k^{(k)} A(f_k, f_2) + A(f_{k+1}, f_2) = 0,$$

$$\ldots \qquad \ldots \qquad \ldots \qquad \ldots \qquad \ldots \qquad \ldots$$

$$A(e_{k+1}, f_k) = \alpha_1^{(k)} A(f_1, f_k) + \alpha_2^{(k)} A(f_2, f_k) + \cdots + \alpha_k^{(k)} A(f_k, f_k) + A(f_{k+1}, f_k) = 0.$$

$$\tag{23}$$

By hypothesis, this inhomogeneous system of equations with coefficients

$$A(f_i, f_j) = a_{ij} \qquad (i, j = 1, 2, \ldots, k)$$

has a nonvanishing determinant, and hence can be solved uniquely; therefore, we can determine the quantities $\alpha_i^{(k)}$ and thereby construct the desired vector e_{k+1}. To determine all the coefficients $\alpha_i^{(k)}$ and all the vectors e_k, we must solve the appropriate system (23) for every k; thus, in all, we must solve $n - 1$ systems of linear equations.

Denote the components of the vector x with respect to the basis $e_1, e_2, \ldots, e_n$ just constructed by $\xi_1, \xi_2, \ldots, \xi_n$ and the components of the vector y with respect to this basis by $\eta_1, \eta_2, \ldots, \eta_n$. In this basis, the bilinear form $A(x, y)$ becomes

$$A(x, y) = \sum_{i=1}^{n} \lambda_i \xi_i \eta_i. \tag{24}$$

To calculate the coefficients λ_i, we argue as follows: Consider the bilinear form $A(x, y)$ only in the subspace

$$L_m = L(e_1, e_2, \ldots, e_m) \quad \text{where } m \leqslant n.$$

The form $A(x, y)$ clearly has the matrix

$$\begin{Vmatrix} a_{11} & a_{12} & \ldots & a_{1m} \\ a_{21} & a_{22} & \ldots & a_{2m} \\ \cdot & \cdot & \ldots & \cdot \\ \cdot & \cdot & \ldots & \cdot \\ a_{m1} & a_{m2} & \ldots & a_{mm} \end{Vmatrix}$$

relative to the basis $f_1, f_2, \ldots, f_m$ of the subspace L_m, and the matrix

$$\begin{Vmatrix} \lambda_1 & 0 & \ldots & 0 \\ 0 & \lambda_2 & \ldots & 0 \\ . & . & \ldots & . \\ 0 & 0 & \ldots & \lambda_m \end{Vmatrix}$$

relative to the basis $e_1, e_2, \ldots, e_m$. As we have seen, the transformation matrix from the basis $f_1, f_2, \ldots, f_m$ to the basis $e_1, e_2, \ldots, e_m$ corresponding to the transformation formulas (21), has determinant 1. By equation (6) of Sec. 40, we must have

$$\det \begin{Vmatrix} a_{11} & a_{12} & \ldots & a_{1m} \\ a_{21} & a_{22} & \ldots & a_{2m} \\ . & . & \ldots & . \\ . & . & \ldots & . \\ a_{m1} & a_{m2} & \ldots & a_{mm} \end{Vmatrix} = \det \begin{Vmatrix} \lambda_1 & 0 & \ldots & 0 \\ 0 & \lambda_2 & \ldots & 0 \\ . & . & \ldots & . \\ . & . & \ldots & . \\ 0 & 0 & \ldots & \lambda_m \end{Vmatrix},$$

or, using the principal minor notation

$$\delta_m = \lambda_1 \lambda_2 \ldots \lambda_m \qquad (m = 1, 2, \ldots, n). \tag{25}$$

It follows immediately from (25) that

$$\lambda_1 = \delta_1 = a_{11}, \quad \lambda_2 = \frac{\delta_2}{\delta_1}, \quad \lambda_3 = \frac{\delta_3}{\delta_2}, \quad \ldots, \quad \lambda_n = \frac{\delta_n}{\delta_{n-1}}. \tag{26}$$

The equations (26) allow us to find the coefficients of the bilinear form in a canonical basis without calculating the basis itself. Moreover, provided that the principal minors $\delta_1, \delta_2, \ldots, \delta_n$ of the matrix $A_{(f)}$ are nonvanishing, our proof also allows us to determine the values of the positive and negative indices of inertia of the bilinear form $A(x, y)$ and therefore of the quadratic form $A(x, x)$.

Problem. Apply Jacobi's method to reduce the bilinear form

$$A(x, y) = \xi_1\eta_1 - \xi_1\eta_2 - \xi_2\eta_1 + \xi_1\eta_3 + \xi_3\eta_1 + 2\xi_2\eta_3 + 2\xi_3\eta_2 + \xi_3\eta_3 + \xi_2\eta_2$$

to canonical form.

Hint. First renumber the variables in such a way that the matrix of the bilinear form $A(x, y)$ is transformed into a form to which Jacobi's method is applicable.

46. Positive Definite Forms

46.1. The following definitions correspond to the definitions introduced for quadratic forms at the end of Sec. 43. The symmetric bilinear form $A(x, y)$ is called *nonsingular* if its rank equals the dimension of the space, i.e., if all the coefficients $\lambda_1, \lambda_2, \ldots, \lambda_n$ are nonzero in the canonical representation (20) of the form $A(x, y)$. If, moreover, all these coefficients are positive, $A(x, y)$ is called *positive definite*. According to Sec. 43, a positive definite bilinear form $A(x, y)$ is characterized by the fact that the corresponding quadratic form $A(x, x)$ takes a *positive* value for every $x \neq 0$.

An important example of a symmetric positive definite bilinear form in the space V_3 is the scalar product of the vectors x and y. In fact, the following relations are immediate consequences of the definition of the scalar product:

$$(x, y) = (y, x),$$
$$(x, x) = |x|^2 > 0 \quad \text{for } x \neq 0.$$

The first of these relations shows that the bilinear form (x, y) is symmetric, and the second shows that the corresponding quadratic form takes a positive value for every $x \neq 0$; consequently, the bilinear form (x, y) is positive definite.

Below, symmetric positive definite bilinear forms will play a particularly important role; by using such forms we shall be able to introduce the concepts of the length of a vector and the angle between vectors in a general linear space (Chapter 7).

46.2. The question arises of how to decide whether a symmetric bilinear form $A(x, y)$ is positive definite by examining its matrix. The following theorem answers this question:

THEOREM 27. *A necessary and sufficient condition for the symmetric matrix $A = \|a_{ik}\|$ to define a positive definite bilinear form $A(x, y)$ is that the descending principal minors*

$$a_{11}, \quad \begin{vmatrix} a_{11} & a_{12} \\ a_{21} & a_{22} \end{vmatrix}, \quad \begin{vmatrix} a_{11} & a_{12} & a_{13} \\ a_{21} & a_{22} & a_{23} \\ a_{31} & a_{32} & a_{33} \end{vmatrix}, \quad \ldots, \det \|a_{ik}\| \qquad (27)$$

of the matrix $\|a_{ik}\|$ all be positive.

Proof. If the principal minors (27) of the matrix A are positive, then by equation (26) of Sec. 45, all the canonical coefficients λ_k of the form $A(x, y)$ in some basis are positive, i.e., $A(x, y)$ is positive definite. Conversely, suppose that the form $A(x, y)$ is positive definite. We wish to show that then the principal minors (27) of the matrix $\|a_{ik}\|$ are positive. In fact, the principal minor

$$M = \begin{vmatrix} a_{11} & a_{12} & \cdots & a_{1m} \\ a_{21} & a_{22} & \cdots & a_{2m} \\ \cdot & \cdot & \cdots & \cdot \\ a_{m1} & a_{m2} & \cdots & a_{mm} \end{vmatrix}$$

defines the matrix $\|a_{ik}\|$ $(i, k = 1, 2, \ldots, m)$ of the bilinear form $A(x, y)$ in the subspace L_m spanned by the first m basis vectors. Since $A(x, y)$ is positive definite on the subspace L_m $[A(x, x) > 0$ for $x \neq 0]$, then in the subspace L_m there exists a canonical basis in which $A(x, y)$ can be written in canonical form with positive coefficients. In particular, relative to this basis, the determinant of $A(x, y)$, which equals the product of the canonical coefficients, is also positive. Bearing in mind the relation between determinants of a bilinear form in different bases [equation (6) of Sec. 40], we see that the determinant of $A(x, y)$ relative to the original basis is also positive. But the determinant of $A(x, y)$ relative to the original basis is just the minor M. Therefore $M > 0$, and the proof of Theorem 27 is complete.

Problem 1. State the conditions under which a symmetric matrix $\|a_{ik}\|$ defines a negative definite bilinear form.

Hint. $\|-a_{ik}\|$ must be the matrix of a positive definite form.

$$Ans. \quad a_{11} < 0, \quad \begin{vmatrix} a_{11} & a_{12} \\ a_{21} & a_{22} \end{vmatrix} > 0, \ldots, (-1)^n \det \|a_{ik}\| > 0.$$

Problem 2. Given a symmetric matrix $A = \|a_{ik}\|$ with the properties

$$a_{11} > 0, \quad \begin{vmatrix} a_{11} & a_{12} \\ a_{21} & a_{22} \end{vmatrix} > 0, \ldots, \det \|a_{ik}\| > 0,$$

show that $a_{nn} > 0$.

Problem 3. In applications of linear algebra to analysis (e.g., in the theory of conditional extrema), it is often required to solve the following problem: Given the matrix $A = \|a_{ik}\|$ of a symmetric bilinear form $A(x, y)$, find out if the form is positive definite in the subspace specified by a system of k independent linear equations

$$\sum_{j=1}^{n} b_{ij}\xi_j = 0 \qquad (i = 1, 2, \ldots k; k < n).$$

Show that a necessary and sufficient condition for this to be the case is that the descending principal minors of orders $2k + 1, 2k + 2, \ldots, k + n$ of the matrix

$$\Delta = (-1)^k \begin{Vmatrix} 0 & 0 & \ldots & 0 & b_{11} & b_{12} & \ldots & b_{1n} \\ 0 & 0 & \ldots & 0 & b_{21} & b_{22} & \ldots & b_{2n} \\ . & . & \ldots & . & . & . & \ldots & . \\ 0 & 0 & \ldots & 0 & b_{k1} & b_{k2} & \ldots & b_{kn} \\ b_{11} & b_{21} & \ldots & b_{k1} & a_{11} & a_{12} & \ldots & a_{1n} \\ b_{12} & b_{22} & \ldots & b_{k2} & a_{21} & a_{22} & \ldots & a_{2n} \\ . & . & \ldots & . & . & . & \ldots & . \\ b_{1n} & b_{2n} & \ldots & b_{kn} & a_{n1} & a_{n2} & \ldots & a_{nn} \end{Vmatrix}$$

be positive, assuming that the rank of the matrix $\|b_{ij}\|$ equals k and that the determinant made up of the first k columns of $\|b_{ij}\|$ is nonvanishing.[1]

---

[1] See the note by R. Y. Shostak, *Uspekhi Matem. Nauk*, vol. 9, no. 2 (1954), p. 199.

*47. Multilinear Forms

In analogy to bilinear forms, we can consider linear functions of a larger number of vectors (three, four, and more); all such functions are called *multilinear forms*. A multilinear form $A(x_1, x_2, \ldots, x_k)$ is called *symmetric* if it does not change if any two of its arguments are interchanged, and it is called *antisymmetric* if it changes sign when any two of its arguments are interchanged. An example of an antisymmetric multilinear form in three vectors x, y and z (a trilinear form) of the space V_3 is the triple product of x, y and z, i.e. $(x, y \times z)$, where $\times$ denotes the vector product. An example of an antisymmetric multilinear form in n vectors

$$x_1 = (a_{11}, a_{12}, \ldots, a_{1n}), x_2 = (a_{21}, a_{22}, \ldots, a_{2n}), \ldots, x_n = (a_{n1}, a_{n2}, \ldots, a_{nn})$$

is the determinant

$$A(x_1, x_2, \ldots, x_n) = \begin{vmatrix} a_{11} & a_{12} & \cdots & a_{1n} \\ a_{21} & a_{22} & \cdots & a_{2n} \\ \cdot & \cdot & \cdots & \cdot \\ a_{n1} & a_{n2} & \cdots & a_{nn} \end{vmatrix}. \tag{28}$$

A somewhat more general example is the product of the determinant (28) with a fixed number C. In fact, we now show that *every antisymmetric multilinear form $A(x_1, x_2, \ldots, x_n)$ in n vectors $x_1, x_2, \ldots, x_n$ of an n-dimensional linear space R with a fixed basis $e_1, e_2, \ldots, e_n$ equals the determinant (28) multiplied by some constant C.*

Let C denote the quantity $A(e_1, e_2, \ldots, e_n)$. Then we can easily calculate the quantity $A(e_{i_1}, e_{i_2}, \ldots, e_{i_n})$ where $i_1, i_2, \ldots, i_n$ are any integers from 1 to n. If two of these numbers are equal, then the quantity $A(e_{i_1}, e_{i_2}, \ldots, e_{i_n})$ vanishes, since on the one hand it does not change when the arguments corresponding to these numbers are interchanged, while on the other hand it has to change sign because of the antisymmetry property. If all the numbers $i_1, i_2, \ldots, i_n$ are different, then by making the same number of interchanges of adjacent arguments as there are inversions in the sequence of indices $i_1, i_2, \ldots, i_n$, we can cause the arguments to be arranged in normal order (cf. the proof of the theorem on the determinant of a matrix product in Sec. 29.5); let the required number of interchanges be N. Then we have

$$A(e_{i_1}, e_{i_2}, \ldots, e_{i_n}) = (-1)^N C.$$

Now let

$$x_i = \sum_{j=1}^{n} a_{ij} e_j \qquad (i = 1, 2, \ldots, n)$$

be an arbitrary system of n vectors of the space R, and write down the multilinear form

$$A(x_1, x_2, \ldots, x_n) = A\left(\sum_{i_1=1}^{n} a_{1i_1}e_{i_1}, \sum_{i_2=1}^{n} a_{2i_2}e_{i_2}, \ldots, \sum_{i_n=1}^{n} a_{ni_n}e_{i_n}\right)$$

$$= \sum_{i_1, i_2, \ldots, i_n=1}^{n} a_{1i_1}a_{2i_2} \ldots a_{ni_n}A(e_{i_1}, e_{i_2}, \ldots, e_{i_n})$$

$$= C \sum_{i_1, i_2, \ldots, i_n=1}^{n} (-1)^N a_{1i_1}a_{2i_2} \ldots a_{ni_n}.$$

Since in each term of the last sum, N denotes the number of inversions in the arrangement of the second subscripts of the elements a_{ij} when the first subscripts are in normal order, it follows that each term is one of the terms in the determinant (28) with the appropriate sign. Therefore, the sum of all the terms equals just the determinant (28). This proves our assertion.

In particular, we have shown that the triple product of three vectors x, y and z of the space V_3, with respect to any basis, can be written as a third-order determinant in the components of x, y and z, with a coefficient equal to the triple product of the basis vectors.

Problem 1. Prove the following theorem: *An antisymmetric multilinear form in $n + 1$ vectors of an n-dimensional space R vanishes identically.*

Problem 2. Prove the following theorem: *An antisymmetric multilinear form in $n - 1$ vectors of an n-dimensional space can be written in any basis as a determinant, whose first $n - 1$ rows consist of the components of the vector arguments and whose last (n'th) row is fixed.*

Problem 3. Show that every antisymmetric bilinear form $A(x, y) \not\equiv 0$ can always be reduced to the canonical form

$$A(x, y) = \sigma_1\tau_2 - \sigma_2\tau_1 + \sigma_3\tau_4 - \sigma_4\tau_3 + \cdots + \sigma_{2k-1}\tau_{2k} - \sigma_{2k}\tau_{2k-1}.$$

Hint. Use the equation $A(e_1, e_2) = 1$ to find the first pair of basis vectors. Then construct the subspace L defined by the equations

$$A(e_1, x) = 0, \qquad A(e_2, x) = 0.$$

If the form $A(x, y)$ does not vanish identically in this subspace, find vectors $e_3, e_4 \in L$ such that $A(e_3, e_4) = 1$, and so on.

7

EUCLIDEAN SPACES

---

48. Introduction

The explanation of a large variety of geometric facts rests to a great extent on the possibility of making measurements, basically measurements of the lengths of straight line segments and the angles between them. So far, we are not in a position to make such measurements in a general linear space; of course, this has the effect of narrowing the scope of our investigations. A natural way to extend these "metric" methods to the case of general linear spaces is to begin with the definition of the scalar product of two vectors which is adopted in analytic geometry (and which is suitable as of now only for ordinary vectors, i.e., elements of the space V_3). This definition reads as follows: *The scalar product of two vectors is the product of the lengths of the vectors and the cosine of the angle between them.* Thus, this definition already rests on the possibility of measuring the lengths of vectors and the angles between them; on the other hand, if we know the scalar product for an arbitrary pair of vectors, we can deduce the lengths of vectors and the angles between them. In fact, the square of the length of a vector equals the scalar product of the vector with itself, while the cosine of the angle between two vectors is just the ratio of their scalar product to the product of their lengths. Therefore, the possibility of measuring lengths and angles (and with it, the whole field of geometry associated with measurements, so-called "metric geometry"), is already implicit in the concept of the scalar product. In the case of a general linear space, the simplest approach is to introduce the concept of the scalar product of two vectors (independently of the lengths of vectors and the angles between them) and then use the scalar product to define these lengths and angles.

We now look for properties of the ordinary scalar product which can be used to construct a similar quantity in a general linear space. As we have already seen in Sec. 46.1, in the space V_3 the scalar product (x, y) is a symmetric positive definite bilinear form in the vectors x and y. Quite generally, we can define such a form in any linear space. Thus, we are led to consider a fixed but arbitrary symmetric positive definite bilinear form $A(x, y)$, defined on a general linear space, which we call the "scalar product" of the vectors x and y. We then use the scalar product to define the length of every vector and the angle between every pair of vectors by the same formulas as those used in the space V_3. Of course, at this point, only further study will show how successful this definition is; however, in the course of this and subsequent chapters, we shall see that with this definition we can in fact extend the methods of metric geometry to general linear spaces, thereby acquiring much more powerful tools for investigating various mathematical objects encountered in algebra and analysis.

The following important fact should be noted: In a given linear space, the initial positive definite bilinear form can be chosen in a variety of different ways. The length of a vector x calculated by using one such form will be different from the length of the same vector x calculated by using another form; a similar remark pertains to the angle between two vectors. Thus, lengths of vectors and the angles between them are not uniquely defined. However, this lack of uniqueness should not disturb us, for there is certainly nothing very surprising about the fact that different numbers will be assigned as the length of the same line segment if we measure the segment in different units. In fact, we can say that the choice of the original symmetric positive definite bilinear form is analogous to the choice of "units" for measuring lengths of vectors and angles between them.

A linear space on which a "unit" symmetric positive definite bilinear form is defined will henceforth be called *Euclidean*. As before, a linear space without a "unit" form will be called *affine*.

49. Definition of a Euclidean Space

A linear space R is said to be *Euclidean*, if (a) there is a rule assigning to every pair of vectors in R a real number called the *scalar product* of the vectors x and y and denoted by (x, y), and if (b) this rule satisfies the following requirements:

I. $(x, y) = (y, x)$ (commutative law);
II. $(x, y + z) = (x, y) + (x, z)$ (distributive law);
III. $(\lambda x, y) = \lambda (x, y)$ for any real number λ;
IV. $(x, x) > 0$ for $x \neq 0$ and $(x, x) = 0$ for $x = 0$.

Taken together, these axioms imply that the scalar product of the vectors x

and y is a bilinear form (Axioms II and III), which is symmetric (Axiom I) and positive definite (Axiom IV). Conversely, *any bilinear form which is symmetric and positive definite can be chosen as the scalar product.*

Example 1. In the space V_3 of free vectors (Sec. 11), the scalar product is defined as in the beginning of Sec. 48, and Axioms I–IV express the familiar properties of the scalar product, proved in vector analysis.

Example 2. In the space T_n (see Sec. 11), we define the scalar product of the vectors $x = (\xi_1, \xi_2, \ldots, \xi_n)$ and $y = (\eta_1, \eta_2, \ldots, \eta_n)$ by the formula

$$(x, y) = \xi_1\eta_1 + \xi_2\eta_2 + \cdots + \xi_n\eta_n. \tag{1}$$

This definition generalizes the familiar expression for the scalar product of three-dimensional vectors in terms of the components of the vectors with respect to an orthogonal coordinate system. The reader can easily verify that Axioms I–IV are satisfied.

We note that formula (1) is not the only way of introducing a scalar product in T_n. A description of all possible ways of introducing a scalar product in an n-dimensional space has essentially been given in Sec. 46.

Example 3. In the space $C(a, b)$ of continuous functions on the interval $a \leqslant t \leqslant b$ (Sec. 11), we define the scalar product of the functions $x(t)$ and $y(t)$ by the formula

$$(x, y) = \int_a^b x(t)y(t)\,dt. \tag{2}$$

By applying the usual rules of integration, it is easy to verify that Axioms I–IV are satisfied. Henceforth, the space $C(a, b)$, with the scalar product defined by (2), will be denoted by $C_2(a, b)$.

> ***Problem.*** Suppose we define the scalar product of two vectors of the space V_3 in the following ways: (a) the product of the lengths of the vectors; (b) the product of the lengths of the vectors and the cube of the cosine of the angle between them; (c) twice the usual definition of the scalar product. Which of these spaces is Euclidean?
>
> *Ans.* (a) is not Euclidean, since Axioms II and III are violated (for $\lambda = -1$); (b) is not Euclidean, since Axiom II is violated; (c) is Euclidean, and the new definition of the scalar product merely corresponds to a change of units along the coordinate axes.

Since the scalar product of the vectors x and y is a bilinear form, equation (2) of Sec. 40 holds, and in the present case takes the form

$$\left(\sum_{i=1}^{k} \alpha_i x_i, \sum_{j=1}^{m} \beta_j y_j \right) = \sum_{i=1}^{k} \sum_{j=1}^{m} \alpha_i \beta_j (x_i, y_j); \tag{3}$$

here $x_1, \ldots, x_k, y_1, \ldots, y_m$ are arbitrary vectors of the Euclidean space R, and $\alpha_1, \ldots, \alpha_k, \beta_1, \ldots, \beta_m$ are arbitrary real numbers.

50. Basic Metric Concepts

Equipped with the scalar product, we now proceed to define the basic metric concepts, i.e., the length of a vector and the angle between two vectors.

50.1. The length of a vector. By the *length* of a vector x in a Euclidean space R, we mean the quantity

$$|x| = +\sqrt{(x, x)}. \tag{4}$$

Example 1. In the space V_3 our definition reduces to the usual definition of the length of a vector.

Example 2. In the space T_n, the length of the vector $x = (\xi_1, \xi_2, \ldots, \xi_n)$ is given by

$$|x| = +\sqrt{\xi_1^2 + \xi_2^2 + \cdots + \xi_n^2}.$$

Example 3. In the space $C_2(a, b)$, the length of the vector $x(t)$ turns out to be

$$|x| = +\sqrt{(x, x)} = +\sqrt{\int_a^b x^2(t)\, dt}.$$

This quantity is sometimes written $\|x(t)\|$ and is called the *norm* of the function $x(t)$ (in order to avoid misleading connotations connected with the phrase "length of a function").

It follows from Axiom IV that every vector x of a Euclidean space R has a length; if $x \neq 0$, the length is positive, and if $x = 0$ (i.e., if x is the zero vector), the length is zero. The equality

$$|\alpha x| = \sqrt{(\alpha x, \alpha x)} = \sqrt{\alpha^2 (x, x)} = |\alpha|\sqrt{(x, x)} = |\alpha|\,|x| \tag{5}$$

shows that *the absolute value of a numerical factor can be factored out of the sign denoting the length of a vector.*

A vector x which has length 1 is said to be *normalized.* Every nonzero vector y can be normalized, i.e., multiplied by a number λ such that the result is a vector of length 1. In fact, solving the equation $|\lambda y| = 1$ for λ, we obtain

$$|\lambda| = \frac{1}{|y|}.$$

A set $F \subset R$ is said to be *bounded*, if the lengths of all the vectors $x \in F$ are bounded by a fixed constant. An example of a bounded set is the unit sphere of the space R, i.e., the set of all vectors $x \in R$ of length not exceeding 1.

50.2. The angle between two vectors. By *the angle between two vectors x and y*, we mean the angle (lying between 0 and 180 degrees) whose cosine is the ratio

$$\frac{(x, y)}{|x|\,|y|}.$$

For ordinary vectors (in the space V_3), our definition agrees with the usual way of writing the angle between two vectors in terms of the scalar product. To be able to apply this definition in a general Euclidean space, we must prove that the ratio has an absolute value no greater than unity, for any vectors x and y. To prove this, consider the vector $\lambda x - y$, where λ is a real number. By Axiom IV, we have

$$(\lambda x - y, \lambda x - y) \geqslant 0 \tag{6}$$

for any λ. Using (3), we can write this inequality in the form

$$\lambda^2(x, x) - 2\lambda(x, y) + (y, y) \geqslant 0. \tag{7}$$

The left-hand side of the inequality is a quadratic trinomial in λ with positive coefficients, which cannot have distinct real roots, since then it would not have the same sign for all λ. Therefore, the discriminant $(x, y)^2 - (x, x)(y, y)$ of the trinomial cannot be positive, and hence $(x, y)^2 \leqslant (x, x)(y, y)$. Taking the square root, we obtain

$$|(x, y)| \leqslant |x|\,|y|, \tag{8}$$

as required.

Suppose now that the equality sign holds in the inequality (8), i.e.,

$$|(x, y)| = |x|\,|y|.$$

Then the expression (7) has a vanishing discriminant and therefore has one real root λ_0. Thus we obtain

$$\lambda_0^2(x, x) - 2\lambda_0(x, y) + (y, y) = (\lambda_0 x - y, \lambda_0 x - y) = 0,$$

whence it follows by Axiom IV that $\lambda_0 x - y = 0$ or $y = \lambda_0 x$. In geometric language, this result can be stated as follows: *If the absolute value of the scalar product of two vectors equals the product of their lengths, then the vectors are collinear.*

The inequality (8) is usually called the *Schwarz inequality* (alternatively the *Cauchy-Schwarz* inequality or the *Buniakovsky inequality*).

Example 1. In the space V_3, the Schwarz inequality is an obvious consequence of the definition of the scalar product as the product of the lengths of two vectors and the cosine of the angle between them.

Example 2. In the space T_n, the Schwarz inequality takes the form

$$\left| \sum_{j=1}^{n} \xi_j \eta_j \right| \leqslant \sqrt{\sum_{j=1}^{n} \xi_j^2} \sqrt{\sum_{j=1}^{n} \eta_j^2}, \tag{9}$$

and is valid for any vectors $x = (\xi_1, \xi_2, \ldots, \xi_n)$ and $y = (\eta_1, \eta_2, \ldots, \eta_n)$, or equivalently, for any two sets of real numbers $\xi_1, \xi_2, \ldots, \xi_n$ and $\eta_1, \eta_2, \ldots, \eta_n$.

Example 3. In the space $C_2(a, b)$, the Schwarz inequality takes the form

$$\left| \int_a^b x(t)y(t)\,dt \right| \leqslant \sqrt{\int_a^b x^2(t)\,dt} \sqrt{\int_a^b y^2(t)\,dt}. \tag{10}$$

Problem 1. Find the angle between opposite sides of the regular tetrahedron.

Hint. Denote by e_1, e_2, e_3 the vectors directed along three sides of the tetrahedron drawn from a common vertex, and express the other sides of the tetrahedron as vectors.   *Ans.* 90°.

Problem 2. Find the angles of the "triangle" formed in the space $C_2(-1, 1)$ by the vectors $x_1(t) = 1$, $x_2(t) = t$, $x_3(t) = 1 - t$.   *Ans.* 90°, 60°, 30°.

50.3. Orthogonality. The vectors x and y are called *orthogonal* if $(x, y) = 0$. If $x \neq 0$ and $y \neq 0$, then according to the general definition of the angle between two vectors, $(x, y) = 0$ means that x and y make an angle of 90° with each other. The zero vector is orthogonal to any vector $x \in R$.

Example 1. In the space T_n, the orthogonality condition for the vectors $x = (\xi_1, \xi_2, \ldots, \xi_n)$ and $y = (\eta_1, \eta_2, \ldots, \eta_n)$ takes the form

$$\xi_1\eta_1 + \xi_2\eta_2 + \cdots + \xi_n\eta_n = 0.$$

For example, the vectors $e_1 = (1, 0, \ldots, 0)$, $e_2 = (0, 1, \ldots, 0), \ldots, e_n = (0, 0, \ldots, 1)$ are orthogonal (in pairs).

Example 2. In the space $C_2(a, b)$, the orthogonality condition for the vectors $x = x(t)$ and $y = y(t)$ takes the form

$$\int_a^b x(t)y(t)\,dt = 0.$$

The reader can easily verify, by calculating the appropriate integrals, that in the space $C_2(-\pi, \pi)$, any two vectors of the "trigonometric system"

$$1,\ \cos t,\ \sin t,\ \cos 2t,\ \sin 2t,\ \ldots,\ \cos nt,\ \sin nt,\ \ldots$$

are orthogonal.

We now deduce some simple propositions associated with the concept of orthogonality.

LEMMA 1. *If the nonzero vectors* $x_1, x_2, \ldots, x_k$ *are orthogonal, then they are linearly independent.*

Proof. Assume that the vectors are linearly dependent; then a relation

$$C_1x_1 + C_2x_2 + \cdots + C_kx_k = 0$$

holds, where $C_1 \neq 0$, for example. Taking the scalar product of this equation with x_1, we obtain $C_1(x_1, x_1) = 0$, since by hypothesis the vectors $x_1, x_2, \ldots, x_k$ are orthogonal.   It follows that $(x_1, x_1) = 0$, i.e., that x_1 is the zero vector, contrary to hypothesis.

The result of this lemma is often used in the following form:

If a sum of orthogonal vectors is zero, then each summand is zero.

LEMMA 2. *If the vectors $y_1, y_2, \ldots, y_k$ are orthogonal to the vector x, then any linear combination $\alpha_1 y_1 + \alpha_2 y_2 + \cdots + \alpha_k y_k$ is also orthogonal to the vector x.*

Proof.   We have

$$(\alpha_1 y_1 + \cdots + \alpha_k y_k, x) = \alpha_1(y_1, x) + \cdots + \alpha_k(y_k, x) = 0.$$

Therefore, the vector $\alpha_1 y_1 + \cdots + \alpha_k y_k$ is orthogonal to the vector x, as asserted.

The set of all linear combinations $\alpha_1 y_1 + \alpha_2 y_2 + \cdots + \alpha_k y_k$ forms a subspace

$$L = L(y_1, y_2, \ldots, y_k),$$

the linear manifold spanned by the vectors $y_1, y_2, \ldots, y_k$ (Sec. 16).   Therefore, if x is orthogonal to the vectors $y_1, y_2, \ldots, y_k$, it is orthogonal to every vector of the subspace L.   In this case, we say that *the vector x is orthogonal to the subspace L.*   In general, if $F \subset R$ is any set of vectors in the Euclidean space R, we say that the vector x is orthogonal to the set F, if x is orthogonal to every vector in F. According to Lemma 2, the set G of all vectors x orthogonal to a set F, is itself a subspace of the space R.   The most common situation is the case where F is a subspace; then the subspace G is called the *orthogonal complement* of the subspace F.

50.4. The Pythagorean theorem and its generalization.   Let the vectors x and y be orthogonal.   Then, by analogy with elementary geometry, we can call the vector $x + y$ the *hypotenuse* of the right triangle determined by the vectors x and y.   Taking the scalar product of $x + y$ with itself, and using the orthogonality of the vectors x and y, we obtain

$$|x + y|^2 = (x + y, x + y) = (x, x) + 2(x, y) + (y, y)$$
$$= (x, x) + (y, y) = |x|^2 + |y|^2.$$

This proves the *Pythagorean theorem* in a general Euclidean space, i.e., *the square of the hypotenuse equals the sum of the squares of the sides.*   It is easy to generalize this theorem to the case of any number of summands.   In fact, let the vectors $x_1, x_2, \ldots, x_k$ be orthogonal and let

$$z = x_1 + x_2 + \cdots + x_k.$$

Then we have

$$|z|^2 = (x_1 + x_2 + \cdots + x_k, x_1 + x_2 + \cdots + x_k)$$
$$= |x_1|^2 + |x_2|^2 + \cdots + |x_k|^2. \quad (11)$$

50.5. The triangle inequalities. If x and y are arbitrary vectors, then by analogy with elementary geometry, it is natural to call $x + y$ the *third side of the triangle determined by the vectors x and y.* Using the Schwarz inequality, we find

$$|x + y|^2 = (x + y, x + y) = (x, x) + 2(x, y) + (y, y)$$

$$\begin{cases} \leqslant |x|^2 + 2|x|\,|y| + |y|^2 = (|x| + |y|)^2, \\ \geqslant |x|^2 - 2|x|\,|y| + |y|^2 = (|x| - |y|)^2, \end{cases}$$

or

$$|x + y| \leqslant |x| + |y|, \tag{12}$$

$$|x + y| \geqslant |\,|x| - |y|\,|. \tag{13}$$

The inequalities (12) and (13) are called the *triangle inequalities.* Geometrically, they mean that *the length of any side of a triangle is no greater than the sum of the lengths of the two other sides and no less than the absolute value of the difference of the lengths of the two other sides.*

Problem. Write down the triangle inequalities in the space $C_2(a, b)$.

We could successively carry over all the theorems of elementary geometry to any Euclidean space, if we so desired; for the time being, we confine ourselves to the facts given above. Later (Sec. 52), we shall prove a general theorem, from which it follows that all the theorems of elementary geometry are valid in any Euclidean space.

51. Orthogonal Bases in an n-Dimensional Euclidean Space

51.1.

THEOREM 28. *In any n-dimensional Euclidean space R, there exists a basis consisting of n nonzero orthogonal vectors.*

Proof. There exists a canonical basis $y_1, y_2, \ldots, y_n$ for the bilinear form (x, y), just as for any other symmetric bilinear form in an n-dimensional space (Sec. 44). The condition $(y_i, y_k) = 0$ $(i \neq k)$, satisfied by the vectors of the canonical basis, is in this case just the condition that the vectors y_i and y_k be orthogonal; thus, the canonical basis $y_1, y_2, \ldots, y_n$ consists of n (pairwise) orthogonal vectors, Q.E.D. (In the next chapter, we shall study practical methods for constructing orthogonal bases.)

From now on, it will be convenient to normalize the vectors $y_1, y_2, \ldots, y_n$ by dividing each of them by its length. We thereby obtain a so-called *orthonormal basis* in the space R.

51.2. Let $e_1, e_2, \ldots, e_n$ be an arbitrary orthonormal basis in an n-dimensional Euclidean space R. Every vector $x \in R$ can be represented in the form

$$x = \xi_1 e_1 + \xi_2 e_2 + \cdots + \xi_n e_n, \tag{14}$$

where $\xi_1, \xi_2, \ldots, \xi_n$ are the components of the vector x with respect to the basis $e_1, e_2, \ldots, e_n$. We shall also call these components *Fourier coefficients* of the vector x with respect to the orthonormal system $e_1, e_2, \ldots, e_n$. (For the origin of this nomenclature, see Sec. 87.) Taking the scalar product of (14) with e_i, we obtain the following expression for the Fourier coefficients:

$$\xi_i = (x, e_i) \qquad (i = 1, 2, \ldots, n). \tag{15}$$

If $y = \eta_1 e_1 + \eta_2 e_2 + \cdots + \eta_n e_n$ is any other vector of the space R, then, according to (3), we have

$$(x, y) = \xi_1 \eta_1 + \xi_2 \eta_2 + \cdots + \xi_n \eta_n. \tag{16}$$

Thus, *in an orthonormal basis, the scalar product of two vectors equals the sum of the products of the components (or Fourier coefficients) of the vectors.* In particular, setting $y = x$, we obtain

$$|x|^2 = (x, x) = \xi_1^2 + \xi_2^2 + \cdots + \xi_n^2. \tag{17}$$

Problem. Find the cosines of the angles between the line $\xi_1 = \xi_2 = \cdots = \xi_n$ and the coordinate axes in the space T_n.

Ans. $\cos \varphi = \dfrac{1}{\sqrt{n}}$.

52. Isomorphism of Euclidean Spaces

The results of Sec. 51 show that as far as its metric properties are concerned, an arbitrary abstract n-dimensional Euclidean space R does not differ from the concrete Euclidean space T_n (Sec. 49). To make this statement more precise, we agree on the following definition:

Two Euclidean spaces R' and R'' are said to be *isomorphic* if a one-to-one correspondence can be established between their elements which has the following properties:

1. If the vectors $x'', y'' \in R''$ correspond to the vectors $x', y' \in R'$, then the vector $x'' + y'' \in R''$ corresponds to the vector $x' + y' \in R'$, and the vector $\alpha x'' \in R''$ corresponds to the vector $\alpha x' \in R'$, for any real α.

2. If the vectors $x'', y'' \in R''$ correspond to the vectors $x', y' \in R'$, then the numbers (x'', y'') and (x', y') are equal.

We now prove the following isomorphism theorem:

THEOREM 29. *Any two finite-dimensional Euclidean spaces R' and R'' with the same dimension are isomorphic.*

Proof. Choose an arbitrary orthonormal basis $e_1', e_2', \ldots, e_n'$ in the space R' and an arbitrary orthonormal basis $e_1'', e_2'', \ldots, e_n''$ in the space R''. With every vector

$$x' = \xi_1 e_1' + \xi_2 e_2' + \cdots + \xi_n e_n' \in R',$$

associate the vector

$$x'' = \xi_1 e_1'' + \xi_2 e_2'' + \cdots + \xi_n e_n'' \in R''.$$

This is obviously a one-to-one correspondence. We now verify the two conditions for an isomorphism. The first condition can be verified, just as in the case of the analogous theorem for affine spaces (Sec. 18), by merely expressing the results of the linear operations in terms of components. The validity of the second condition follows at once from the fact that in both spaces the scalar product is given by the same formula (16). This completes the proof of Theorem 29.

In particular, the three-dimensional spaces V_3 and T_3 are isomorphic. Moreover, the space V_3 is isomorphic to any three-dimensional subspace of any Euclidean space R. (It is obvious that any subspace of a Euclidean space is itself a Euclidean space with the same scalar product as that defined on the whole space.) This implies the previously mentioned fact to the effect that every metric theorem of elementary geometry (i.e., every theorem pertaining to the space V_3 or its two-dimensional subspace V_2) is automatically satisfied in any Euclidean space R. For example, by using the isomorphism theorem just proved, the Schwarz inequality and the Pythagorean theorem proved in Sec. 50 could have been derived directly from the fact that they are valid in elementary geometry.

53. The Norm of a Linear Operator

In the next three sections, we shall study linear operators defined on a Euclidean space R. The presence of a metric in the space R allows us to associate with every linear operator A a nonnegative number $\|A\|$, called the *norm* of the operator. In fact, consider the numerical function $F(x) = |Ax|$, defined for all vectors $x \in R$. By the *norm* of the operator A is meant the least upper bound or supremum (if it exists) of the values of $F(x)$ for unit vectors, i.e.,

$$\|A\| = \sup_{|x|=1} |Ax|. \tag{18}$$

Every vector x_0 for which the least upper bound (18) is achieved is called a *maximal vector* of the operator A. We now show that the quantity $\|A\|$ exists for any linear operator defined on an n-dimensional Euclidean space. (In an infinite-dimensional space, the case $\|A\| = \infty$ can occur.) The proof goes as follows: The length of the vector Ax is obviously a continuous function of the components $\eta_1, \eta_2, \ldots, \eta_n$ of the vector Ax. Since each of these components is a continuous function of the components $\xi_1, \xi_2, \ldots, \xi_n$ of the vector x, $|Ax|$ itself is a continuous function of the components $\xi_1, \xi_2, \ldots, \xi_n$ of the vector x. Since the sphere $|x| = 1$ is a bounded and closed subset of n-dimensional space, it follows from a familiar theorem of analysis[1] that the continuous function $|Ax|$ is bounded on this sphere. Since every bounded set of real numbers has a least upper bound, the number $\|A\|$ exists. Moreover, by another familiar theorem,[2] there is a point x_0 on the sphere $|x| = 1$ at which the continuous function $|Ax|$ achieves its least upper bound $\|A\|$; the vector corresponding to x_0 is a maximal vector of the operator A.

Example 1. The norm of the zero operator is obviously zero. Conversely, if $\|A\| = 0$, then A carries every normalized vector x_0 into zero; since every vector x is collinear with some normalized vector x_0, $Ax = 0$ for any x. Therefore, $\|A\| = 0$ implies that $A = 0$.

Example 2. The identity operator E has norm 1, since $|Ex| = |x|$ for any vector x.

Example 3. In an n-dimensional Euclidean space, let the diagonal operator A be defined by the relations

$$Ae_i = \lambda_i e_i \qquad (i = 1, 2, \ldots, n),$$

where $e_1, e_2, \ldots, e_n$ is an orthonormal basis. Then the norm of the operator A equals the largest of the numbers $|\lambda_i|$ $(i = 1, 2, \ldots, n)$. To see this, let the largest of the numbers $|\lambda_i|$ be $|\lambda_1|$, say. Then, for any normalized vector

$$x = \sum_{i=1}^{n} \xi_i e_i,$$

we have

$$|Ax|^2 = (Ax, Ax) = \left(A \sum_{i=1}^{n} \xi_i e_i, A \sum_{i=1}^{n} \xi_i e_i\right) = \sum_{i=1}^{n} \lambda_i^2 \xi_i^2 \leqslant \lambda_1^2 \sum_{i=1}^{n} \xi_i^2 = \lambda_1^2,$$

---

[1] See, e.g.: A. N. Kolmogorov and S. V. Fomin, *Elements of the Theory of Functions and Functional Analysis* (translated by L. F. Boron), Graylock Press, Rochester (1957), Theorem 2, p. 63. (See also Corollary to Theorem 1, p. 58.)

[2] See, e.g.: A. N. Kolmogorov and S. V. Fomin, *op. cit.*, Theorem 3, p. 63.

which implies that $|Ax| \leqslant |\lambda_1|$ for $|x| = 1$. On the other hand, setting $x = e_1$, we obtain

$$|Ax| = |Ae_1| = |\lambda_1 e_1| = |\lambda_1|.$$

It follows that

$$\sup_{|x|=1} |Ax| = |\lambda_1|,$$

as asserted. At the same time, we have shown that the vector e_1 is a maximal vector of the operator A.

Example 4. The norm of a linear operator which is given by the matrix $A = \|a_i^{(j)}\|$ relative to the orthonormal basis $e_1, e_2, \ldots, e_n$ satisfies the inequalities

$$\max_j \sum_{i=1}^{n} (a_i^{(j)})^2 \leqslant \|A\|^2 \leqslant \sum_{i=1}^{n} \sum_{j=1}^{n} (a_i^{(j)})^2. \tag{19}$$

Proof. By the definition of the norm of an operator, we have

$$|Ae_j| \leqslant \|A\|$$

for any $j = 1, 2, \ldots, n$. But, since

$$Ae_j = \sum_{i=1}^{n} a_i^{(j)} e_i, \qquad |Ae_j|^2 = \sum_{i=1}^{n} (a_i^{(j)})^2,$$

we obtain

$$\sum_{i=1}^{n} (a_i^{(j)})^2 \leqslant \|A\|^2$$

for any $j = 1, 2, \ldots, n$. This gives the left-hand side of the inequality (19). Moreover, if

$$x_0 = \sum_{j=1}^{n} \xi_j e_j$$

is a maximal vector of A, then by the Schwarz inequality

$$\|A\|^2 = |Ax_0|^2 = \left| \sum_{j=1}^{n} \xi_j Ae_j \right|^2 \leqslant \left\{ \sum_{j=1}^{n} |\xi_j| \, |Ae_j| \right\}^2$$

$$\leqslant \sum_{j=1}^{n} |\xi_j|^2 \sum_{j=1}^{n} |Ae_j|^2 = \sum_{j=1}^{n} |Ae_j|^2 = \sum_{j=1}^{n} \left| \sum_{i=1}^{n} a_i^{(j)} e_i \right|^2 = \sum_{j=1}^{n} \sum_{i=1}^{n} (a_i^{(j)})^2,$$

which gives the right-hand side of the inequality (19).

Problem. State the conditions under which the equality signs hold in the inequality (19).

Ans. The first inequality becomes an equality if one of the basis vectors is maximal; the second inequality becomes an equality if all the vectors Ae_j $(j = 1, 2, \ldots, n)$ are collinear.

We now show that *the inequality*

$$|Ax| \leqslant \|A\| \, |x| \tag{20}$$

holds for any $x \in R$ *and any linear operator* A *with finite norm* $\|A\|$. First, we note that the inequality (20) holds for any unit vector by the definition of the norm of the operator A. If x is an arbitrary nonzero vector [the inequality (20) obviously holds for a zero vector], then $x/|x|$ is a unit vector and therefore

$$\left| A \frac{x}{|x|} \right| \leqslant \|A\|.$$

But since A is a linear operator, we have

$$\frac{1}{|x|} |Ax| = \left| A \frac{x}{|x|} \right| \leqslant \|A\|.$$

Multiplying this inequality by $|x|$, we obtain the desired inequality (20).

    Problem. Verify the inequalities

$$\|A + B\| \leqslant \|A\| + \|B\| \quad \text{and} \quad \|AB\| \leqslant \|A\| \, \|B\|$$

for any two linear operators A and B.

54. Orthogonal Matrices and Isometric Operators

54.1. In Chapter 5 we studied the transformation formulas governing changes of coordinates in an n-dimensional linear space in going from one basis to another. In a Euclidean space, the most important situation is a transformation from one orthonormal basis

$$\{e\} = \{e_1, e_2, \ldots, e_n\}$$

to another orthonormal basis

$$\{f\} = \{f_1, f_2, \ldots, f_n\}.$$

We now examine the form of the transformation formulas in this case. Expressing the vectors of the basis $\{f\}$ with respect to the basis $\{e\}$, we obtain the equations

$$\begin{aligned}
f_1 &= q_1^{(1)}e_1 + q_2^{(1)}e_2 + \cdots + q_n^{(1)}e_n, \\
f_2 &= q_1^{(2)}e_1 + q_2^{(2)}e_2 + \cdots + q_n^{(2)}e_n, \\
& \quad \cdots \qquad \cdots \qquad \cdots \qquad \cdots \\
f_n &= q_1^{(n)}e_1 + q_2^{(n)}e_2 + \cdots + q_n^{(n)}e_n,
\end{aligned} \tag{21}$$

with coefficient matrix $Q = \|q_i^{(j)}\|$. Since the vectors $\{f\}$ are orthonormal, using equation (16), we obtain

$$(f_i, f_j) = \sum_{k=1}^{n} q_k^{(i)} q_k^{(j)} = \begin{cases} 0 \text{ for } i \neq j, \\ 1 \text{ for } i = j = 1, 2, \ldots, n. \end{cases} \tag{22}$$

Every matrix $Q = \|q_i^{(j)}\|$ with the properties (22) is called *orthogonal*. If we take an arbitrary orthogonal matrix $Q = \|q_i^{(j)}\|$, then the vectors $\{f\}$ defined in terms of the vectors $\{e\}$ by the equations (21) are orthonormal. Therefore, *every orthogonal matrix is the transformation matrix from one orthonormal basis to another orthonormal basis.* In particular, *an orthogonal matrix Q is always nonsingular,* i.e., $\det Q \neq 0$.

The elements $q_i^{(j)}$ of an orthogonal matrix Q have a direct geometric meaning: Taking the scalar product of e_i with the jth equation (21), we obtain

$$q_i^{(j)} = (f_j, e_i) = \cos (\widehat{f_j, e_i}),$$

i.e., *the numbers $q_i^{(j)}$ are the cosines of the angles between the new and the old basis vectors.*

54.2. Consider the linear operator Q which carries each vector e_i into a new vector f_i in accordance with the formulas (21). The operator Q has the important property that *it does not change the metric.* In other words, the scalar product of the new vectors Qx and Qy is the same as the scalar product of the old vectors x and y. To see this, let

$$x = \sum_{i=1}^{n} \xi_i e_i, \qquad y = \sum_{j=1}^{n} \eta_j e_j.$$

Then

$$(Qx, Qy) = \sum_{i,j=1}^{n} \xi_i \eta_j (Qe_i, Qe_j) = \sum_{i,j=1}^{n} \xi_i \eta_j (f_i, f_j) = \sum_{i=1}^{n} \xi_i \eta_i = (x, y),$$

since $(f_i, f_j) = 0$ if $i \neq j$ and 1 if $i = j$. Every linear operator which does not change the metric of a space, i.e., which satisfies the condition

$$(Qx, Qy) = (x, y)$$

for any pair of vectors, is called an *isometric operator.* Thus, we have just seen that the operator which transforms one orthonormal basis into another is *isometric*. It is obvious that the converse is true, i.e., every isometric operator carries any orthonormal basis into another orthonormal basis.

Problem. A linear operator Q preserves the length of every vector. Show that Q is isometric.

Hint. First method: The angles of a triangle are uniquely determined by its sides. Second method: The symmetric bilinear form (Qx, Qy) is uniquely determined by the quadratic form (Qx, Qx).

54.3. We now construct the matrix which is the inverse of a given orthogonal matrix Q. Solving the system (21) with respect to the vectors $e_1, e_2,$ $\ldots, e_n$, we obtain

$$e_1 = p_1^{(1)}f_1 + p_2^{(1)}f_2 + \cdots + p_n^{(1)}f_n,$$
$$e_2 = p_1^{(2)}f_1 + p_2^{(2)}f_2 + \cdots + p_n^{(2)}f_n,$$
$$\cdots \quad \cdots \quad \cdots \quad \cdots$$
$$e_n = p_1^{(n)}f_1 + p_2^{(n)}f_2 + \cdots + p_n^{(n)}f_n.$$

Since these formulas again describe a transformation from one orthonormal basis to another, the numbers $p_i^{(j)}$ must satisfy the same relations (22) as the numbers $q_i^{(j)}$, i.e., *the matrix of the inverse transformation is also orthogonal.* Moreover, we clearly have

$$q_i^{(j)} = (f_j, e_i), \tag{23}$$
$$p_i^{(j)} = (e_j, f_i), \tag{24}$$

so that $p_i^{(j)} = q_i^{(i)}$. Thus, *the inverse of the orthogonal matrix Q is just the transpose of Q.* Actually, this fact is an immediate consequence of equation (22), which can be written in the form

$$QQ' = E \tag{25}$$

or

$$Q' = Q^{-1}.$$

In particular, it follows from (25) that

$$\det Q \det Q' = (\det Q)^2 = 1. \tag{26}$$

Thus, *the determinant of an orthogonal matrix can only take the values ± 1.*

54.4. Finally, we write down the transformation formulas for the components of a vector x in going from one orthonormal basis $\{e\}$ to another orthonormal basis $\{f\}$. Let $\xi_1, \xi_2, \ldots, \xi_n$ be the coordinates of the vector x with respect to the basis $\{e\}$, and let $\eta_1, \eta_2, \ldots, \eta_n$ be its components with respect to the basis $\{f\}$. According to Sec. 35, the matrix of the transformation from the components $\xi_1, \xi_2, \ldots, \xi_n$ to the components $\eta_1, \eta_2, \ldots, \eta_n$ is the matrix $(A^{-1})'$, where A is the transformation matrix from the basis $\{e\}$ to the basis $\{f\}$. Since in the present case, the transformation matrix is an orthogonal matrix Q, we have $Q^{-1} = Q'$, from which it follows that $(Q^{-1})' = Q$. Therefore, the transformation formulas from the components $\xi_1, \xi_2, \ldots, \xi_n$ to the components $\eta_1, \eta_2, \ldots, \eta_n$ are written in terms of the same matrix Q as used in (21) to describe the transformation from the basis $\{e\}$ to the basis $\{f\}$, i.e.

$$\eta_1 = q_1^{(1)}\xi_1 + q_2^{(1)}\xi_2 + \cdots + q_n^{(1)}\xi_n,$$
$$\eta_2 = q_1^{(2)}\xi_1 + q_2^{(2)}\xi_2 + \cdots + q_n^{(2)}\xi_n,$$
$$\cdots \quad \cdots \quad \cdots \quad \cdots \tag{27}$$
$$\eta_n = q_1^{(n)}\xi_1 + q_2^{(n)}\xi_2 + \cdots + q_n^{(n)}\xi_n.$$

Such a coordinate transformation in a Euclidean space is called an *isometric transformation*. Since $Q^{-1} = Q'$, the formulas describing the inverse transformation are obtained by using the transposed matrix, i.e.,

$$\begin{aligned}
\xi_1 &= q_1^{(1)}\eta_1 + q_1^{(2)}\eta_2 + \cdots + q_1^{(n)}\eta_n, \\
\xi_2 &= q_2^{(1)}\eta_1 + q_2^{(2)}\eta_2 + \cdots + q_2^{(n)}\eta_n, \\
&\quad \cdots \qquad \cdots \qquad \cdots \qquad \cdots \\
\xi_n &= q_n^{(1)}\eta_1 + q_n^{(2)}\eta_2 + \cdots + q_n^{(n)}\eta_n.
\end{aligned} \tag{28}$$

Problem 1. Let A be an orthogonal matrix. Show that $A_{ik} = a_{ik} \det A$ is the cofactor of the element a_{ik} of A.

Problem 2. Show that the product of two orthogonal matrices is again an orthogonal matrix.

Problem 3. Show that for $n = 2$, every orthogonal matrix with determinant $+1$ has the form

$$\left\| \begin{array}{cc} \cos \varphi & -\sin \varphi \\ \sin \varphi & \cos \varphi \end{array} \right\|,$$

i.e., is the matrix of a rotation.

Problem 4. In an n-dimensional Euclidean space R, a linear operator K_t, depending on a parameter t ($t_0 \leqslant t \leqslant t_1$) is called a *rotation* if it satisfies the following conditions:

1) $K_{t_0} = E$ (the identity operator);

2) $K_t(x)$ depends continuously on t for every x, i.e., as $\Delta t \to 0$, $|K_{t+\Delta t}x - K_t x| \to 0$ for every $x \in R$;

3) For any x, y and any fixed t ($t_0 \leqslant t \leqslant t_1$), $(K_t x, K_t y) = (x, y)$.

Show that $\det K_t = 1$.

Hint. $\det K_t$ is a continuous function of t.

Problem 5. (Continuation) Show that the product of two rotations K_t ($t_0 \leqslant t \leqslant t_1$) and K_s ($s_0 \leqslant s \leqslant s_1$) is also a rotation.

Problem 6. (Continuation) Let $R' \subset R$ be a subspace of dimension no greater than $n - 2$, and let x and y be unit vectors orthogonal to R'. Construct a rotation which carries x into y but has no effect on R'.

Hint. Construct an orthonormal basis $e_1, e_2, \ldots, e_n$ such that $e_1 = x$ and e_2 lies in the plane determined by the vectors x and y. Then consider the operator K_t with matrix

$$K_t = \left\| \begin{array}{cccccc} \cos t & -\sin t & 0 & 0 & \ldots & 0 \\ \sin t & \cos t & 0 & 0 & \ldots & 0 \\ 0 & 0 & 1 & 0 & \ldots & 0 \\ 0 & 0 & 0 & 1 & \ldots & 0 \\ \cdot & \cdot & \cdot & \cdot & \ldots & \cdot \\ \cdot & \cdot & \cdot & \cdot & \ldots & \cdot \\ 0 & 0 & 0 & 0 & \ldots & 1 \end{array} \right\|.$$

Problem 7. (Continuation) Show that every isometric operator Q with $\det Q = 1$ is the result of some rotation.

Hint. Let $e_1, e_2, \ldots, e_n$ be an orthonormal basis and let $f_i = Qe_i$ $(i = 1, 2, \ldots, n)$. Consider the following rotations:

1) $e_1 \rightarrow f_1$, with e_2 going into some $e_2^{(2)}$;

2) $e_2^{(2)} \rightarrow f_2$, with f_1 fixed and e_3 going into some $e_3^{(2)}$;

3) $e_3^{(2)} \rightarrow f_3$, with f_1 and f_2 fixed, and so on.

Then use Problem 5.

Problem 8. (Continuation) Show that every isometric operator Q with $\det Q = -1$ is the product of a rotation operator and an operator corresponding to reflection in a hyperplane of dimension $n - 1$.

Problem 9. Show that the sum of the squares of all the minors of order k appearing in k fixed rows of an orthogonal matrix equals 1. Show that the sum of the products of all the minors of order k appearing in one group of k rows with the corresponding minors in another group of k rows equals 0.

Hint. Apply Problem 4 of Sec. 29.5.

Problem 10. An operator A which preserves the orthogonality of any pair of vectors x, y, i.e., such that $(x, y) = 0$ implies $(Ax, Ay) = 0$, is called an *isogonal* operator. Isometric operators and similarity operators ($Ax = \lambda x$ for any x) are isogonal, and so is the product of any similarity operator and any isometric operator. Show that every isogonal operator is the product of a similarity operator and an isometric operator.

Hint. A given isogonal operator C transforms the orthonormal basis $e_1, e_2, \ldots, e_n$ into an orthogonal basis

$$f_1' = \alpha_1 f_1, \quad f_2' = \alpha_2 f_2, \quad \ldots, \quad f_n' = \alpha_n f_n,$$

where $f_1, f_2, \ldots, f_n$ are normalized. Let Q be the isometric operator carrying the vectors $f_1, f_2, \ldots, f_n$ into $e_1, e_2, \ldots, e_n$. Then the matrix of the isogonal operator QC is diagonal. Show that the condition $\alpha_i \neq \alpha_j$ allows one to construct a pair of orthogonal vectors which are carried into nonorthogonal vectors by the operator QC.

55. The Relation between Linear Operators and Bilinear Forms. The Adjoint Operator

55.1. Let A be a linear operator defined on a Euclidean space R. For any two vectors x and y of the space R, we can construct the quantity $A(x, y) = (x, Ay)$. It is easily verified that this numerical function of the two vectors x and y is a bilinear form. In fact, because of the definitions of linear operator and scalar product (Secs. 26, 49), the following relations, which together imply that $A(x, y)$ is a bilinear form, are valid:

$$A(x_1 + x_2, y) = (x_1 + x_2, Ay) = (x_1, Ay) + (x_2, Ay)$$
$$= A(x_1, y) + A(x_2, y),$$
$$A(\alpha x, y) = (\alpha x, Ay) = \alpha(x, Ay) = \alpha A(x, y),$$
$$A(x, y_1 + y_2) = (x, A(y_1 + y_2)) = (x, Ay_1 + Ay_2)$$
$$= (x, Ay_1) + (x, Ay_2) = A(x, y_1) + A(x, y_2),$$
$$A(x, \alpha y) = (x, A(\alpha y)) = (x, \alpha Ay) = \alpha(x, Ay) = \alpha A(x, y).$$

Suppose now that R is an n-dimensional Euclidean space. Let $A = \|a_i^{(j)}\|$ be the matrix of the operator A relative to some orthonormal basis $\{e\} = \{e_1, e_2, \ldots, e_n\}$, and construct the matrix of the bilinear form relative to this basis; according to Sec. 40.2, the element a_{ij} of this matrix is given by

$$a_{ij} = A(e_i, e_j) = (e_i, Ae_j).$$

Moreover, according to Sec. 27

$$Ae_j = \sum_{k=1}^{n} a_k^{(j)} e_k,$$

so that

$$a_{ij} = \left(e_i, \sum_{k=1}^{n} a_k^{(j)} e_k\right) = a_i^{(j)}.$$

Thus, the matrix of the operator A in the basis $\{e\}$ coincides with the matrix of the form (x, Ay) in the same basis.

Conversely, suppose that a bilinear form $A(x, y)$ is given in an n-dimensional Euclidean space R. Then *there exists a linear operator A for which the relation*

$$A(x, y) = (x, Ay)$$

holds for any x and y in R. To prove this assertion, choose an orthonormal basis $e_1, e_2, \ldots, e_n$ in the space R and construct the linear operator A whose matrix in this basis coincides with the matrix of the bilinear form $A(x, y)$ in this basis. Then construct the bilinear form (x, Ay). By what has already been proved, the matrix of (x, Ay) in the basis $\{e\}$ coincides with the matrix of the operator A, and therefore with the matrix of the form $A(x, y)$. But then

$$A(x, y) \equiv (x, Ay),$$

i.e., both forms have the same value for any x and y, Q.E.D.

It was no accident that we stated and proved this property for a *finite*-dimensional Euclidean space; in an *infinite*-dimensional Euclidean space, it is in general no longer true.   Moreover, in an n-dimensional affine space, it is no longer permissible to identify a bilinear form and a linear operator on the grounds that they have the same matrix A in some basis.   In fact, if we go over to a new basis by using the transformation matrix C, the matrix of the bilinear form transforms into $C'AC$ (Sec. 40.4), while the matrix of the linear operator transforms into $C^{-1}AC$ (Sec. 38), and these two results are in general different.   In a Euclidean space, the transformation from one orthonormal basis to another is accomplished by using an orthogonal matrix C.   For an orthogonal matrix, $C' = C^{-1}$ (Sec. 54) and consequently, the results of transforming the matrix of the bilinear form and the matrix of the operator are the same.

55.2. With any bilinear form $A(x, y)$ defined on an n-dimensional Euclidean space R, we can associate another linear operator A^* defined by the condition

$$A(x, y) = (A^*x, y), \tag{29}$$

which is to hold for any pair of vectors $x, y \in R$.   We now show that an operator A^* satisfying this condition actually exists.   Let $e_1, e_2, \ldots, e_n$ be an orthonormal basis of the space R and set

$$A^*e_i = \sum_{j=1}^{n} a_j^{*(i)}e_j,$$

where the numbers $a_j^{*(i)}$ are yet to be determined.   Substituting $x = e_i$, $y = e_j$ in the relation (29), we find

$$A(e_i, e_j) = a_{ij} = (A^*e_i, e_j) = a_j^{*(i)}.$$

This uniquely determines the numbers $a_j^{*(i)}$; clearly, the matrix $\|a_j^{*(i)}\|$ is the transpose of the matrix of the form $A(x, y)$.   The operator A^* which has the matrix $\|a_j^{*(i)}\|$ relative to the basis $e_1, e_2, \ldots, e_n$ satisfies (29) for $x = e_i$, $y = e_j$.   Moreover, (29) is satisfied for any pair x and y, since two bilinear forms which take the same values for the basis vectors are identically equal.

We might just as well have constructed A^* starting with the operator A instead of the form $A(x, y)$.   In this case, we have the following result:

THEOREM 30.   *With every linear operator A defined on an n-dimensional Euclidean space R, we can uniquely associate an operator A^* which is defined on the same space and satisfies the relation*

$$(A^*x, y) = (x, Ay)$$

for any vectors x and y.   In any orthonormal basis of the space R, the matrix of the operator A^* is the transpose of the matrix of the operator A.

The operator A^* is called the *adjoint* of the operator A.   In particular, if $A^* = A$, the operator A is called *self-adjoint*.   In other words, the operator A is self-adjoint if the bilinear form $A(x,y)$ corresponding to the operator A is symmetric, i.e.,

$$(Ax, y) = (x, Ay). \tag{30}$$

An operator A which satisfies (30) for any two vectors x and y is therefore also called *symmetric*. Equation (30) can also serve as the definition of a self-adjoint operator in an infinite-dimensional space, where we can no longer explicitly construct the adjoint operator. By Theorem 30, the matrix of a symmetric operator in any orthonormal basis coincides with its own transpose, i.e., is a symmetric matrix. Conversely, every operator A which has a symmetric matrix in some orthonormal basis is a symmetric operator.

Problem 1. Prove the formulas

$$(A^*)^* = A,$$
$$(A + B)^* = A^* + B^*,$$
$$(\lambda A)^* = \lambda A^*,$$
$$(AB)^* = B^*A^*.$$

Problem 2. Two operators A and B are given in an n-dimensional affine space R. Suppose it is known that the matrices of these operators are transposes of each other in some basis. Will this property be preserved in every other basis? *Ans.* No.

Problem 3. Find the adjoints of the following operators:

a) An operator A defined on an n-dimensional space, which carries every vector e_i of an orthonormal basis into $\lambda_i e_i$, where λ_i is some number $(i = 1, 2, \ldots, n)$;

b) An isometric operator Q. *Ans.* $A^* = A,\ Q^* = Q^{-1}$.

Problem 4. Show that the operators $A_1 = A^*A$, $A_2 = AA^*$, and $A_3 = A + A^*$ are self-adjoint, for any operator A whatsoever. Is the operator $A_4 = A - A^*$ self-adjoint, if $A \neq A^*$? *Ans.* No, since $A_4^* = -A_4$.

Problem 5. The quantity

$$[A] = \sup_{|x| \leqslant 1,\, |y| \leqslant 1} |A(x, y)|$$

is called the *norm* of the bilinear form $A(x, y)$. Show that if the operator A corresponds to the bilinear form $A(x, y)$, then $\|A\| = [A]$.

Problem 6. Show that the operators A and A^* have the same norm.

Problem 7. Show that if x_0 is a maximal vector of the operator A, then

$$\frac{1}{\|A\|} Ax_0$$

is a maximal vector of the operator A^*.

Problem 8. Show that if $N(A)$ and $T(A)$ are the null space and range of the operator A, then the orthogonal complements of these subspaces are the range and null space, respectively, of the operator A^*.

Problem 9. Let $\mathscr{R}'$ be a left ideal in the algebra $\mathscr{R}$ of all linear operators defined on a Euclidean space R (see Sec. 32). Show that the set $\mathscr{R}''$ of all operators which are adjoint to the operators in $\mathscr{R}'$ is a right ideal of the algebra $\mathscr{R}$.

Problem 10. By introducing a metric in the linear space R, derive the second half of Theorem 25 of Sec. 32 from the first half.

Hint. Use the results of Problems 8 and 9.

8

ORTHOGONALIZATION
AND THE MEASUREMENT
OF VOLUME

In the preceding chapter, we studied orthogonal coordinate systems in an n-dimensional Euclidean space (i.e., coordinate systems defined by orthonormal bases), and we saw how useful such systems are for solving problems of a metric nature. In the present chapter, we shall develop some practical methods for constructing orthogonal systems of vectors, and at the same time, we shall study some related geometric problems.

56. Perpendiculars

56.1. Let R' be a subspace of the Euclidean space R, and let f be a vector which is in general not an element of R'. We pose the problem of representing f in the form

$$f = g + h,\tag{1}$$

where the vector g belongs to the subspace R' and the vector h is orthogonal to R'. The vector g appearing in the expansion (1) is called the *projection of f onto the subspace R'* and the vector h is called the *perpendicular dropped from the end of f onto the subspace R'*. This terminology calls to mind certain familiar geometric associations, but it is not intended to do more than just suggest these associations. (Since the concept of the "end of a vector" plays no role in our axiomatics, it is inappropriate to look for any logical content in this terminology.)

We now explicitly construct the representation (1), assuming that R' is finite-dimensional, say k-dimensional. To do this, we introduce an orthonormal basis $e_1, e_2, \ldots, e_k$ in the subspace R', and write the vector g in the form

$$g = \alpha_1 e_1 + \alpha_2 e_2 + \cdots + \alpha_k e_k,\tag{2}$$

where the numbers $\alpha_1, \alpha_2, \ldots, \alpha_k$ are yet to be determined. The vector $h = f - g$ must be orthogonal to the subspace R'; for this, it is necessary and sufficient that the relations

$$(h, e_i) = (f - g, e_i) = 0 \qquad (i = 1, 2, \ldots, k) \tag{3}$$

hold. Substituting the expression (2) into these relations, we obtain

$$(f - g, e_i) = (f - \alpha_1 e_1 - \alpha_2 e_2 - \cdots - \alpha_k e_k, e_i)$$
$$= (f, e_i) - \alpha_i(e_i, e_i) = (f, e_i) - \alpha_i,$$

since the vectors $e_1, e_2, \ldots, e_k$ are orthonormal, by hypothesis. Therefore, the vector h is orthogonal to the subspace R' if and only if the coefficients α_i appearing in the expression (2) satisfy the relations

$$\alpha_i = (f, e_i) \qquad (i = 1, 2, \ldots, k).$$

This proves both the existence and the uniqueness of the expansion (1) in the case where the subspace R' is finite-dimensional. We defer consideration of the general case until Sec. 86.

56.2. Applying the Pythagorean theorem to the expansion (1), we obtain

$$|f|^2 = |g|^2 + |h|^2,$$

which implies the inequality

$$0 \leqslant |h| \leqslant |f|; \tag{4}$$

this inequality expresses the geometric fact that *the length of a perpendicular does not exceed the length of the line segment from which it is dropped.* The first equality sign holds if $|h| = 0$; this means that $f = g + 0$, i.e., f is an element of the subspace R'. The second equality sign holds if $|h| = |f|$; according to the Pythagorean theorem, this means that $g = 0$ or $f = 0 + h$, i.e., f is orthogonal to the subspace R'. In any other configuration of f, the length of h is less than the length of f. Thus, to reiterate, $|h| = 0$ means that f belongs to R', and $|h| = |f|$ means that f is orthogonal to R'.

Applying the Pythagorean theorem once again to the decomposition (2), we obtain

$$|g|^2 = \alpha_1^2 + \alpha_2^2 + \cdots + \alpha_k^2 = \sum_{j=1}^{k} \alpha_j^2,$$

so that

$$|f|^2 = |h|^2 + \sum_{j=1}^{k} \alpha_j^2.$$

In particular, for any (finite) orthonormal system $e_1, e_2, \ldots, e_k$ and any vector f, we obtain the inequality

$$\sum_{j=1}^{n} \alpha_j^2 \leqslant |f|^2,$$

called *Bessel's inequality*. The geometric meaning of this inequality is clear: The square of the length of the vector f is not less than the sum of the squares of its projections onto any k mutually orthogonal directions.

56.3. In the applications, one sometimes needs an explicit solution of the problem of dropping a perpendicular onto a subspace R', given some basis

$$\{b\} = \{b_1, b_2, \ldots, b_k\}$$

in R' (in general, not an orthonormal basis). To solve this problem, we expand g, the "foot of the perpendicular," with respect to the basis $\{b\}$, i.e., we write

$$g = \beta_1 b_1 + \beta_2 b_2 + \cdots + \beta_k b_k.$$

We then impose on the vector $h = f - g$ the condition that it be orthogonal to all the vectors $b_1, b_2, \ldots, b_k$, thereby obtaining the following system of equations:

$$(h, b_1) = (f - g, b_1) = (f, b_1) - \beta_1(b_1, b_1) - \beta_2(b_2, b_1) - \cdots - \beta_k(b_k, b_1) = 0,$$
$$(h, b_2) = (f - g, b_2) = (f, b_2) - \beta_1(b_1, b_2) - \beta_2(b_2, b_2) - \cdots - \beta_k(b_k, b_2) = 0,$$
$$\ \cdot \qquad \cdots \qquad\qquad \cdots\cdots \qquad\qquad \cdots \qquad \cdots$$
$$(h, b_k) = (f - g, b_k) = (f, b_k) - \beta_1(b_1, b_k) - \beta_2(b_2, b_k) - \cdots - \beta_k(b_k, b_k) = 0,$$

with determinant

$$D = \begin{vmatrix} (b_1, b_1) & (b_2, b_1) & \ldots & (b_k, b_1) \\ (b_1, b_2) & (b_2, b_2) & \ldots & (b_k, b_2) \\ \cdot & \cdot & \cdots & \cdot \\ \cdot & \cdot & \cdots & \cdot \\ \cdot & \cdot & \cdots & \cdot \\ (b_1, b_k) & (b_2, b_k) & \ldots & (b_k, b_k) \end{vmatrix}.$$

We have already proved that the solution of this problem exists and is unique. Therefore, we can conclude that the determinant D must be *nonvanishing*. Solving the system by Cramer's rule, we obtain an expression for the coefficients β_j ($j = 1, 2, \ldots, k$):

$$\beta_j = \frac{1}{D} \begin{vmatrix} (b_1, b_1) & (b_2, b_1) & \ldots & (b_{j-1}, b_1) & (f, b_1) & (b_{j+1}, b_1) & \ldots & (b_k, b_1) \\ (b_1, b_2) & (b_2, b_2) & \ldots & (b_{j-1}, b_2) & (f, b_2) & (b_{j+1}, b_2) & \ldots & (b_k, b_2) \\ \cdot & \cdot & \cdots & \cdot & \cdot & \cdot & \cdots & \cdot \\ \cdot & \cdot & \cdots & \cdot & \cdot & \cdot & \cdots & \cdot \\ \cdot & \cdot & \cdots & \cdot & \cdot & \cdot & \cdots & \cdot \\ (b_1, b_k) & (b_2, b_k) & \ldots & (b_{j-1}, b_k) & (f, b_k) & (b_{j+1}, b_k) & \ldots & (b_k, b_k) \end{vmatrix}.$$

Problem 1. Write the vector $f = (5, 2, -2, 2)$ of the space T_4 as the sum of a vector g lying in the subspace R' spanned by the vectors

$$b_1 = (2, 1, 1, -1), \qquad b_2 = (1, 1, 3, 0)$$

and another vector h orthogonal to this subspace.

Ans. $g = (3, 1, -1, -2)$, $h = (2, 1, -1, 4)$.

Problem 2. Solve the same problem where now $f = (-3, 5, 9, 3)$ and R' is spanned by the vectors

$$b_1 = (1, 1, 1, 1), \qquad b_2 = (2, -1, 1, 1), \qquad b_3 = (2, -7, -1, -1).$$
$$Ans. \quad g = (1, 7, 3, 3), h = (-4, -2, 6, 0).$$

Problem 3. Prove that of all the vectors in the subspace R', the vector g (the projection of f onto R') makes the smallest angle with f.

Problem 4. Prove that if the vector g_0 in the subspace R' is orthogonal to g, the projection of f onto R', then g_0 is orthogonal to f itself.

Hint. Take the scalar product of (1) with the vector g_0.

56.4. The problem of dropping a perpendicular can be posed not only for a subspace, but also for a hyperplane. In the case of a hyperplane, the formulation is as follows: Suppose that in a Euclidean space R, we are given a vector f and a hyperplane R'', generated by parallel displacement of the subspace R'. We wish to show that there exists a unique decomposition

$$f = g + h, \tag{5}$$

where the vector g belongs to the hyperplane R'' and the vector h is orthogonal to the subspace R'. (Saying that g belongs to the hyperplane R'' means geometrically that the endpoint of g lies in the hyperplane R'', while its initial point, as usual, is at the origin of coordinates. One must not imagine that the whole vector g lies in the hyperplane R''!) The geometric meaning of the decomposition (5) is illustrated in Figure 1(a). In the decomposition (5), the summands are in general no longer orthogonal.

The problem is now easily reduced to the previous problem of dropping a perpendicular onto a subspace. In fact, if we fix any vector in the hyperplane R'' and subtract it from both sides of (5), we obtain the problem of representing the vector $f - f_0$ as a sum of vectors $g - f_0$ and h, the first of which belongs to the subspace R' and the second of which is orthogonal to R' [see Figure 1(b)]. By our previous result, this representation exists; therefore, the representation (5) also exists. It remains only to prove the uniqueness of the representation (5). If there were two such representations

$$f = g_1 + h_1 = g_2 + h_2,$$

then we would have

$$0 = (g_1 - g_2) + (h_1 - h_2).$$

Here $g_1 - g_2$ belongs to the subspace R' and $h_1 - h_2$ is orthogonal to R'. It follows that $g_1 - g_2 = h_1 - h_2 = 0$, Q.E.D.

Problem. Show that the perpendicular dropped from the origin of coordinates onto a hyperplane H has the smallest length of all the vectors joining the origin with H.

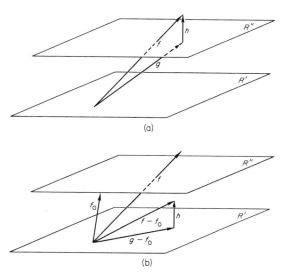

FIGURE 1

57. The General Orthogonalization Theorem

57.1. The following theorem is of fundamental importance in constructing orthogonal systems in a Euclidean space:

THEOREM 31. (*Orthogonalization theorem*) *Let* $x_1, x_2, \ldots, x_k, \ldots$ *be a finite or infinite sequence of vectors in a Euclidean space R, and let*

$$L_k = L(x_1, x_2, \ldots, x_k)$$

denote the linear manifold spanned by the first k of these vectors. *Then there exists a system of vectors* $y_1, y_2, \ldots, y_k, \ldots$ *which has the following properties:*

1. *For any integer k, the linear manifold* L'_k *spanned by the vectors* $y_1, y_2, \ldots, y_k$ *coincides with the linear manifold* L_k;

2. *For any integer k, the vector* y_{k+1} *is orthogonal to* L_k.

Proof. We shall prove the theorem by induction, i.e., assuming that k vectors $y_1, y_2, \ldots, y_k$ have been constructed which satisfy the conditions of the theorem, we shall construct a vector y_{k+1} such that the vectors $y_1, y_2, \ldots, y_k, y_{k+1}$ also satisfy the conditions of the theorem. First we note that if we set $y_1 = x_1$, the condition $L'_1 = L_1$ is obviously satisfied. Now the subspace L_k is finite-dimensional. Therefore, according to Sec. 56, there exists a decomposition

$$x_{k+1} = g_k + h_k, \tag{6}$$

where g_k is an element of L_k and h_k is orthogonal to L_k. Setting $y_{k+1} = h_k$, we verify that the conditions of the theorem are satisfied for this choice of y_{k+1}. By the induction hypothesis, the subspace L_k contains the vectors $y_1, y_2, \ldots, y_k$; therefore, the larger subspace L_{k+1} also contains these vectors. Moreover, it follows from (6) that L_{k+1} contains the vector $h_k = y_{k+1}$. Therefore, the subspace L_{k+1} contains all the vectors $y_1, y_2, \ldots, y_k, y_{k+1}$ and hence also contains the linear manifold L'_{k+1} spanned by these vectors. Conversely, the subspace L'_{k+1} contains the vectors $x_1, x_2, \ldots, x_k$, and moreover by (6), L'_{k+1} contains the vector x_{k+1} as well. It follows that L'_{k+1} contains the whole subspace L_{k+1}. Therefore,

$$L'_{k+1} = L_{k+1},$$

and the first assertion of the theorem is proved. The second assertion is an obvious consequence of the construction of the vector $y_{k+1} = h_k$. Thus, the induction is justified, and the proof of the theorem is complete.

57.2. In the present case, the inequality (4) takes the form

$$0 \leqslant |y_{k+1}| \leqslant |x_{k+1}|. \tag{7}$$

As shown in Sec. 56.2, the equality $|y_{k+1}| = 0$ means that the vector x_{k+1} belongs to the subspace L_k, and is therefore a linear combination of the vectors $x_1, x_2, \ldots, x_k$. The opposite equality $|y_{k+1}| = |x_{k+1}|$ means that the vector x_{k+1} is orthogonal to the subspace L_k, and hence is orthogonal to each of the vectors $x_1, x_2, \ldots, x_k$.

57.3. *Remark. Every system of vectors $z_1, z_2, \ldots, z_k, \ldots$ satisfying the conditions of the orthogonalization theorem, coincides to within numerical factors with the system $y_1, y_2, \ldots, y_k, \ldots$ constructed in proving the theorem.* In fact, the vector z_{k+1} must belong to the subspace L_{k+1}, and at the same time z_{k+1} must be orthogonal to the subspace L_k. The first of these conditions implies the existence of an expansion

$$z_{k+1} = C_1 y_1 + C_2 y_2 + \cdots + C_k y_k + C_{k+1} y_{k+1} = \tilde{y}_k + C_{k+1} y_{k+1},$$

where $\tilde{y}_k = C_1 y_1 + C_2 y_2 + \cdots + C_k y_k \in L_k$, and $C_{k+1} y_{k+1}$ is orthogonal to L_k. The second condition implies that $\tilde{y}_k = 0$ and hence that

$$z_{k+1} = C_{k+1} y_{k+1},$$

as required.

Problem 1. Given the system of vectors $x_1 = \mathbf{i}$, $x_2 = 2\mathbf{i}$, $x_3 = 3\mathbf{i}$, $x_4 = 4\mathbf{i} - 2\mathbf{j}$, $x_5 = -\mathbf{i} + 10\mathbf{j}$, $x_6 = \mathbf{i} + \mathbf{j} + 5\mathbf{k}$ in V_3, construct the vectors $y_1, y_2, \ldots, y_6$ discussed in the orthogonalization theorem.

         Ans. $y_1 = \mathbf{i}, y_2 = y_3 = 0, y_4 = -2\mathbf{j}, y_5 = 0, y_6 = 5\mathbf{k}$.

Problem 2. Using the method of the orthogonalization theorem, construct an orthogonal basis in the three-dimensional subspace of the space T_4 spanned by the vectors $(1, 2, 1, 3)$, $(4, 1, 1, 1)$, $(3, 1, 1, 0)$.

Ans. $(1, 2, 1, 3)$, $(10, -1, 1, -3)$, $(19, -87, -61, 72)$.

Problem 3. As shown in Sec. 51, every orthogonal basis in an n-dimensional space R_n is a canonical basis for the bilinear form (x, y). Let $x_1, x_2, \ldots, x_n$ be an arbitrary basis of the space R_n. Starting with this basis, construct a canonical basis for the bilinear form (x, y) by Jacobi's method (Sec. 45), and show that it coincides with the basis obtained by the orthogonalization process.

Hint. Use Sec. 56.3.

Problem 4. Suppose $F = \{x_1, x_2, \ldots, x_k\}$ and $G = \{y_1, y_2, \ldots, y_k\}$ are two finite sets of vectors in a Euclidean space R_n. Show that a necessary and sufficient condition for there to exist an isometric operator Q taking every vector x_i into the corresponding vector y_i $(i = 1, 2, \ldots, k)$ is that the relations

$$(x_i, x_j) = (y_i, y_j) \qquad (i, j = 1, 2, \ldots, k)$$

hold.

Hint. Applying the orthogonalization process to the given systems, obtain the orthonormal systems $e_1, e_2, \ldots$ and $f_1, f_2, \ldots$ Using Sec. 56.3, show that the formulas expressing the vectors $x_1, x_2, \ldots, x_k$ in terms of $e_1, e_2, \ldots$ are the same formulas as those expressing the vectors $y_1, y_2, \ldots, y_k$ in terms of $f_1, f_2, \ldots$ Then define Q as the operator which maps the system $e_1, e_2, \ldots$ into the system $f_1, f_2, \ldots$

Problem 5. *The angles between two subspaces.* Let R' and R'' be two subspaces of a Euclidean space R. Let the unit vector e' vary over the unit sphere of the space R', and let the unit vector e'' vary (independently of e') over the unit sphere of the space R''. For some pair of vectors $e' = e'_1$, $e'' = e''_1$, the angle between e' and e'' achieves a minimum, which we denote by φ_1. Now let e' vary over its unit sphere while remaining orthogonal to e'_1, and let e'' vary over its unit sphere while remaining orthogonal to e''_1. With these constraints, the angle between e' and e'' achieves a minimum $\varphi_2 \geqslant \varphi_1$ for some pair $e' = e'_2$, $e'' = e''_2$. Then let e' vary over its unit sphere while remaining orthogonal to e'_1 and e'_2, and let e'' vary over its unit sphere while remaining orthogonal to e''_1 and e''_2. In this way, we get a new minimum angle $\varphi_3 \geqslant \varphi_2$ and a new pair e'_3 and e''_3. Continuing this process, we obtain a set of angles $\varphi_1, \varphi_2, \ldots, \varphi_k$, the number of which equals the smaller of the dimensions of R' and R''. The angles $\varphi_1, \varphi_2, \ldots, \varphi_k$ are called the *angles between the subspaces R' and R''.* Prove the following facts:

a) The angles $\varphi_1, \varphi_2, \ldots, \varphi_k$ are defined uniquely and do not depend on the choice of the vectors $e'_1, e''_1, e'_2, e''_2, \ldots$ if these vectors are not uniquely defined by the construction;

b) The angles $\varphi_1, \varphi_2, \ldots, \varphi_k$ determine the subspaces R' and R'' to within their spatial orientation, i.e., if there are two pairs of subspaces R', R'' and S', S'', such that the angles between R' and R'' are the same as the angles between S' and S'', then there exists an isometric operator which simultaneously carries S' into R' and S'' into R''.

c) Given any preassigned angles $\varphi_1 \leqslant \varphi_2 \leqslant \ldots \leqslant \varphi_k \leqslant \pi/2$, we can construct a pair of spaces R' and R'', such that $\varphi_1, \varphi_2, \ldots, \varphi_k$ are the angles between R' and R''.

Hint. b) Consider the finite systems $e_1', e_1'', e_2', e_2'', \ldots, e_k', e_k''$ and $f_1', f_1'', f_2', f_2'', \ldots, f_k', f_k''$, obtained while determining the angles between the subspaces R', R'', and S', S'', respectively. According to the construction, we have

$$(e_i', e_i'') = (f_i', f_i'') = \cos \varphi_i \qquad (i = 1, 2, \ldots, k),$$
$$(e_i', e_j') = (f_i', f_j') = 0,$$
$$(e_i'', e_j'') = (f_i'', f_j'') = 0 \qquad (i \neq j).$$

Show that we also have

$$(e_i', e_j'') = (f_i', f_j'') = 0,$$

by using Problem 3 of Sec. 56. Then use the result of Problem 4.

Problem 6. Given two subspaces R' and R'', let $m(R', R'')$ denote the maximum length of the perpendiculars dropped onto R'' from the ends of the unit vectors $e' \in R'$; define the quantity $m(R'', R')$ similarly. The quantity

$$\theta = \max\{m(R', R''), m(R'', R')\}$$

is called the *spread* of the subspaces R' and R''. Show that the subspaces R' and R'' have the same dimension if $\theta < 1$. (M. A. Krasnosyelski and M. G. Krein.)

Hint. Assuming that the dimension of R'' is greater than the dimension of R', consider the vector $e'' \in R''$ which is orthogonal to the projection of R' onto R''; then use Problem 4 of Sec. 56.

*58. Legendre Polynomials

In the Euclidean space $C_2(-1, +1)$, apply the orthogonalization theorem to the system of functions

$$x_0(t) = 1, \ x_1(t) = t, \ \ldots, \ x_k(t) = t^k, \ \ldots.$$

In this case, the subspace $L_k = L(1, t, \ldots, t^k)$ is obviously the set of all polynomials in t of degree $n \leqslant k$. For any k, the function $x_k(t)$ is linearly independent of the functions with smaller indices (see Sec. 12). Therefore, by Sec. 56.2, the functions $y_0(t), y_1(t), \ldots$ obtained by orthogonalization are all nonzero. By its very construction, $y_k(t)$ must be a polynomial in t of degree k. In particular, direct calculation using the method of the orthogonalization theorem, yields

$$y_0(t) = 1, \ y_1(t) = t, \ y_2(t) = t^2 - \tfrac{1}{3}, \ y_3(t) = t^3 - \tfrac{3}{5}t, \text{ etc.}$$

These polynomials were introduced in 1785 by the French mathematician Legendre, in connection with certain problems in potential theory, but it was not until some 30 years later, in 1814, that the general formula for the Legendre polynomials was found by Rodrigues, who showed that the polynomial $y_n(t)$ is given by

$$p_n(t) = \frac{d^n}{dt^n} [(t^2 - 1)^n] \qquad (n = 0, 1, 2, \ldots), \tag{8}$$

to within a numerical factor. We now prove this formula, using the remark of Sec. 57.3, i.e., we shall show that the polynomial $p_n(t)$ satisfies the conditions of the orthogonalization theorem, whence it will follow from the remark in question that $p_n(t)$ must equal $C_n y_n(t)$ for every n, as required.

58.1. *The linear manifold spanned by the vectors $p_0(t)$, $p_1(t), \ldots, p_n(t)$ coincides with the set of all polynomials of degree no greater than n.* In fact, by (8), the polynomial $p_k(t)$ is clearly a polynomial in t of degree k; in particular

$$\begin{aligned}
&p_0(t) = a_{00}, \\
&p_1(t) = a_{10} + a_{11}t, \\
&p_2(t) = a_{20} + a_{21}t + a_{22}t^2, \\
& \quad \ldots\ldots \\
&p_k(t) = a_{k0} + a_{k1}t + \cdots + a_{kk}t^k, \\
& \quad \ldots\ldots \\
&p_n(t) = a_{n0} + a_{n1}t + \cdots + a_{nk}t^k + \cdots + a_{nn}t^n,
\end{aligned} \tag{9}$$

where the leading coefficients $a_{00}, a_{11}, \ldots, a_{nn}$ are nonzero. Thus, all the polynomials $p_0(t)$, $p_1(t)$, $\ldots$, $p_n(t)$ are elements of the linear manifold spanned by the functions $1, t, \ldots, t^n$, which is obviously just the set L_n of all polynomials in t of degree no greater than n. Conversely, the functions $1, t, \ldots, t^n$ can be expressed as linear combinations of $p_0(t)$, $p_1(t)$, $\ldots$, $p_n(t)$, since the matrix of the linear transformation (9) has the nonvanishing determinant $a_{00}a_{11} \ldots a_{nn}$. Therefore, the linear manifold $L(p_0(t), p_1(t), \ldots, p_n(t))$ coincides with the linear manifold $L(1, t, \ldots, t^n)$ and hence coincides with the set L_n, as required.

58.2. *The vector $p_n(t)$ is orthogonal to the subspace L_{n-1}.* It is sufficient to verify that the polynomial $p_n(t)$ is orthogonal, in the sense of the space $C_2(-1, +1)$, to the functions $1, t, \ldots, t^{n-1}$. To show this, we first prove the following lemma:

LEMMA (*On n-fold zeros*).   *Suppose that the function $f(t)$ can be represented in the form*

$$f(t) = (t - t_0)^n \varphi(t), \tag{10}$$

where $\varphi(t_0) \neq 0$ (if this is the case, we say that the function $f(t)$ has an *n*-fold zero for $t = t_0$). *Then, if it is assumed that* $f(t)$ *and* $\varphi(t)$ *have* n *continuous derivatives, the following relations hold:*

$$f(t_0) = 0, f'(t_0) = 0, \ldots, f^{(n-1)}(t_0) = 0, f^{(n)}(t_0) \neq 0.$$

Proof. Differentiating (10), we obtain

$$\begin{aligned} f'(t) &= n(t - t_0)^{n-1}\varphi(t) + (t - t_0)^n\varphi'(t) \\ &= (t - t_0)^{n-1}[n\varphi(t) + (t - t_0)\varphi'(t)] \\ &= (t - t_0)^{n-1}\varphi_1(t), \end{aligned}$$

where

$$\varphi_1(t) = n\varphi(t) + (t - t_0)\varphi'(t);$$

in particular,

$$\varphi_1(t_0) = n\varphi(t_0) \neq 0.$$

Similarly, we obtain

$$f''(t) = (t - t_0)^{n-2}\varphi_2(t), \quad \text{where } \varphi_2(t_0) \neq 0,$$
$$\ldots\ldots$$
$$f^{(n-1)}(t) = (t - t_0)\varphi_{n-1}(t), \quad \text{where } \varphi_{n-1}(t_0) \neq 0,$$
$$f^{(n)}(t) = \varphi_n(t), \quad \text{where } \varphi_n(t_0) \neq 0.$$

Substituting $t = t_0$ in all these expressions, we obtain the desired result.

In particular, the function

$$(t^2 - 1)^n = (t - 1)^n(t + 1)^n$$

has an *n*-fold zero at the points $t = \pm 1$. Therefore, for $t = \pm 1$, the quantity $[(t^2 - 1)^n]^{(k)}$ vanishes for $k < n$ and is different from zero for $k = n$.

We now show that the function

$$p_n(t) = [(t^2 - 1)^n]^{(n)}$$

is orthogonal to the functions $1, t, \ldots, t^{n-1}$, by calculating the scalar product of t^k and $p_n(t)$ for $k < n$. Integrating by parts, we obtain

$$\begin{aligned} (t^k, p_n(t)) &= \int_{-1}^{+1} t^k[(t^2 - 1)^n]^{(n)} dt \\ &= t^k[(t^2 - 1)^n]^{(n-1)} \Big|_{-1}^{+1} - k\int_{-1}^{+1} t^{k-1}[(t^2 - 1)^n]^{(n-1)} dt. \end{aligned}$$

By the lemma on *n*-fold zeros, the first term in the right-hand side vanishes. We integrate the second term by parts again, and continue this process until the exponent of t becomes zero. The result is

$$\begin{aligned} (t^k, p_n(t)) &= -kt^{k-1}[(t^2 - 1)^n]^{(n-2)}\Big|_{-1}^{+1} + k(k-1)\int_{-1}^{+1} t^{k-2}[(t^2 - 1)^n]^{(n-2)} dt \\ &= \cdots = \pm k!\int_{-1}^{+1} [(t^2 - 1)^n]^{(n-k)} dt = \pm k![(t^2 - 1)^n]^{(n-k-1)}\Big|_{-1}^{+1}, \end{aligned}$$

which vanishes, as required. Thus, we have finally proved that for every n, the polynomial $y_n(t)$ is the same as the polynomial

$$p_n(t) = [(t^2 - 1)^n]^{(n)},$$

except for a numerical factor.

We now calculate $p_n(1)$, by applying to the function

$$(t^2 - 1)^n = (t + 1)^n(t - 1)^n$$

Leibniz' formula for n-fold differentiation of a product. The result is

$$[(t + 1)^n(t - 1)^n]^{(n)} = (t + 1)^n[(t - 1)^n]^{(n)} + \binom{n}{1}[(t + 1)^n]'[(t - 1)^n]^{(n-1)} + \cdots$$
$$= (t + 1)^n n! + \binom{n}{1}n(t + 1)^{n-1}n(n - 1) \ldots 2(t - 1) + \cdots,$$

where $\binom{n}{k} = n!/k!(n - k)!$. If we substitute $t = 1$, all the terms of this sum vanish from the second term on. Thus, we find

$$p_n(1) = 2^n n!$$

For numerical purposes, it is convenient to make the values of our orthogonal functions equal 1 for $t = 1$. To achieve this, we have to multiply by the factor $1/2^n n!$. In fact, it is actually these normalized polynomials which are called the *Legendre polynomials*, i.e., the Legendre polynomial of degree n, denoted by $P_n(t)$, is given by the formula

$$P_n(t) = \frac{1}{2^n n!} [(t^2 - 1)^n]^{(n)}.$$

Problem 1. Find the leading coefficient A_n of the Legendre polynomial $P_n(t)$.

$$\text{Ans.} \quad A_n = \frac{(2n)!}{2^n(n!)^2}.$$

Problem 2. Show that $P_n(t)$ is an even function for even n, and an odd function for odd n. In particular, find $P_n(-1)$.

$$\text{Ans.} \quad P_n(-1) = (-1)^n.$$

Problem 3. Show that if the polynomial $tP_{n-1}(t)$ is expanded in terms of the Legendre polynomials, i.e.,

$$tP_{n-1}(t) = a_0P_0(t) + a_1P_1(t) + \cdots + a_nP_n(t),$$

then the coefficients $a_0, a_1, \ldots, a_{n-3}$ and a_{n-1} are zero.

Hint. Express the coefficients as scalar products.

Problem 4. Find the coefficients a_{n-2} and a_n of the expansion of the polynomial $tP_{n-1}(t)$ given in the preceding problem, thereby obtaining the recurrence formula

$$nP_n(t) = (2n - 1)tP_{n-1}(t) - (n - 1)P_{n-2}(t).$$

Hint. Use the results of Problems 1 and 2.

Problem 5. Find the polynomial

$$Q(t) = t^n + b_1 t^{n-1} + \cdots + b_{n-1} t + b_n$$

for which the integral

$$\int_{-1}^{+1} Q^2(t)\, dt$$

has the smallest value.

Hint. Expand $Q(t)$ in Legendre polynomials.

Ans. $\quad Q(t) = \dfrac{1}{A_n} P_n(t)$.

Problem 6. Show that the polynomial

$$Q_n(t) = P'_{n+1}(t) - P'_{n-1}(t)$$

is the Legendre polynomial $P_n(t)$ multiplied by a factor; find this factor.

Hint. Use the remark of Sec. 57.3 and the result of Problem 1.

Ans. $\quad Q_n(t) = (2n + 1)P_n(t)$.

Problem 7. Find the norm of the Legendre polynomial $P_n(t)$.

Ans. $\quad \|P_n(t)\|^2 = 2/(2n + 1)$.

59. The Gram Determinant

By the *Gram determinant* is meant the determinant of the form

$$G(x_1, x_2, \ldots, x_k) = \begin{vmatrix} (x_1, x_1) & (x_1, x_2) & \ldots & (x_1, x_k) \\ (x_2, x_1) & (x_2, x_2) & \ldots & (x_2, x_k) \\ \cdot & \cdot & \ldots & \cdot \\ (x_k, x_1) & (x_k, x_2) & \ldots & (x_k, x_k) \end{vmatrix},$$

where $x_1, x_2, \ldots, x_k$ are arbitrary vectors of the Euclidean space R. In Sec. 56.3, we saw that in the case of linearly independent vectors $x_1, x_2, \ldots, x_k$, the determinant $G(x_1, x_2, \ldots, x_k)$ is nonvanishing. We now calculate the value of $G(x_1, x_2, \ldots, x_k)$ by applying the orthogonalization process to the vectors $x_1, x_2, \ldots, x_k$. We begin by setting $y_1 = x_1$ and requiring that the vector

$$y_2 = \alpha_1 y_1 + x_2$$

be orthogonal to y_1. We replace the vector x_1 by y_1 everywhere in the determinant $G(x_1, x_2, \ldots, x_k)$; then we multiply the first column of $G(x_1, x_2, \ldots, x_k)$ by α_1 (associating α_1 with the second factors of the scalar products) and add it to the second column. Next, we multiply the first row of the determinant by α_1 (associating α_1 with the first factors of the scalar products) and add it to the second row. As a result, the vector y_2 will appear at every place in the determinant where x_2 appeared formerly. Now let

$$y_3 = \beta_1 y_1 + \beta_2 y_2 + x_3$$

be orthogonal to y_1 and y_2. Multiply the first column by β_1 and the second column by β_2 and add them to the third column; then carry out the same operations on the rows. As a result, x_3 will be replaced by y_3 everywhere in $G(x_1, x_2, \ldots, x_k)$. We can continue this process until we arrive at the last column (and row). Since these operations do not change the value of the determinant, we finally obtain

$$G(x_1, x_2, \ldots, x_k) = \begin{vmatrix} (y_1, y_1) & 0 & \ldots & 0 \\ 0 & (y_2, y_2) & \ldots & 0 \\ \cdot & \cdot & \ldots & \cdot \\ \cdot & \cdot & \ldots & \cdot \\ \cdot & \cdot & \ldots & \cdot \\ 0 & 0 & \ldots & (y_k, y_k) \end{vmatrix} \tag{11}$$

$$= (y_1, y_1)(y_2, y_2) \ldots (y_k, y_k).$$

Moreover, by the result of Sec. 57.2, we have the inequality

$$0 \leqslant G(x_1, x_2, \ldots, x_k) \leqslant (x_1, x_1)(x_2, x_2) \ldots (x_k, x_k). \tag{12}$$

We now examine the conditions under which the quantity $G(x_1, x_1, \ldots, x_k)$ can take the values 0 or $(x_1, x_1)(x_2 \, x_2) \ldots (x_k, x_k)$. It follows from the form (11) of the Gram determinant that it vanishes if and only if one of the vectors $y_1, y_2, \ldots, y_k$ vanishes. But according to Sec. 57.2, this implies that the vectors $x_1, x_2, \ldots, x_k$ are linearly dependent. Moreover, according to (11) and Sec. 57.2, the second equality sign holds in the inequality (12) only in the case where the vectors $x_1, x_2, \ldots, x_k$ are already orthogonal. Thus, we have proved the following theorem:

THEOREM 32. (*Theorem on the Gram determinant*) *The Gram determinant of the vectors* $x_1, x_2, \ldots, x_k$ *vanishes if the vectors are linearly dependent and is positive if they are linearly independent. It equals the product of the squares of the lengths of the vectors* $x_1, x_2, \ldots, x_k$ *if they are orthogonal and is less than this quantity otherwise.*

60. The Volume of a k-Dimensional Hyperparallelepiped

As is well known from elementary geometry, the area of a parallelogram equals the product of a base and the corresponding altitude. If the parallelogram is determined by two vectors x_1 and x_2, then for the base we can take the length of the vector x_1 and for the altitude we can take the length of the perpendicular dropped from the end of the vector x_2 onto the line containing the vector x_1. Similarly, the volume of the parallelepiped determined by the vectors x_1, x_2 and x_3 equals the product of the area of a base and the corresponding altitude; for the area of the base we choose the area of the parallelogram determined by the vectors x_1 and x_2, and for the altitude we take the length of the perpendicular dropped from the end of the vector x_3 onto the plane of the vectors x_1 and x_2.

These considerations make the following a very natural inductive definition of the volume of a k-dimensional hyperparallelepiped in an n-dimensional Euclidean space: Suppose we are given a system of vectors $x_1, x_2, \ldots, x_k$ in a Euclidean space R. Denote by h_j the perpendicular dropped from the end of the vector x_{j+1} onto the subspace

$$L(x_1, x_2, \ldots, x_j) \qquad (j = 1, 2, \ldots, k - 1).$$

Then introduce the following notation:

$V_1 = |x_1|$ (a one-dimensional volume, i.e., the length of the vector x_1),

$V_2 = V_1|h_1|$ (a two-dimensional volume, i.e., the area of the parallelogram determined by the vectors x_1, x_2),

$V_3 = V_2|h_2|$ (a three-dimensional volume, i.e., the volume of the parallelepiped determined by the vectors x_1, x_2, x_3),

. . . . . .

$V_k = V_{k-1}|h_{k-1}|$ (a k-dimensional volume, i.e., the volume of the hyperparallelepiped determined by the vectors $x_1, x_2, \ldots, x_k$).

Obviously, the volume V_k can be written in the form

$$V_k \equiv V[x_1, x_2, \ldots, x_k] = |x_1|\,|h_1|\,\ldots\,|h_{k-1}|.$$

Using equation (11) of Sec. 59, we can express the quantity V_k in terms of the vectors $x_1, x_2, \ldots, x_k$ as follows:

$$V_k^2 = \begin{vmatrix} (x_1, x_1) & (x_1, x_2) & \ldots & (x_1, x_k) \\ (x_2, x_1) & (x_2, x_2) & \ldots & (x_2, x_k) \\ \ldots & \ldots & \ldots, & \ldots \\ (x_k, x_1) & (x_k, x_2) & \ldots & (x_k, x_k) \end{vmatrix}.$$

Thus, *the Gram determinant of the k vectors $x_1, x_2, \ldots, x_k$ equals the square of the volume of the k-dimensional hyperparallelepiped determined by these vectors.* Let

$$\xi_i^{(j)} \qquad (j = 1, 2, \ldots, k;\ i = 1, 2, \ldots, n)$$

be the components of the vector x_j with respect to an orthonormal basis $e_1, e_2, \ldots, e_n$. Expressing the scalar products in terms of the components of the vectors involved, we obtain the following formula for V_k^2:

$$V_k^2 = \begin{vmatrix} \xi_1^{(1)}\xi_1^{(1)} + \xi_2^{(1)}\xi_2^{(1)} + \cdots + \xi_n^{(1)}\xi_n^{(1)} & \ldots & \xi_1^{(1)}\xi_1^{(k)} + \xi_2^{(1)}\xi_2^{(k)} + \cdots + \xi_n^{(1)}\xi_n^{(k)} \\ \xi_1^{(2)}\xi_1^{(1)} + \xi_2^{(2)}\xi_2^{(1)} + \cdots + \xi_n^{(2)}\xi_n^{(1)} & \ldots & \xi_1^{(2)}\xi_1^{(k)} + \xi_2^{(2)}\xi_2^{(k)} + \cdots + \xi_n^{(2)}\xi_n^{(k)} \\ \ldots & \ldots & \ldots \\ \xi_1^{(k)}\xi_1^{(1)} + \xi_2^{(k)}\xi_2^{(1)} + \cdots + \xi_n^{(k)}\xi_n^{(1)} & \ldots & \xi_1^{(k)}\xi_1^{(k)} + \xi_2^{(k)}\xi_2^{(k)} + \cdots + \xi_n^{(k)}\xi_n^{(k)} \end{vmatrix}.$$

We now use an argument similar to that used in Sec. 29.5. Every column of the determinant just written is the sum of n "elementary columns" with elements of the form $\xi_i^{(j)}\xi_i^{(\alpha)}$, where the indices α and i are fixed in each elementary column, while j ranges from 1 to k. Therefore, the whole determinant equals the sum of n^k "elementary determinants" consisting only of elementary columns. In each elementary column, the factor $\xi_i^{(\alpha)}$ is constant and can therefore be factored out of the elementary determinant. As a result, each elementary determinant takes the form

$$
\xi_{i_1}^{(1)}\xi_{i_2}^{(2)} \ldots \xi_{i_k}^{(k)}
\begin{vmatrix}
\xi_{i_1}^{(1)} & \xi_{i_2}^{(1)} & \ldots & \xi_{i_k}^{(1)} \\
\xi_{i_1}^{(2)} & \xi_{i_2}^{(2)} & \ldots & \xi_{i_k}^{(2)} \\
\cdot & \cdot & \ldots & \cdot \\
\xi_{i_1}^{(k)} & \xi_{i_2}^{(k)} & \ldots & \xi_{i_k}^{(k)}
\end{vmatrix},
\tag{13}
$$

where $i_1, i_2, \ldots, i_k$ are numbers from 1 to n. If some of these numbers are the same, then the corresponding elementary determinant obviously vanishes. Thus, we need only consider the case where $i_1, i_2, \ldots, i_k$ are all different. In the entire sum, we group together those terms of the form (13) which have the same indices $i_1, i_2, \ldots, i_k$, but arranged in different orders; we denote the sum of all such terms by

$$
M^2[i_1', i_2', \ldots, i_k'],
$$

where $i_1', i_2', \ldots, i_k'$ denote the numbers $i_1, i_2, \ldots, i_k$ arranged in increasing order. An argument similar to that used in Sec. 29.5 then leads to the following result: In the $n \times k$ matrix

$$
\|\xi_i^{(j)}\| \qquad (i = 1, 2, \ldots, n; j = 1, 2, \ldots, k),
$$

the quantity $M^2[i_1', i_2', \ldots, i_k']$ is the square of the minor of order k formed by the columns with indices $i_1', i_2', \ldots, i_k'$. The sum of *all* the terms (13) equals the sum of the squares of all the minors of order k of the matrix $\|\xi_i^{(j)}\|$. In other words, the square of the volume of the k-dimensional hyperparallelepiped determined by the vectors $x_1, x_2, \ldots, x_k$ equals the sum of the squares of all the minors of order k in the matrix consisting of the components of the vectors $x_1, x_2, \ldots, x_k$ with respect to any orthonormal basis $e_1, e_2, \ldots, e_n$.

In the case $k = n$, the matrix $\|\xi_i^{(j)}\|$ has only one minor of order k, equal to the determinant of the matrix $\|\xi_i^{(j)}\|$. Therefore, *the volume of the n-dimensional hyperparallelepiped determined by the vectors $x_1, x_2, \ldots, x_n$ equals the absolute value of the determinant formed from the components of the vectors $x_1, x_2, \ldots, x_n$ with respect to any orthonormal basis.*

Problem 1. Derive the first result of Problem 9 of Sec. 54 from geometric considerations.

Hint. An isometric operator preserves k-dimensional volumes.

Problem 2. Let A be any linear operator defined on an n-dimensional Euclidean space R. Show that the ratio

$$k(A) = \frac{V[Ax_1, Ax_2, \ldots, Ax_n]}{V[x_1, x_2, \ldots, x_n]}$$

is a constant (i.e., is independent of the choice of the vectors $x_1, x_2, \ldots, x_n$), and find the value of $k(A)$, the so-called "distortion coefficient."

Ans. $k(A) = |\det A|$.

Problem 3. Show that the relation $k(AB) = k(A)k(B)$ holds for any two linear operators.

Problem 4. Let Q be a linear operator defined on an n-dimensional Euclidean space R $(n \geqslant 3)$. Suppose that Q does not change the area of any parallelogram, i.e.,

$$V[x, y] = V[Qx, Qy].$$

Show that Q is an isometric operator.

Hint. It is sufficient to show that Q is an isogonal operator (Problem 10 of Sec. 54). Assuming that there is a right angle which is not transformed into a right angle, construct a parallelogram whose area changes as a result of applying the operator Q.

Problem 5. Let Q be a linear operator defined on an n-dimensional Euclidean space R, and suppose that Q does not change the volume of any k-dimensional hyperparallelepiped $(k < n)$. Show that Q is isometric. (M. A. Krasnosyelski.)

Hint. Generalize the construction of Problem 4.

Remark. For $k = n$, the assertion of Problem 5 is not valid, since then any operator Q with $\det Q = \pm 1$ will satisfy the condition of the problem.

Problem 6. Let $y_1, y_2, \ldots, y_m$ be the orthogonal projections of the vectors $x_1, x_2, \ldots, x_m$ onto some subspace. Show that the volume of the hyperparallelepiped determined by the vectors $y_1, y_2, \ldots, y_m$ does not exceed the volume of the hyperparallelepiped determined by the vectors $x_1, x_2, \ldots, x_m$.

Problem 7. (Continuation) In Problem 6, assume that both the vectors $x_1, x_2, \ldots, x_m$ and the vectors $y_1, y_2, \ldots, y_m$ are linearly independent. Show that the formula

$$V[y_1, y_2, \ldots, y_m] = V[x_1, x_2, \ldots, x_m] \cos \alpha_1 \cos \alpha_2 \cdots \cos \alpha_m$$

holds, where $\alpha_1, \alpha_2, \ldots, \alpha_m$ are the angles between the subspaces

$$L(x_1, x_2, \ldots, x_m) = L_1 \quad \text{and} \quad L(y_1, y_2, \ldots, y_m) = L_2$$

(Problem 5 of Sec. 57).

Hint. In the subspaces L_1 and L_2, let $e_1, e_2, \ldots, e_m$ and $f_1, f_2, \ldots, f_m$ be the bases obtained while constructing the angles $\alpha_1, \alpha_2, \ldots, \alpha_m$. In the space R construct a basis $e_1, e_2, \ldots, e_m, e_{m+1}, \ldots, e_n$ which begins with the vectors obtained by orthogonalizing the vectors $e_1, e_2, \ldots, e_m, f_1, f_2, \ldots, f_m$. Expand the vectors $x_1, x_2, \ldots, x_m$ and $y_1, y_2, \ldots, y_m$ with respect to this basis. Show that the matrices corresponding to these expansions each have only one minor of order m, if we disregard minors which are known to vanish. Then use the expression for the volume of a hyperparallelepiped in terms of the minors of the corresponding matrix.

Problem 8. By a *k-vector* we shall mean a set of k vectors of the Euclidean space R. Two k-vectors $\{x_1, x_2, \ldots, x_k\}$ and $\{y_1, y_2, \ldots, y_k\}$ are said to be *equal* if

1) The volume $V[x_1, x_2, \ldots, x_k]$ equals the volume $V[y_1, y_2, \ldots, y_k]$,

2) The linear manifold $L(x_1, x_2, \ldots, x_k)$ coincides with the linear manifold $L(y_1, y_2, \ldots, y_k)$,

3) The systems $x_1, x_2, \ldots, x_k$ and $y_1, y_2 \ldots, y_k$ have the same orientation (i.e., the   operator in the space $L(x_1, x_2, \ldots, x_k)$ carrying the system $x_1, x_2, \ldots, x_k$ into the system $y_1, y_2, \ldots, y_k$ has a positive determinant).

Show that a k-vector $\{x_1, x_2, \ldots, x_k\}$ in an n-dimensional space R is uniquely determined if we know the values of all the minors of order k of the $n \times k$ matrix
$$\| \xi_i^{(j)} \| \qquad (i = 1, 2, \ldots, n; j = 1, 2, \ldots, k),$$
which is formed from the components of the vectors $x_1, x_2, \ldots, x_k$ with respect to any orthonormal basis $e_1, e_2, \ldots, e_n$ of R.

Hint. Use Problem 2 of Sec. 19 and Problem 1 of Sec. 29.5.

Problem 9. If the k-vector $\{x_1, x_2, \ldots, x_k\}$ equals the k-vector $\{y_1, y_2, \ldots, y_k\}$ (Problem 8), show that the minors of order k of the matrix formed from the components of the vectors $x_1, x_2, \ldots, x_k$ equal the corresponding minors of the matrix formed from the components of the vectors $y_1, y_2, \ldots, y_k$.

Hint. Verify this assertion in the special basis whose first k vectors belong to the subspace $L(x_1, x_2, \ldots, x_k)$. To go over to the general case, use Problem 1 of Sec. 29.5 (show that det $\| a_i^{(j)} \| = 1$).

Problem 10. By the *angles* between two k-vectors $\{x_1, x_2, \ldots, x_k\}$ and $\{y_1, y_2, \ldots, y_k\}$, we mean the set of angles between the subspaces
$$L_1 = L(x_1, x_2, \ldots, x_k) \quad \text{and} \quad L_2 = L(y_1, y_2, \ldots, y_k)$$
(Problem 5 of Sec. 57), which are subject, however, to one supplementary condition, namely that the vectors $e_1, e_2, \ldots, e_k$ chosen in the subspace L_1 (when constructing the angles) have the same orientation as the vectors $x_1, x_2, \ldots, x_k$ (this condition plays a role only in constructing the last vector e_k), and similarly for the subspace L_2. Show that the angles $\beta_1, \beta_2, \ldots, \beta_k$ between the k-vectors and the angles $\alpha_1, \alpha_2, \ldots, \alpha_k$ between the corresponding subspaces are connected by the following relations:
$$\alpha_j = \beta_j \, (j < k), \quad \alpha_k = \beta_k \quad \text{or} \quad \alpha_k = \pi - \beta_k.$$

Problem 11. By the *scalar product* of two k-vectors $X = \{x_1, x_2, \ldots, x_k\}$ and $Y = \{y_1, y_2, \ldots, y_k\}$, specified by the matrices X and Y relative to some orthonormal basis of the space R, we mean the sum of all the products of the minors of order k of the matrix X with the corresponding minors of the matrix Y. Show that this scalar product equals

$$V[x_1, x_2, \ldots, x_k] \, V[y_1, y_2, \ldots, y_k] \cos \beta_1 \cos \beta_2 \ldots \cos \beta_k,$$

where $\beta_1, \beta_2, \ldots, \beta_k$ are the angles between the k-vectors X and Y.

Hint. Choose a basis in the space R like that chosen in Problem 7, and verify that the formula is valid in this basis. Then go over to the general case in the way indicated in Problem 9.

Problem 12. Show that the scalar product of two k-vectors

$$X = \{x_1, x_2, \ldots, x_k\} \quad \text{and} \quad Y = \{y_1, y_2, \ldots, y_k\}$$

can be written in the form

$$\{X, Y\} = \begin{vmatrix} (x_1, y_1) & (x_1, y_2) & \ldots & (x_1, y_k) \\ (x_2, y_1) & (x_2, y_2) & \ldots & (x_2, y_k) \\ \cdot & \cdot & \ldots & \cdot \\ (x_k, y_1) & (x_k, y_2) & \ldots & (x_k, y_k) \end{vmatrix}.$$

61. Hadamard's Inequality

Using the results of the preceding section, we can obtain an important estimate for the absolute value of an arbitrary determinant

$$D = \begin{vmatrix} \xi_{11} & \xi_{12} & \ldots & \xi_{1k} \\ \xi_{21} & \xi_{22} & \ldots & \xi_{2k} \\ \cdot & \cdot & \ldots & \cdot \\ \xi_{k1} & \xi_{k2} & \ldots & \xi_{kk} \end{vmatrix}$$

of order k. If we interpret the numbers $\xi_{i1}, \xi_{i2}, \ldots, \xi_{ik}$ $(i = 1, 2, \ldots, k)$ as the components of a vector x_i with respect to an orthonormal basis in a k-dimensional Euclidean space, then the last result of Sec. 60 allows us to interpret the absolute value of the determinant D as the volume of the k-dimensional hyperparallelepiped determined by the vectors $x_1, x_2, \ldots, x_k$. Then, using the expression for this volume in terms of the Gram determinant, we have

$$D^2 = G(x_1, x_2, \ldots, x_k).$$

Applying Theorem 32, we obtain

$$D^2 \leqslant (x_1, x_1)(x_2, x_2) \ldots (x_k, x_k) = \prod_{i=1}^{k} \sum_{j=1}^{k} \xi_{ij}^2,$$

an inequality known as *Hadamard's inequality*. Moreover, we note that according to Theorem 32, the equality holds if and only if the vectors $x_1, x_2, \ldots, x_k$ are pairwise orthogonal.

The geometric meaning of Hadamard's inequality is transparent:

The volume of a hyperparallelepiped does not exceed the product of the lengths of its sides; it equals this product if and only if its sides are orthogonal.

Problem 1. Let $x_1, x_2, \ldots, x_k, y, z$ be vectors in a Euclidean space R. Prove the inequality

$$\frac{V[x_1, x_2, \ldots, x_k, y, z]}{V[x_1, x_2, \ldots, x_k, y]} \leqslant \frac{V[x_1, x_2, \ldots, x_k, z]}{V[x_1, x_2, \ldots, x_k]}. \tag{14}$$

Hint. This is a question of comparing the altitudes of two hyperparallele-pipeds.

Problem 2. Let $x_1, x_2, \ldots, x_m$ be vectors in a Euclidean space R. Prove the inequality

$$V[x_1, x_2, \ldots, x_m] \leqslant \prod_{k=1}^{m} \{V[x_1, \ldots, x_{k-1}, x_{k+1}, \ldots, x_m]\}^{1/(m-1)}. \tag{15}$$

What is the geometric meaning of this inequality?
Hint. The inequalities

$$\frac{V[x_1, x_2, \ldots, x_m]}{V[x_1, \ldots, x_{k-1}, x_{k+1}, \ldots, x_m]} \leqslant \frac{V[x_1, x_2, \ldots, x_k]}{V[x_1, x_2, \ldots, x_{k-1}]} \qquad (k = 1, 2, \ldots, m)$$

are easily obtained from the inequality (14). Multiply them all together for $k = 1, 2, \ldots, m$, make appropriate cancellations, and then take the $(m - 1)$th root. The geometric meaning of the inequality is the following: The volume of an m-dimensional hyperparallelepiped does not exceed the product of the $(m - 1)$th roots of the volumes of its $(m - 1)$-dimensional "faces."

Problem 3. (Continuation) Prove the following inequalities, which are more precise than Hadamard's inequality:

$$V[x_1, x_2, \ldots, x_m]$$

$$\leqslant \prod_{k=1}^{m} \{V[x_1, \ldots, x_{k-1}, x_{k+1}, \ldots, x_m]\}^{1/(m-1)}$$

$$\leqslant \prod_{1 \leqslant k < l \leqslant m} \{V[x_1, \ldots, x_{k-1}, x_{k+1}, \ldots x_{l-1}, x_{l+1}, \ldots x_m]\}^{1 \cdot 2/(m-1)(m-2)}$$

$$\leqslant \cdots \leqslant \prod_{1 \leqslant s_1 < s_2 < \ldots < s_r \leqslant m} \{V[x_{s_1}, x_{s_2}, \ldots, x_{s_r}]\}^{1 \cdot 2 \ldots (n-r)/(m-1)(m-2) \ldots r}$$

$$\leqslant \cdots \leqslant \prod_{1 \leqslant s_1 < s_2 \leqslant m} \{V[x_{s_1}, x_{s_2}]\}^{1/(m-1)} \leqslant \prod_{s=1}^{m} |x_s|.$$

(M. K. Faguet.)

Hint. Write the inequality (15) for $x_{s_1}, x_{s_2}, \ldots, x_{s_r}$ where $s_1, s_2, \ldots, s_r$ take all permissible values, and then multiply these inequalities together.

Problem 4. If $|a_{ik}| \leqslant M$, then according to Hadamard's inequality, we have

$$\det \|a_{ik}\| \leqslant M^n n^{n/2}.$$

Show that this estimate cannot be improved for $n = 2^m$.

Hint. We have to construct a hyperparallelepiped in a 2^m-dimensional space such that the projections of its edges onto each axis have absolute values no greater than M and such that its volume is exactly $M^n n^{n/2}$. For $M = 1$, the matrix A_m of the components of the 2^m-dimensional vectors determining this hyperparallelepiped are given by the following recurrence formula:

$$A_m = \begin{Vmatrix} A_{m-1} & A_{m-1} \\ A_{m-1} & -A_{m-1} \end{Vmatrix}, \qquad A_1 = \begin{Vmatrix} 1 & 1 \\ 1 & -1 \end{Vmatrix}.$$

Comment. For $n \neq 2^m$, the estimate $M^n n^{n/2}$ can be improved.

62. Incompatible Systems of Linear Equations and the Method of Least Squares

Suppose we are given an incompatible system of linear equations

$$\begin{aligned}
a_{11}x_1 + a_{12}x_2 + \cdots + a_{1m}x_m &= b_1, \\
a_{21}x_1 + a_{22}x_2 + \cdots + a_{2m}x_m &= b_2, \\
\cdots \qquad \cdots \qquad \cdots \qquad \cdots & \\
a_{n1}x_1 + a_{n2}x_2 + \cdots + a_{nm}x_m &= b_n.
\end{aligned} \qquad (16)$$

Since this system is incompatible, it cannot be solved, i.e., we cannot find numbers $c_1, c_2, \ldots, c_m$ which satisfy all the equations of the system when substituted for the unknowns $x_1, x_2, \ldots, x_m$. Thus, if we substitute the numbers $\xi_1, \xi_2, \ldots, \xi_m$ for the unknowns $x_1, x_2, \ldots, x_m$ in the left-hand side of the system (16), we obtain numbers $\gamma_1, \gamma_2, \ldots, \gamma_n$ which differ from the numbers $b_1, b_2, \ldots, b_n$. This suggests the following problem: *Find the numbers $\xi_1, \xi_2, \ldots, \xi_m$ such that the resulting numbers $\gamma_1, \gamma_2, \ldots, \gamma_n$ have the smallest possible "mean square deviation"*

$$\delta^2 = \sum_{j=1}^{n} (\gamma_j - b_j)^2 \qquad (17)$$

from the given numbers $b_1, b_2, \ldots, b_n$, and find the corresponding minimum value of δ^2.

An example of a situation where this problem arises in practice is the following: Suppose we have to determine the coefficients ξ_j in the linear relation

$$b = \xi_1 a_1 + \xi_2 a_2 + \cdots + \xi_m a_m$$

connecting the quantity b and the quantities $a_1, a_2, \ldots, a_m$, when we are given the results of measurements of the a_j ($j = 1, 2, \ldots, m$) and the corresponding values of b. If the ith measurement gives us the value a_{ij} for the quantity a_j and the value b_i for the quantity b, then we must write

$$\xi_1 a_{i1} + \xi_2 a_{i2} + \cdots + \xi_m a_{im} = b_i. \tag{18}$$

If we make n measurements, we obtain a system of n equations of the form (18), i.e., a system of the form (16). As a result of unavoidable measurement errors, this system will generally be incompatible, and then the problem of finding the coefficients $\xi_1, \xi_2, \ldots, \xi_m$ does not reduce to the problem of solving the system (16). This suggests determining the coefficients ξ_j in such a way that every equation is at least approximately valid and the total error is as small as possible. If we take as a measure of the error the mean square deviation of the quantities

$$\gamma_j = \sum_{i=1}^{m} a_{ij}\xi_i$$

from the known quantities b_j, i.e., if we take the formula (17) as a measure of the error, then we arrive at the problem formulated at the beginning of this section. Moreover, in this case, it is also useful to know the quantity δ^2, since this helps to estimate the accuracy of the measurements.

We can immediately solve the problem just stated, if we interpret it geometrically. Consider the m vectors $a_1, a_2, \ldots, a_m$ whose components form the columns of the system (16), i.e.

$$a_1 = (a_{11}, a_{21}, \ldots, a_{n1}), \ldots, a_m = (a_{1m}, a_{2m}, \ldots, a_{nm}).$$

If we form the linear combination $\xi_1 a_1 + \xi_2 a_2 + \cdots + \xi_m a_m$, we obtain the vector $\gamma = (\gamma_1, \gamma_2, \ldots, \gamma_n)$. Our problem is to determine the numbers $\xi_1, \xi_2, \ldots, \xi_m$ in such a way that the vector γ has the smallest possible deviation *in norm* from the given vector $b = (b_1, b_2, \ldots, b_n)$. Now the set of all linear combinations of the vectors $a_1, a_2, \ldots, a_m$ forms a subspace $L(a_1, a_2, \ldots, a_m)$, and the projection of the vector b onto the subspace L is the vector in L which is the closest to b. Therefore, the numbers $\xi_1, \xi_2, \ldots, \xi_m$ have to be chosen in such a way that the linear combination

$$\xi_1 a_1 + \xi_2 a_2 + \cdots + \xi_m a_m$$

reduces to the projection of b onto L. But, as we know, the solution of this problem is given by the last equation in Sec. 56.3, i.e.,

$$\xi_j = \frac{1}{D} \begin{vmatrix} (a_1, a_1) & \cdots & (a_{j-1}, a_1) & (b, a_1) & (a_{j+1}, a_1) & \cdots & (a_m, a_1) \\ \cdot & \cdots & \cdot & \cdot & \cdot & \cdots & \cdot \\ \cdot & \cdots & \cdot & \cdot & \cdot & \cdots & \cdot \\ \cdot & \cdots & \cdot & \cdot & \cdot & \cdots & \cdot \\ (a_1, a_m) & \cdots & (a_{j-1}, a_m) & (b, a_m) & (a_{j+1}, a_m) & \cdots & (a_m, a_m) \end{vmatrix},$$

where D is the Gram determinant $G(a_1, a_2, \ldots, a_m)$.

The results of Sec. 60 also allow us to evaluate the deviation δ itself.   In fact, δ is just the altitude of the $(m + 1)$-dimensional hyperparallelepiped determined by the vectors $a_1, a_2, \ldots, a_m, b$, and is therefore equal to the following ratio of volumes:

$$\frac{V[a_1, a_2, \ldots, a_m, b]}{V[a_1, a_2, \ldots, a_m]}.$$

Using the Gram determinant to write each of these volumes, we finally obtain

$$\delta^2 = \frac{G(a_1, a_2, \ldots, a_m, b)}{G(a_1, a_2, \ldots, a_m)}.$$

Thus, the problem posed at the beginning of this section is now completely solved.

In numerical analysis, the following problem is often encountered (interpolation with the least mean square error): *Given a function $f_0(t)$ defined on the interval $a \leqslant t \leqslant b$, find the polynomial $P(t)$ of degree k $(k < n)$ for which the mean square deviation from the function $f_0(t)$, defined by*

$$\delta^2(f_0, P) = \sum_{j=0}^{n} [f_0(t_j) - P(t_j)]^2$$

is the smallest.   Here $t_0, t_1, \ldots, t_n$ are certain fixed points of the interval $a \leqslant t \leqslant b$.   Using geometric considerations, M. A. Krasnosyelski has given the following simple solution of the problem: Introduce a Euclidean space R consisting of functions $f(t)$ considered *only* at the points $t_0, t_1, \ldots, t_n$. Define the scalar product by

$$(f, g) = \sum_{j=0}^{n} f(t_j)g(t_j).$$

Then the problem reduces to finding the projection of the vector $f_0(t)$ on the subspace of all polynomials of degree not exceeding k.   The coefficients of the desired polynomial

$$P_0(t) = \xi_0 + \xi_1 t + \cdots + \xi_k t^k$$

are given by the same formulas as in the problem analyzed previously, i.e.,

$$\xi_j = \frac{1}{D} \begin{vmatrix} (1, 1) & (t, 1) & \ldots & (t^{j-1}, 1) & (f_0, 1) & (t^{j+1}, 1) & \ldots & (t^k, 1) \\ (1, t) & (t, t) & \ldots & (t^{j-1}, t) & (f_0, t) & (t^{j+1}, t) & \ldots & (t^k, t) \\ \cdot & \cdot & \ldots & \cdot & \cdot & \cdot & \ldots & \cdot \\ \cdot & \cdot & \ldots & \cdot & \cdot & \cdot & \ldots & \cdot \\ \cdot & & \ldots & & & & \ldots & \\ (1, t^k) & (t, t^k) & \ldots & (t^{j-1}, t^k) & (f_0, t^k) & (t^{j+1}, t^k) & \ldots & (t^k, t^k) \end{vmatrix},$$

where D is the Gram determinant $G(1, t, \ldots, t^k)$.   The least square deviation itself is given by the formula

$$\delta^2(f_0, P) = \frac{G(1, t, \ldots, t^k, P)}{G(1, t, \ldots, t^k)}.$$

9

INVARIANT SUBSPACES AND EIGENVECTORS

63. Invariant Subspaces

Let A be a linear operator defined on a linear (affine) space R. The subspace R' of the linear space R is said to be *invariant* with respect to the operator A if $x \in R'$ implies that $Ax \in R'$. In particular, the trivial subspaces, i.e., the whole space and the subspace whose only element is 0, are invariant with respect to any linear operator. We now examine from this point of view the linear operators given in the examples of Sec. 26.

1. *Every subspace is invariant* for the operators of Examples 1, 2, and 3.

2. The rotation operator in the plane (Example 4) *has no nontrivial invariant subspaces.*

3. The projection operator (Example 5) has the following invariant subspaces (among others); the subspace R' of vectors

$$x = \sum_{k=1}^{m} \xi_k e_k$$

which remain unchanged, and the subspace R'' of vectors

$$y = \sum_{k=m+1}^{n} \xi_k e_k$$

which are carried into 0.

4. Every subspace spanned by some of the basis vectors $e_1, e_2, \ldots, e_n$ is invariant with respect to a diagonal operator (Example 6).

5. For the operator corresponding to multiplication by t in the space $C(a, b)$ of Example 7, the set of all functions which vanish on a subinterval Δ of the interval $a \leqslant t \leqslant b$ is an invariant (infinite-dimensional) subspace.

6. The linear combinations of n fixed functions of the form $e^{k_1 t}$, $e^{k_2 t}$, ..., $e^{k_n t}$ form an n-dimensional invariant subspace of the differentiation operator (Example 9).

Problem 1. What feature characterizes the matrix of an operator A, defined on an n-dimensional space R, if the first k basis vectors are chosen from a k-dimensional invariant subspace?

Ans. In the first k columns of the matrix, all the elements from row $k + 1$ to row n vanish.

Problem 2. Show that if A is a nonsingular linear operator defined on an n-dimensional linear space, then every subspace invariant with respect to A is also invariant with respect to A^{-1}.

Problem 3. If the space R can be represented as the direct sum[1] of two subspaces R_1 and R_2 which are invariant with respect to the operator A, and if the first k vectors of the basis $e_1, \ldots, e_k, e_{k+1}, \ldots, e_n$ belong to R_1, while the remaining $n - k$ vectors belong to R_2, then the matrix of the operator A has the form

$$A = \left\| \begin{matrix} A_1 & 0 \\ 0 & A_2 \end{matrix} \right\|,$$

where A_1 is a square matrix of order k and A_2 is a square matrix of order $n - k$. What is the geometric meaning of the matrices A_1 and A_2?

Ans. The matrix A_1 is the matrix of the operator A regarded as an operator acting only on the subspace R_1 and specified relative to the basis $e_1, \ldots, e_k$; A_2 has a similar meaning.

Problem 4. Show that if the annihilating polynomial $P(t)$ of the operator A can be written as a product of two factors, i.e., $P(t) = P_1(t)P_2(t)$, then the matrix of the operator A can be written in the form

$$A = \left\| \begin{matrix} A_1 & 0 \\ 0 & A_2 \end{matrix} \right\|,$$

relative to some basis, where the matrices A_1 and A_2 are annihilated by the polynomials $P_1(t)$ and $P_2(t)$, respectively.

Hint. There exist polynomials $S_1(t)$ and $S_2(t)$ satisfying the relation

$$P_1(t)S_1(t) + P_2(t)S_2(t) \equiv 1.$$

---

[1] R is said to be the *direct sum* of the subspaces R_1 and R_2 if the sum of R_1 and R_2 is R, while the intersection of R_1 and R_2 is the zero vector (cf. Sec. 15.1).

Let R_1 be the subspace of all the vectors $x \in R$ for which $P_1(A)x = 0$, and let R_2 be the subspace of all the vectors $y \in R$ for which $P_2(A)y = 0$. Using the above relation, show that the intersection of R_1 and R_2 is zero, while the sum of R_1 and R_2 is the whole space R. Then use the result of Problem 1.

Problem 5. Show that if the subspace R_1 of the n-dimensional Euclidean space R is invariant with respect to the operator A, then the orthogonal complement of R_1 is invariant with respect to the adjoint operator A^*.

Problem 6. Let A be an isometric operator defined on a Euclidean space R, and let A satisfy the condition of Problem 4. Show that the basis used to construct the matrix of A can be chosen to be *orthonormal*.

Hint. By using the expansion

$$z = P_1(A)S_1(A)z + P_2(A)S_2(A)z$$

and the invariance of the orthogonal complement of the subspace R_1 (Problem 4), show that every vector z orthogonal to R_1 is an element of the subspace R_2.

Problem 7. Show that if the polynomial $[P(t)]^k$ is an annihilating polynomial of the isometric operator A, then so is the polynomial $P(t)$.

Hint. Consider the orthogonal complement Z of the invariant (with respect to A) subspace H of all vectors x for which $P(A)x = 0$. The subspace Z is also invariant with respect to the operator A, and therefore with respect $[P(A)]^{k-1}$. But if $z \in Z$, then $[P(A)]^{k-1}z \in H$, so that

$$[P(A)]^{k-1}z = 0.$$

From this, we find that $[P(t)]^{k-1}$ is an annihilating polynomial of the operator A.

Problem 8. Show that if the annihilating polynomial of the operator A is a polynomial of the second degree, then any vector x of the space R lies in a plane or straight line which is invariant with respect to A.

Hint. The vectors A, Ax and A^2x are linearly dependent.

Problem 9. Given any isometric operator A, show that the space R can be represented as a direct sum of an invariant plane (or straight line) and an orthogonal invariant subspace.

Hint. Every real polynomial, including the annihilating polynomials of the operator A, can be factored into relatively prime real factors of the form $[P(t)]^k$, where $P(t)$ is a polynomial of degree no greater than 2. Then use Problems 4 to 8.

Problem 10. Given any isometric operator A defined on a linear space R, show that an orthonormal basis can be chosen relative to which the matrix of the operator A has the form

$$
A = \begin{Vmatrix}
\cos\alpha_1 & -\sin\alpha_1 & & & & & & & \\
\sin\alpha_1 & \cos\alpha_1 & & & & & & & \\
& & \cos\alpha_2 & -\sin\alpha_2 & & & & & \\
& & \sin\alpha_2 & \cos\alpha_2 & & & & & \\
& & & & \ddots & & & & \\
& & & & & -1 & & & \\
& & & & & & -1 & & \\
& & & & & & & \ddots & \\
& & & & & & & & 1 \\
& & & & & & & & & 1 \\
& & & & & & & & & & \ddots
\end{Vmatrix},
$$

where all the unwritten elements of the matrix vanish.

Hint. First find the matrix of the isometric operator A regarded as operating in a two-dimensional or one-dimensional subspace. Then use the result of Problem 8.

64. Eigenvectors and Eigenvalues

64.1. A special role is played by the one-dimensional invariant subspaces of the operator A; they are also called *invariant directions* (or *eigenrays*). Every (nonzero) vector belonging to a one-dimensional invariant direction of the operator A, is called an *eigenvector* of the operator A. In other words, the vector $x \neq 0$ is called an *eigenvector* of the operator A, if A carries the vector x into a collinear vector, i.e.,

$$Ax = \lambda x. \tag{1}$$

The real number λ appearing in (1) is called the *eigenvalue* (or *characteristic value*) *of the operator A corresponding to the eigenvector x.* In this regard, we again examine the examples of Sec. 26:

1. In Examples 1, 2, and 3, every nonzero vector of the space is an eigenvector, and the corresponding eigenvalues are 0, 1, λ.

2. The rotation operator (Example 4) has no eigenvectors.

3. The projection operator (Example 5) has eigenvectors of the form

$$x = \sum_{k=1}^{m} \xi_k e_k \quad \text{and} \quad y = \sum_{k=m+1}^{n} \xi_k e_k,$$

with corresponding eigenvalues 1 and 0. It can be verified that the projection operator has no other eigenvectors.

4. The diagonal operator (Example 6), by its very definition, has the eigenvectors $e_1, e_2, \ldots, e_n$ with corresponding eigenvalues $\lambda_1, \lambda_2, \ldots, \lambda_n$.

5. The operator corresponding to multiplication by t (Example 7) has no eigenvectors in the space $C(a, b)$, since there is no continuous function $x(t) \not\equiv 0$ which satisfies the equation

$$tx(t) \equiv \lambda x(t).$$

6. We shall discuss the eigenvectors of the Fredholm operator (Example 8) later (Chapter 12).

7. To find the eigenvectors of the differentiation operator (Example 9), we have to solve the equation $x'(t) = \lambda x(t)$, which gives $x(t) = Ce^{\lambda t}$. The result is an infinite set of eigenvectors with different eigenvalues.

64.2. We now prove two simple properties of eigenvectors:

LEMMA 1. *The eigenvectors $x_1, x_2, \ldots, x_m$ of an operator A corresponding to distinct eigenvalues $\lambda_1, \lambda_2, \ldots, \lambda_m$ are linearly independent.*

Proof. We prove this assertion by induction on the integer m. Obviously, the lemma is true for $m = 1$. Assuming that the lemma is true for any $m - 1$ eigenvectors of the operator A, we show that it remains true for any m eigenvectors of A. In fact, assume the contrary, i.e., that there is a linear relation

$$C_1 x_1 + C_2 x_2 + \cdots + C_m x_m = 0$$

between m eigenvectors of the operator A, with $C_1 \neq 0$, say. Applying the operator A to this relation gives

$$C_1 \lambda_1 x_1 + C_2 \lambda_2 x_2 + \cdots + C_m \lambda_m x_m = 0.$$

Multiplying the first equation by λ_m and subtracting it from the second equation, we obtain

$$C_1(\lambda_1 - \lambda_m)x_1 + C_2(\lambda_2 - \lambda_m)x_2 + \cdots + C_{m-1}(\lambda_{m-1} - \lambda_m)x_{m-1} = 0,$$

which, by the induction hypothesis, implies that all the coefficients

$$C_1(\lambda_1 - \lambda_m), \ldots, C_{m-1}(\lambda_{m-1} - \lambda_m)$$

vanish. However, $C_1(\lambda_1 - \lambda_m) = 0$ contradicts the conditions $C_1 \neq 0$, $\lambda_1 \neq \lambda_m$. Therefore, our assumption that there is linear dependence between $x_1, x_2, \ldots, x_m$ is false, Q.E.D. In particular, *a linear operator A defined on an m-dimensional space cannot have more than m eigenvectors with different eigenvalues.*

LEMMA 2. *The eigenvectors of a linear operator A which correspond to a given eigenvalue λ span a subspace $R^{(\lambda)} \subset R$.*

Proof. If $Ax_1 = \lambda x_1$, $Ax_2 = \lambda x_2$, then

$$A(\alpha x_1 + \beta x_2) = \alpha A x_1 + \beta A x_2 = \alpha \lambda x_1 + \beta \lambda x_2 = \lambda(\alpha x_1 + \beta x_2),$$

Q.E.D. The subspace $R^{(\lambda)}$ is called the *eigenspace* (or *characteristic subspace*) *of the operator A, corresponding to the eigenvalue λ.*

Problem 1. Show that if the linear operators A and B commute (i.e., $AB = BA$), then every eigenspace of the operator A is an invariant subspace of the operator B.

Problem 2. Suppose that the operator A has a set of eigenspaces whose sum is the whole space R and which are such that every eigenspace is invariant with respect to the operator B. Show that A and B commute.

Problem 3. Let x and y be eigenvectors of the operator A which correspond to *different* eigenvalues. Show that $\alpha x + \beta y$ ($\alpha \neq 0$, $\beta \neq 0$) is certainly not an eigenvector of A.

Problem 4. If every vector of the space R is an eigenvector of the operator A, show that $A = \lambda E$ (where λ is a real number and E is the identity transformation).
Hint. Use the result of Problem 3.

Problem 5. Show that if the linear operator A commutes with *all* linear operators defined on the space, then $A = \lambda E$.
Hint. Suitably selecting an operator B and using Problem 1, reduce the solution to Problem 4.

Problem 6. Let the linear operator A have the eigenvector e_0 with eigenvalue λ_0. Show that e_0 is an eigenvector of the operator A^2 corresponding to the eigenvalue λ_0^2.

Problem 7. Even if a linear operator A has no eigenvectors, the operator A^2 may have eigenvectors (e.g., the operator in the plane corresponding to rotation through $90°$). Show that if the operator A^2 has an eigenvector with a *nonnegative* eigenvalue $\lambda = \mu^2$, then the operator A also has an eigenvector.
Hint. Use the factorization of the operator $A^2 - \mu^2 E$.

65. Calculation of Eigenvectors and Eigenvalues in a Finite-Dimensional Space

Let $e_1, e_2, \ldots, e_n$ be a basis for the n-dimensional space R_n, and let A be a linear operator defined on R. Let the vector

$$x = \sum_{k=1}^{n} \xi_k e_k$$

be an eigenvector of the operator A, so that $Ax = \lambda x$ for some real λ. Using equation (7) of Sec. 27, we can rewrite this equation in coordinate form as follows:

$$\lambda \xi_1 = a_1^{(1)} \xi_1 + a_1^{(2)} \xi_2 + \cdots + a_1^{(n)} \xi_n,$$
$$\lambda \xi_2 = a_2^{(1)} \xi_1 + a_2^{(2)} \xi_2 + \cdots + a_2^{(n)} \xi_n,$$
$$\lambda \xi_n = a_n^{(1)} \xi_1 + a_n^{(2)} \xi_2 + \cdots + a_n^{(n)} \xi_n,$$

or

$$(a_1^{(1)} - \lambda)\xi_1 + a_1^{(2)} \xi_2 + \cdots + a_1^{(n)} \xi_n = 0,$$
$$a_2^{(1)} \xi_1 + (a_2^{(2)} - \lambda)\xi_2 + \cdots + a_2^{(n)} \xi_n = 0, \qquad (2)$$
$$a_n^{(1)} \xi_1 + a_n^{(2)} \xi_2 + \cdots + (a_n^{(n)} - \lambda)\xi_n = 0.$$

This homogeneous system of equations in the unknowns $\xi_1, \xi_2, \ldots, \xi_n$ has a nontrivial solution if and only if its determinant vanishes, i.e.,

$$\Delta(\lambda) \equiv \begin{vmatrix} a_1^{(1)} - \lambda & a_1^{(2)} & \cdots & a_1^{(n)} \\ a_2^{(1)} & a_2^{(2)} - \lambda & \cdots & a_2^{(n)} \\ \cdot & \cdot & \cdots & \cdot \\ a_n^{(1)} & a_n^{(2)} & \cdots & a_n^{(n)} - \lambda \end{vmatrix} = 0. \qquad (3)$$

The polynomial of degree n in λ appearing in (3) is already familiar to us as the *characteristic polynomial* of the operator A (Sec. 38.3). To each of its real roots λ_0 there corresponds an eigenvector obtained by substituting λ_0 for λ in (2) and then solving the resulting compatible system for the quantities $\xi_1, \xi_2, \ldots, \xi_n$. We now study the various possibilities which can occur in solving the characteristic equation (3).

    1. *The case of no real roots.* If all the roots of the equations $\Delta(\lambda) = 0$ are imaginary, then the linear operator A obviously has no eigenvectors at all. For example, as already noted, the rotation operator in the plane $\dot{V}_2$, corresponding to rotation through the angle

$$\varphi_0 \neq m\pi \qquad (m = 0, \pm 1, \pm 2, \ldots),$$

has no eigenvectors. This fact, which is geometrically obvious, is easily proved algebraically. Indeed, for the rotation operator, equation (3) takes the form

$$\begin{vmatrix} \cos \varphi_0 - \lambda & -\sin \varphi_0 \\ \sin \varphi_0 & \cos \varphi_0 - \lambda \end{vmatrix} = 0,$$

which becomes

$$1 - 2\lambda \cos \varphi_0 + \lambda^2 = 0,$$

after calculating the determinant. If

$$\varphi_0 \neq m\pi \qquad (m = 0, \pm 1, \pm 2, \ldots),$$

this last equation has no real roots.

2. *The case of n distinct roots.* If all n roots $\lambda_1, \lambda_2, \ldots, \lambda_n$ of the equation $\Delta(\lambda) = 0$ are real and different, we can find n different eigenvectors of the operator A, by solving the system (2) for $\lambda = \lambda_1, \lambda_2, \ldots, \lambda_n$ in turn. By Lemma 1 of Sec. 64.2 the eigenvectors $f_1, f_2, \ldots, f_n$ so obtained are linearly independent. Choosing them as a new basis, we can construct the matrix of the operator A relative to this basis. Since

$$Af_1 = \lambda_1 f_1,$$
$$Af_2 = \lambda_2 f_2,$$
$$\cdots\cdots$$
$$Af_n = \lambda_n f_n,$$

the matrix $A_{(f)}$ has the form

$$\begin{Vmatrix} \lambda_1 & 0 & \ldots & 0 \\ 0 & \lambda_2 & \ldots & 0 \\ \cdot & \cdot & \ldots & \cdot \\ 0 & 0 & \ldots & \lambda_n \end{Vmatrix}.$$

Using the definition of a diagonal operator (Example 6 of Sec. 26), we can formulate this result as follows:

Every linear operator defined on an n-dimensional space, whose characteristic polynomial has n distinct real roots, is diagonal; the matrix of the operator relative to the basis consisting of its eigenvectors is diagonal, and the diagonal elements are the eigenvalues of the operator.

3. *The case of multiple real roots.* Let $\lambda = \lambda_0$ be a real root of multiplicity $r \geqslant 1$ of the characteristic equation (3). The following question arises: What is the dimension of the corresponding eigenspace $R^{(\lambda_0)}$, or in other words, how many linearly independent solutions does the system

(2) have for $\lambda = \lambda_0$? Let the dimension of this eigenspace be m, and choose a basis in the space R whose first m vectors lie in $R^{(\lambda_0)}$. Then, in the matrix of the operator A relative to this basis, all the elements of the first m columns vanish, except the elements along the principal diagonal, which are all equal to λ_0. Therefore, the determinant of the matrix $A - \lambda E$ contains the factor $(\lambda_0 - \lambda)^m$, and the multiplicity of the root λ_0 of the characteristic polynomial $\det(A - \lambda E)$ will not be less than m, i.e., the inequality $r \geqslant m$ holds. This result can be stated as follows:

The dimension of the eigenspace of the operator A which corresponds to the root λ_0 of the characteristic polynomial, does not exceed the multiplicity of the root λ_0.

In particular, *simple* roots of the characteristic polynomial correspond to *one-dimensional* eigenspaces.

In Examples 1, 2, 3, 5, and 6 of Sec. 26, it is easily verified that the dimensions of the eigenspaces are the same as the multiplicities of the corresponding roots of the characteristic polynomials. However, it can happen that the dimension of an eigenspace is actually less than the multiplicity of the corresponding root of the characteristic polynomial. As an example, consider the operator A in V_2 specified by the matrix

$$A_{(e)} = \left\| \begin{matrix} \lambda_0 & 0 \\ \mu & \lambda_0 \end{matrix} \right\|,$$

with arbitrary $\mu \neq 0$. The characteristic polynomial is $(\lambda_0 - \lambda)^2$ and has a double root $\lambda = \lambda_0$. In this case, the system (2) takes the form

$$0\xi_1 + 0\xi_2 = 0,$$
$$\mu\xi_1 + 0\xi_2 = 0,$$

which has the unique solution

$$\xi_1 = 0, \; \xi_2 = 1,$$

except for a numerical factor. Thus, the eigenspace of the operator A corresponding to the eigenvalue $\lambda = \lambda_0$ has the dimension 1, which is less than the multiplicity of the root λ_0.

Problem 1. Find the eigenvalues and the eigenvectors of the operators given by the following matrices:

a) $\left\| \begin{matrix} 2 & -1 & -1 \\ 0 & -1 & 0 \\ 0 & 2 & 1 \end{matrix} \right\|;$
b) $\left\| \begin{matrix} -1 & -2 & 2 \\ 0 & 1 & 0 \\ 0 & 0 & 1 \end{matrix} \right\|;$

c) $\left\| \begin{matrix} 2 & -1 & 0 \\ 0 & 1 & -1 \\ 0 & 1 & 3 \end{matrix} \right\|;$
d) $\left\| \begin{matrix} 0 & 0 & 1 & -1 \\ -1 & 0 & 1 & -1 \\ 0 & 0 & 0 & 0 \\ 0 & 0 & 0 & 1 \end{matrix} \right\|.$

Ans. a) $\lambda_1 = 2, f_1 = (1, 0, 0)$; $\lambda_2 = 1, f_2 = (1, 0, 1)$; $\lambda_3 = -1, f_3 = (0, 1, -1)$;

   b) $\lambda_1 = -1, f_1 = (1, 0, 0)$; $\lambda_2 = \lambda_3 = 1, f_2 = (1, 0, 1), f_3 = (0, 1, -1)$;

   c) $\lambda_1 = 2, f_1 = (1, 0, 0)$;

   d) $\lambda_1 = 1, f_1 = (1, 0, 0, -1)$; $\lambda_2 = 0, f_2 = (0, 1, 0, 0)$.

Problem 2. Let the operator A defined on the n-dimensional space R have a k-dimensional invariant subspace R'. Then, temporarily regarding A as defined only on the subspace R', we can construct a characteristic polynomial of degree k for A. Show that this polynomial is a factor of the characteristic polynomial of the operator A defined on the whole space R.

Hint. Use Problem 1 of Sec. 63 and the invariance of the characteristic polynomial (Sec. 38.3).

Problem 3. Find all the invariant subspaces of a diagonal operator with n distinct diagonal elements, and show that there are 2^n such subspaces.

Hint. We must prove that there are k eigenvectors in every k-dimensional invariant subspace. To show this, use the result of Problem 2.

66. Eigenvectors of Symmetric Operators

66.1. We now investigate the eigenvectors of an important class of linear operators acting on a Euclidean space R, namely, *symmetric operators* (Sec. 55.2). We are interested in this class for the following reasons: In the first place, operators of this type play an important role in mathematical physics (see ·Chapter ·12); secondly, the theory of the eigenvectors of these operators can be given in definitive form.

It will be recalled that an operator A defined on a Euclidean space R is called symmetric if the corresponding bilinear form (x, Ay) is symmetric, i.e., if the relation

$$(Ax, y) = (x, Ay) \qquad (4)$$

holds for any vectors x and y in the space R. As we saw in Sec. 55.2, in an n-dimensional Euclidean space, the matrix A of a symmetric operator relative to any orthonormal basis coincides with its own transpose, i.e., A is a symmetric matrix. Conversely, every operator A which has a symmetric matrix relative to some orthonormal basis is a symmetric operator.

66.2. We begin by proving some simple lemmas:

LEMMA 1. *The eigenvectors of a symmetric operator A which correspond to different eigenvalues are orthogonal.*

Proof. Consider the two equations

$$Ax = \lambda x,$$
$$Ay = \mu y,$$

where $\lambda \neq \mu$. Take the scalar product of the first equation with y, and take the scalar product of the second equation with x; then subtract the second equation from the first. The result is

$$(Ax, y) - (x, Ay) = (\lambda - \mu)(x, y).$$

The left-hand side of this equation vanishes, because of the symmetry of the operator A. Since $\lambda \neq \mu$, this implies that $(x, y) = 0$, Q.E.D.

LEMMA 2. *Let R' be a subspace of the Euclidean space R which is invariant with respect to the symmetric operator A. Then the orthogonal complement R'' of the subspace R' (Sec. 50.3) is also invariant with respect to A.*

Proof. Let x be any vector of the subspace R', and let y be any vector of the subspace R''. By hypothesis, we have $(Ax, y) = 0$. By the symmetry of the operator A, this implies that $(x, Ay) = 0$, i.e., the vector Ay is orthogonal to any vector $x \in R'$. Therefore, $Ay \in R''$ for any $y \in R''$, Q.E.D.

LEMMA 3. *If $|e| = 1$ and A is a symmetric operator, then*

$$|Ae|^2 \leqslant |A^2e|,$$

where the equality sign holds if and only if e is an eigenvector of A^2 with the eigenvalue

$$\lambda = |Ae|^2.$$

Proof. By the symmetry of the operator and the Schwarz inequality:

$$|Ae|^2 = (Ae, Ae) = (A^2e, e) \leqslant |A^2e|\,|e| = |A^2e|. \tag{5}$$

The Schwarz inequality reduces to an equality if and only if the two vectors appearing in it are collinear (Sec. 50.2). Therefore, the equality holds in (5) if and only if

$$A^2e = \lambda e,$$

so that e is an eigenvector of the operator A^2. Substituting this expression in (5), we find λ:

$$(A^2e, e) = (\lambda e, e) = \lambda = |Ae|^2.$$

This completes the proof of Lemma 3.

Before proving the next lemma, we recall from Sec. 53 that the vector e_0, $(|e_0| = 1)$, is said to be a maximal vector of the operator A if $|Ae_0| = \|A\|$.

LEMMA 4. *If e_0 is a maximal vector of the symmetric operator A, then e_0 is an eigenvector of the operator A^2, corresponding to the eigenvalue $\|A\|^2$.*

Proof. By Lemma 3 and by the definition of the norm of a linear operator, we have

$$\|A\|^2 = |Ae_0|^2 \leqslant |A^2e_0| \leqslant \|A\| \, |Ae_0| \leqslant \|A\|^2,$$

so that

$$|Ae_0|^2 = |A^2e_0| = \|A\|^2.$$

By Lemma 3, e_0 is an eigenvector of the operator A^2, corresponding to the eigenvalue

$$|Ae_0|^2 = \|A\|^2,$$

Q.E.D.

LEMMA 5. *If the symmetric operator A has a maximal vector, then A also has an eigenvector with eigenvalue* $+\|A\|$ *or* $-\|A\|$.

Proof. Let x_0 be a maximal vector of the operator A. By Lemma 4, x_0 is an eigenvector of the operator A^2 with eigenvalue $\|A\|^2$, i.e.,

$$A^2x_0 = \|A\|^2x_0.$$

Denoting $\|A\|$ by μ, we can write this relation as

$$(A - \mu E)(A + \mu E)x_0 = 0.$$

Assume that $z_0 = (A + \mu E)x_0 \neq 0$. Then

$$(A - \mu E)z_0 = 0,$$

or equivalently

$$Az_0 = \mu z_0,$$

which implies that z_0 is an eigenvector of the operator with the eigenvalue $\mu = \|A\|$. On the other hand, if $(A + \mu E)x_0 = 0$, then

$$Ax_0 = -\mu x_0,$$

so that x_0 is an eigenvector of the operator A with eigenvalue $\mu = -\|A\|$. This proves Lemma 5.

We note that in Lemmas 1 to 5, we did not assume that the space R is finite-dimensional. However, the next theorem is based in an essential way on the assumption that the space is finite-dimensional.

THEOREM 33. (*Symmetric operator theorem*) *A symmetric operator A, defined on an n-dimensional space R, has n (pairwise) orthogonal eigenvectors.*

Proof. The operator A, like any operator defined on an n-dimensional space, has a maximal vector (Sec. 53). Applying Lemma 5, we find that the operator A has an eigenvector e_1 with eigenvalue λ_1 equal to $\|A\|$ or

$-\|A\|$. Consider the orthogonal complement R' (with respect to the space R) of the one-dimensional subspace R_1 spanned by the vector e_1; R' is an $(n-1)$-dimensional subspace of the space R, and by Lemma 2, R' is invariant with respect to the operator A. We can now regard the operator A as being defined only on the subspace R'. By the argument above, we can find a new eigenvector e_2 of A in the space R'. The absolute value of the corresponding eigenvalue λ_2 equals the least upper bound of the function $\|Ax\|$ on the unit sphere of the space R', and hence does not exceed $|\lambda_1| = \|A\|$.

The subspace R_2 spanned by the vectors e_1 and e_2 is invariant with respect to the operator A, since

$$A(\alpha_1 e_1 + \alpha_2 e_2) = \alpha_1 A e_1 + \alpha_2 A e_2 = \alpha_1 \lambda_1 e_1 + \alpha_2 \lambda_2 e_2.$$

Therefore, by Lemma 2, the orthogonal complement R'' of the space R_2 is also invariant with respect to A. Repeating the argument just given, we construct in R'' a third eigenvector e_3 with eigenvalue λ_3, where $|\lambda_3| \leqslant |\lambda_2| \leqslant |\lambda_1|$. Continuing this process, we successively construct a system of n (pairwise) orthogonal eigenvectors $e_1, e_2, \ldots, e_n$ whose corresponding eigenvalues $\lambda_1, \lambda_2, \ldots, \lambda_n$ are arranged in order of decreasing absolute value, i.e., $|\lambda_1| \geqslant |\lambda_2| \geqslant \cdots \geqslant |\lambda_n|$. This completes the proof of Theorem 33.

COROLLARY 1. *Every symmetric operator A is diagonal, i.e., there exists a basis (which is in fact orthogonal) in which the matrix of A has diagonal form.*

COROLLARY 2. *If the matrix $\|a_i^{(j)}\|$ is symmetric, then its characteristic equation* (3) *(cf. Sec. 65) has no complex roots. Moreover, each real root of multiplicity r of* (3) *corresponds to precisely r linearly independent solutions of the system* (2).

Proof. By Sec. 65.3, to each real root λ corresponds a set of no more than r linearly independent solutions of the system (2), where r is the multiplicity of λ. If we assume that less than r linearly independent solutions of (2) correspond to some λ or that there exist complex roots of the characteristic equation (3), then the total number of linearly independent solutions of the systems of the form (2) corresponding to all the real roots of (3) is certainly less than n. But we have just proved that a symmetric operator with matrix $\|a_i^{(j)}\|$ must have n linearly independent eigenvectors. Therefore, neither of these two possibilities can occur, which proves the corollary.

The following theorem shows that the result of Theorem 33 cannot be generalized to nonsymmetric operators.

THEOREM 34. *If the linear operator A defined on an n-dimensional Euclidean space R has n orthogonal eigenvectors, then A is symmetric.*

Proof. We take the n orthonormal eigenvectors of the operator A as a basis for the space R. Then, the matrix of the operator A has diagonal form relative to this basis and is hence a symmetric matrix. By Sec. 55.2, this implies that A is a symmetric operator, Q.E.D.

Problem 1. A symmetric operator, defined on an n-dimensional Euclidean space R, is said to be *positive definite* if all its eigenvalues $\lambda_1, \lambda_2, \ldots, \lambda_n$ are positive. Given a positive definite symmetric operator A, show that another positive definite symmetric operator B can always be found, for which $B^2 = A$ (B is the "square root" of the operator A).

Hint. Construct the operator B by the formula

$$Be_i = + \sqrt{\lambda_i}e_i \qquad (i = 1, 2, \ldots, n),$$

where $e_1, e_2, \ldots, e_n$ is the canonical basis of the operator A.

Problem 2. Take the square root of the operator A, given in an orthonormal basis by the matrix

$$A = \begin{Vmatrix} 13 & 14 & 4 \\ 14 & 24 & 18 \\ 4 & 18 & 29 \end{Vmatrix}.$$

Hint. First transform the basis so that the matrix of A has diagonal form.

$$\text{Ans.} \qquad \sqrt{A} = \begin{Vmatrix} 3 & 3 & 0 \\ 2 & 4 & 2 \\ 0 & 2 & 5 \end{Vmatrix}.$$

Problem 3. Show that the square of the norm of any operator A defined on an n-dimensional Euclidean space equals the maximum eigenvalue of the (symmetric) operator A^*A.

Hint. Use an argument similar to that used in the proof of Lemma 3, but without assuming that A is symmetric. Also use Problem 7 of Sec. 55.

Problem 4. Suppose that it is known that a linear operator A is the product of a symmetric operator S and an isometric operator Q, i.e., $A = SQ$. Show that $S^2 = AA^*$.

Problem 5. Show that *every* operator A with $\det A \neq 0$ can be represented as the product of a symmetric and an isometric operator.

Hint. Show that the operator AA^* is symmetric and positive definite (Problem 1), so that we can find a symmetric positive definite operator S for which $S^2 = AA^*$. Then, construct the operator Q by the formula $Q = S^{-1}A$ and show that Q is isometric.

Problem 6. (Continuation) Prove the uniqueness of the representation of the operator A as the product SQ.

Hint. Use the result of Problem 4.

Problem 7. (Continuation) Investigate the existence and uniqueness of the representation $A = SQ$, under the assumption that $\det A = 0$.

Ans. Such a representation exists, but it is not unique.

Problem 8. Show that the symmetric operators A and B commute if and only if they have a common system of n orthogonal eigenvectors.

Hint. Use Problems 1 and 2 of Sec. 64.

67. Examples of Symmetric Operators on Infinite-Dimensional Spaces

In this section, we study some concrete operators which have applications in mathematical physics.

67.1. Consider the operator corresponding to multiplication by t in the Euclidean space $C_2(a, b)$. This operator is symmetric, since for any $x(t)$ and $y(t)$ in $C_2(a, b)$, we have

$$(tx, y) = \int_a^b tx(t)y(t)\,dt = \int_a^b x(t)ty(t)\,dt = (x, ty).$$

We note that the operator corresponding to multiplication by t has no eigenvectors (Sec. 64). Thus, the basic symmetric operator theorem (Sec. 66) cannot be extended to infinite-dimensional spaces.

67.2. Consider the Fredholm integral operator (Example 8 of Sec. 26). If the kernel $K(t, s)$ of this operator is symmetric, i.e., if $K(t, s) = K(s, t)$, then the Fredholm operator is symmetric; in fact, we have

$$(Ax, y) = \int_a^b \int_a^b K(t, s)x(s)y(t)\,dt\,ds = \int_a^b \int_a^b K(s, t)x(s)y(t)\,dt\,ds$$

$$= \int_a^b x(s)\left[\int_a^b K(s, t)y(t)\,dt\right]ds = (x, Ay)$$

for any $x(t)$, $y(t)$ in $C_2(a, b)$. Unlike the operator corresponding to multiplication by t, the symmetric Fredholm operator has eigenvectors, which we shall study later, in Chapter 12.

*67.3. The Sturm-Liouville differential operator. This is the name given to the operator in the space $C_2(a, b)$, which is defined by the formula

$$Lx(t) = \frac{d}{dt}\left[p(t)\frac{dx(t)}{dt}\right] + q(t)x(t),$$

where $p(t)$ and $q(t)$ are given functions. The domain of definition of the operator L is not the whole space $C_2(a, b)$, but rather a part of $C_2(a, b)$, consisting only of twice differentiable functions; this point will be made more precise later. It is obvious that the operator L is linear. We now write down the bilinear forms (Lx, y) and (x, Ly), which we then evaluate, using integration by parts:

$$(Lx, y) = \int_a^b \left\{ \frac{d}{dt}\left[p\frac{dx}{dt}\right] + qx \right\} y \, dt = py\frac{dx}{dt}\Big|_a^b - p\int_a^b \frac{dx}{dt}\frac{dy}{dt}\,dt + \int_a^b qxy\,dt,$$

$$(x, Ly) = \int_a^b \left\{ \frac{d}{dt}\left[p\frac{dy}{dt}\right] + qy \right\} x \, dt = px\frac{dy}{dt}\Big|_a^b - p\int_a^b \frac{dy}{dt}\frac{dx}{dt}\,dt + \int_a^b qyx\,dt.$$

We see that the relation $(Lx, y) = (x, Ly)$ holds, and hence the operator L is symmetric, provided that the condition

$$p\left(y\frac{dx}{dt} - x\frac{dy}{dt}\right)\Big|_a^b = 0 \tag{6}$$

holds for any two functions $x(t)$ and $y(t)$ in the domain of definition Ω of the operator L. For example, the condition (6) is met if

$$x(a) = x(b) = 0 \tag{7}$$

for any function $x(t) \in \Omega$. The relations

$$x(a) + h_1 x'(a) = 0, \qquad x(b) + h_2 x'(b) = 0, \tag{8}$$

$$x(a) = x(b), \qquad x'(a) = x'(b), \tag{9}$$

etc., give other kinds of conditions which can be imposed on the functions $x(t) \in \Omega$. In general, any linear homogeneous conditions which guarantee the validity of (6) are suitable;[2] any such conditions determine a certain domain of definition Ω of the operator L, in which L is a symmetric operator. For example, if $p(t) \equiv 1$ and $q(t) \equiv 0$, we get the operator

$$Lx(t) = x''(t), \tag{10}$$

which is symmetric if the condition (9) is met, for example. To find the eigenfunctions of this operator in the appropriate linear space; we must solve the equation

$$x''(t) = \lambda x(t), \tag{11}$$

where we choose only the solutions satisfying the condition (9). For simplicity, we assume that $a = -\pi$, $b = \pi$. Then, (11) has the obvious solutions

$$x_n = A_n \cos nt + B_n \sin nt,$$

where the corresponding values of λ are $0, 1^2, 2^2, \ldots, n^2$, etc., and the constants A_n and B_n are arbitrary.

---

[2] The condition (6) itself cannot be taken as a definition of the domain Ω, since it involves two functions $x(t)$ and $y(t)$, whereas the definition of Ω must involve only one function.

The Legendre polynomials (Sec. 58) are also eigenfunctions of a Sturm-Liouville operator. To find this operator, we set

$$y_n(t) = (t^2 - 1)^n$$

and differentiate with respect to t, obtaining

$$y_n'(t) = 2nt(t^2 - 1)^{n-1},$$

so that

$$(t^2 - 1)y_n'(t) = 2nty_n(t).$$

Applying Leibniz' formula, we differentiate this equation $n + 1$ times with respect to t; the result is

$$(t^2 - 1)y_n^{(n+2)}(t) + 2t(n + 1)y_n^{(n+1)}(t) + 2\frac{(n + 1)n}{1 \cdot 2} y_n^{(n)}(t)$$
$$= 2nty_n^{(n+1)}(t) + 2n(n + 1)y_n^{(n)}(t).$$

Replacing $y_n^{(n)}(t)$ by $p_n(t)$, we can transform this equation into the form

$$[(t^2 - 1)p_n'(t)]' = n(n + 1)p_n(t),$$

which, when multiplied by $1/(2^n n!)$, yields

$$\frac{d}{dt}\left[(t^2 - 1)\frac{d}{dt}P_n(t)\right] = n(n + 1)P_n(t) \tag{12}$$

(cf. Sec. 58). Equation (12) shows that the Legendre polynomial $P_n(t)$ is the eigenfunction of the Sturm-Liouville operator

$$Lx(t) = \frac{d}{dt}\left[(t^2 - 1)\frac{d}{dt}x(t)\right], \tag{13}$$

corresponding to the eigenvalue $\lambda = n(n + 1)$. The Sturm-Liouville operator (13), defined on the interval $-1 \leqslant t \leqslant +1$, requires no supplementary symmetry conditions, because (6) is automatically satisfied for all twice differentiable functions $x(t) \in C_2(-1, 1)$, since $p(t) = t^2 - 1$ vanishes for $t = \pm 1$.

Application of Lemma 6 of Sec. 66 proves once again that the trigonometric functions and the Legendre polynomials are orthogonal (for different n). Subsequently, in Chapter 12, we shall study further the properties of Sturm-Liouville operators.

10

QUADRATIC FORMS
IN A EUCLIDEAN SPACE

68. The Basic Theorem on Quadratic Forms

We begin with the following theorem concerning symmetric bilinear forms in a Euclidean space:

THEOREM 35. *Every symmetric bilinear form defined on an n-dimensional Euclidean space R has a canonical basis consisting of orthogonal vectors.*

Proof. Consider the linear operator A corresponding to the given symmetric bilinear form $A(x, y)$ (see Sec. 55). The operator A is also symmetric. According to the symmetric operator theorem (Theorem 33), the space R has an orthonormal basis consisting of the eigenvectors of the operator A, and the matrix of A is diagonal relative to this basis. Since this matrix is also the matrix of the bilinear form $A(x, y)$, the orthonormal basis just found is a canonical basis of $A(x, y)$, Q.E.D.

We now apply this result to the study of quadratic forms. Suppose that we are given a quadratic form

$$A(x, x) = \sum_{i,j=1}^{n} a_{ij}\xi_i\xi_j \qquad (a_{ij} = a_{ji}). \tag{1}$$

We shall regard the numbers $\xi_1, \xi_2, \ldots, \xi_n$ as the components of a vector x in an n-dimensional Euclidean space R, with a scalar product defined by the formula

$$(x, y) = \sum_{i=1}^{n} \xi_i\eta_i,$$

where $y = (\eta_1, \eta_2, \ldots, \eta_n)$. The basis

$$e_1 = (1, 0, \ldots, 0), \quad e_2 = (0, 1, \ldots, 0), \quad \ldots, \quad e_n = (0, 0, \ldots, 1)$$

is an orthonormal basis in R, and clearly

$$x = \sum_{i=1}^{n} \xi_i e_i, \qquad y = \sum_{i=1}^{n} \eta_i e_i.$$

Next, consider the bilinear form

$$A(x, y) = \sum_{i, j=1}^{n} a_{ij} \xi_i \eta_j$$

corresponding to the quadratic form (1). By Theorem 35, this form has an orthonormal basis $f_1, f_2, \ldots, f_n$. If the components of the vectors x and y are $\tau_1, \tau_2, \ldots, \tau_n$ and $\theta_1, \theta_2, \ldots, \theta_n$, respectively, relative to this basis, then we can write the bilinear form $A(x, y)$ as

$$A(x, y) = \sum_{i=1}^{n} \lambda_i \tau_i \theta_i,$$

and the quadratic form $A(x, x)$ as

$$A(x, x) = \sum_{i=1}^{n} \lambda_i \tau_i^2. \tag{2}$$

The transition from the basis $e_1, e_2, \ldots, e_n$ to the basis $f_1, f_2, \ldots, f_n$ is accomplished by using the formula

$$f_j = \sum_{i=1}^{n} q_i^{(j)} e_i \qquad (j = 1, 2, \ldots, n),$$

where $Q = \|q_i^{(j)}\|$ is an orthogonal matrix (Sec. 54.1). According to equation (28) of Sec. 54, the relation between the components $\tau_1, \tau_2, \ldots, \tau_n$ and $\xi_1, \xi_2, \ldots, \xi_n$ is given by the system of equations

$$\xi_j = \sum_{i=1}^{n} q_j^{(i)} \tau_i \qquad (j = 1, 2, \ldots, n), \tag{3}$$

obtained by using the transposed matrix Q'. These considerations give us the following important theorem:

THEOREM 36. (*Theorem on quadratic forms in a Euclidean space.*) *Every quadratic form* (1) *can be reduced to the canonical form* (2) *by using an isometric coordinate transformation* (3).

The sequence of operations which must be performed in order to construct the transformation formulas (3) and the canonical form (2) can be deduced from the results of Secs. 65 and 66. We now give this sequence of operations in final form:

1. Using the quadratic form (1), construct the symmetric matrix $A = \|a_i^{(j)}\|$, where $a_i^{(j)} = a_{ij}$.

2. Form the characteristic polynomial $\Delta(\lambda) = \det(A - \lambda E)$ and find its roots. By Corollary 2 of Theorem 33, this polynomial has n real roots (not necessarily distinct).

3. Knowing the roots of the polynomial $\Delta(\lambda)$, we can already write the quadratic form (1) in canonical form; in particular, we can determine its positive and negative indices of inertia.

4. Substitute the root λ_1 into the system (2) of Sec. 65. (For the given root λ_1, the system must have a number of linearly independent solutions just equal to the multiplicity of the root λ_1.) Then find these linearly independent solutions by using the rules for solving homogeneous systems of linear equations.

5. If the multiplicity of the root λ_1 is greater than unity, orthogonalize the linearly independent solutions just obtained, by using the method of Sec. 57.

6. Carrying out the indicated operations for every root, we finally obtain a system of n orthogonal vectors. We then normalize them by dividing each vector by its length. The resulting vectors

$$f_1 = (q_1^{(1)}, q_2^{(1)}, \ldots, q_n^{(1)}),$$
$$f_2 = (q_1^{(2)}, q_2^{(2)}, \ldots, q_n^{(2)}),$$
$$\cdots\cdots\cdots$$
$$f_n = (q_1^{(n)}, q_2^{(n)}, \ldots, q_n^{(n)}),$$

form an orthonormal system.

7. Using the numbers $q_i^{(j)}$, we can write the transformation formulas (3).

8. In order to express the new components $\tau_1, \tau_2, \ldots, \tau_n$ in terms of the old components $\xi_1, \xi_2, \ldots, \xi_n$, we write

$$\tau_i = \sum_{j=1}^{n} q_j^{(i)} \xi_j \qquad (i = 1, 2, \ldots, n),$$

recalling that the inverse of the orthogonal matrix Q is the transposed matrix Q'.

Problem. Transform the following quadratic forms to canonical form by using orthogonal transformations:

a) $2\xi_1^2 + \xi_2^2 - 4\xi_1\xi_2 - 4\xi_2\xi_3$;

b) $2\xi_1^2 + 5\xi_2^2 + 5\xi_3^2 + 4\xi_1\xi_2 - 4\xi_1\xi_3 - 8\xi_2\xi_3$;

c) $2\xi_1^2 + 2\xi_2^2 + 2\xi_3^2 + 2\xi_4^2 - 4\xi_1\xi_2 + 2\xi_1\xi_4 + 2\xi_2\xi_3 - 4\xi_3\xi_4;$

d) $2\xi_1\xi_2 + 2\xi_1\xi_3 - 2\xi_1\xi_4 - 2\xi_2\xi_3 + 2\xi_2\xi_4 + 2\xi_3\xi_4.$

Ans.   a) $4\eta_1^2 + \eta_2^2 - 2\eta_3^2;$ $\eta_1 = \dfrac{2}{3}\xi_1 - \dfrac{2}{3}\xi_2 + \dfrac{1}{3}\xi_3,$

$$\eta_2 = \frac{2}{3}\xi_1 + \frac{1}{3}\xi_2 - \frac{2}{3}\xi_2,$$

$$\eta_3 = \frac{1}{3}\xi_1 + \frac{2}{3}\xi_2 + \frac{2}{3}\xi_3;$$

b) $10\eta_1^2 + \eta_2^2 + \eta_3^2;$ $\eta_1 = \dfrac{1}{3}\xi_1 + \dfrac{2}{3}\xi_2 - \dfrac{2}{3}\xi_3,$

$$\eta_2 = \frac{2}{\sqrt{5}}\xi_1 - \frac{1}{\sqrt{5}}\xi_2,$$

$$\eta_3 = \frac{2}{3\sqrt{5}}\xi_1 + \frac{4}{3\sqrt{5}}\xi_2 + \frac{\sqrt{5}}{3}\xi_3;$$

c) $\eta_1^2 - \eta_2^2 + 3\eta_3^2 + 5\eta_4^2;$ $\eta_1 = \dfrac{1}{2}\xi_1 + \dfrac{1}{2}\xi_2 + \dfrac{1}{2}\xi_3 + \dfrac{1}{2}\xi_4,$

$$\eta_2 = \frac{1}{2}\xi_1 + \frac{1}{2}\xi_2 - \frac{1}{2}\xi_3 - \frac{1}{2}\xi_4,$$

$$\eta_3 = \frac{1}{2}\xi_1 - \frac{1}{2}\xi_2 + \frac{1}{2}\xi_3 - \frac{1}{2}\xi_4,$$

$$\eta_4 = \frac{1}{2}\xi_1 - \frac{1}{2}\xi_2 - \frac{1}{2}\xi_3 + \frac{1}{2}\xi_4;$$

d) $\eta_1^2 + \eta_2^2 + \eta_3^2 - 3\eta_4^2;$ $\eta_1 = \dfrac{\sqrt{2}}{2}\xi_1 + \dfrac{\sqrt{2}}{2}\xi_2,$

$$\eta_2 = \frac{\sqrt{2}}{2}\xi_3 + \frac{\sqrt{2}}{2}\xi_4,$$

$$\eta_3 = \frac{1}{2}\xi_1 - \frac{1}{2}\xi_2 + \frac{1}{2}\xi_3 - \frac{1}{2}\xi_4,$$

$$\eta_4 = \frac{1}{2}\xi_1 - \frac{1}{2}\xi_2 - \frac{1}{2}\xi_3 + \frac{1}{2}\xi_4.$$

69. Uniqueness Questions for Quadratic Forms

In Sec. 43, we saw that neither the canonical form nor the canonical basis of a quadratic form is uniquely defined in an affine space; in general, any preassigned vector can be included in the canonical basis of the quadratic form. The situation is quite different in a Euclidean space, provided that only orthonormal bases are considered. The point is that the matrix of the quadratic form and the matrix of the corresponding symmetric linear operator transform in the same way, as we have already noted (Sec. 55.1); thus, a canonical basis for the quadratic form is simultaneously a basis consisting of the eigenvectors of the symmetric operator, and the coefficients of the quadratic form relative to a canonical basis (the "canonical coefficients") coincide with the eigenvalues of the operator. But the eigenvalues of the operator A are the roots of the equation $\det(A - \lambda E) = 0$; this equation does not depend on the choice of a basis, and is an invariant of the operator A. Therefore, the set of canonical coefficients of the form $A(x,x)$ is uniquely defined. As concerns the canonical basis of the quadratic form $A(x, x)$, it is defined with the same arbitrariness as in the definition of the complete orthonormal set of eigenvectors of the operator A, i.e., apart from permutations of the eigenvectors, we can multiply any of them by -1, or more generally, we can subject them to any isometric transformation in the eigenspace corresponding to a fixed eigenvalue λ.

70. Extremal Properties of Quadratic Forms

Let $A(x, x)$ be a quadratic form in a Euclidean space R. We examine its values on the unit sphere $(x, x) = 1$ of the space R, and inquire at what points of the unit sphere the values of $A(x, x)$ are stationary. It will be recalled that by definition, a differentiable numerical function $f(x)$, defined at the points of a surface U, takes a *stationary value* at the point $x_0 \in U$ if the derivative of the function $f(x)$ along any direction on the surface U vanishes at the point x_0. In particular, the function $f(x)$ is stationary at the points where it has a maximum or a minimum.

The problem of determining the stationary values of a quadratic form on the unit sphere is a problem involving conditional extrema; one method of solving the problem is to use Lagrange's method,[1] as follows: We construct an orthonormal basis in the space R and denote the components of the vector x relative to this basis by $\xi_1, \xi_2, \ldots, \xi_n$. In this coordinate system, our quadratic form becomes

$$A(x, x) = \sum_{i,j=1}^{n} a_{ij}\xi_i\xi_j,$$

and the condition $(x, x) = 1$ becomes

$$\sum_{i=1}^{n} \xi_i^2 = 1.$$

---

[1] See, e.g.: R. Courant, *Differential and Integral Calculus* (translated by E. J. McShane), New York: Interscience Publishers, Inc., vol. 2 (1956), p. 170.

Using Lagrange's method, we construct the function

$$F(\xi_1, \xi_2, \ldots, \xi_n) = \sum_{i,j=1}^{n} a_{ij}\xi_i\xi_j - \lambda \sum_{i=1}^{n} \xi_i^2,$$

and equate to zero its partial derivatives with respect to ξ_i $(i = 1, 2, \ldots, n)$, recalling that $a_{ij} = a_{ji}$. The result is:

$$2 \sum_{j=1}^{n} a_{ij}\xi_j - 2\lambda\xi_i = 0 \qquad (i = 1, 2, \ldots, n).$$

After dividing by 2, we obtain the familiar system (Sec. 65)

$$\begin{aligned}
(a_{11} - \lambda)\xi_1 + a_{12}\xi_2 + \cdots + a_{1n}\xi_n &= 0, \\
a_{21}\xi_1 + (a_{22} - \lambda)\xi_2 + \cdots + a_{2n}\xi_n &= 0, \\
\cdots \quad \cdots \quad \cdots \quad \cdots \quad \\
a_{n1}\xi_1 + a_{n2}\xi_2 + \cdots + (a_{nn} - \lambda)\xi_n &= 0,
\end{aligned}$$

which serves to define the eigenvectors of the symmetric operator corresponding to the quadratic form $A(x, x)$. This implies the following result:

The quadratic form $A(x, x)$ takes stationary values for those vectors of the unit sphere which are eigenvectors of the symmetric operator A corresponding to the form $A(x, x)$.

We now calculate the values which the form takes at its stationary points. To do this, we introduce the corresponding symmetric operator A, and write the quadratic form as

$$A(x, x) = (Ax, x).$$

Suppose that $A(x, x)$ takes a stationary value for the vector e_i. Since we have just shown that e_i is an eigenvector of the operator A, i.e., $Ae_i = \lambda_i e_i$, we have

$$A(e_i, e_i) = (Ae_i, e_i) = \lambda_i(e_i, e_i) = \lambda_i.$$

Therefore, the stationary value of the form $A(x, x)$ at $x = e_i$ equals the corresponding eigenvalue of the operator A. Since the eigenvalues of the operator A are the same as the canonical coefficients of the form $A(x, x)$, we can conclude that *the stationary values of the form $A(x, x)$ and its canonical coefficients coincide.* In particular, the maximum of the form $A(x, x)$ on the unit sphere is equal to its largest canonical coefficient, and the minimum of $A(x, x)$ on the unit sphere is equal to its smallest canonical coefficient.

Problem 1. Prove the last assertion by direct calculation.

Problem 2. What are the stationary values of the quadratic form

$$A(x, x) = x_1^2 + \tfrac{1}{2} x_2^2 + \tfrac{1}{3} x_3^2, \qquad x = (x_1, x_2, x_3),$$

on the sphere $|x| = 1$, and of what type are they?

Ans. A maximum for $x = (\pm 1, 0, 0)$, where $A(x, x) = 1$;
A minimum for $x = (0, 0, \pm 1)$, where $A(x, x) = \tfrac{1}{3}$;
A *minimax* for $x = (0, \pm 1, 0)$, where $A(x, x) = \tfrac{1}{2}$, i.e., the function $A(x, x)$ increases if we go along the unit sphere in one direction from the point x, and decreases if we go in the other direction.

*71. Quadratic Forms on Subspaces

71.1. We can study both quadratic forms and bilinear forms on a k-dimensional subspace $R_k \subset R_n$ rather than on the whole n-dimensional space R_n, and we can look for an orthonormal canonical basis in R_k. Let the quadratic form $A(x, x)$ have the canonical form

$$A(x, x) = \lambda_1 \xi_1^2 + \lambda_2 \xi_2^2 + \cdots + \lambda_n \xi_n^2 \qquad (4)$$

in the whole space R_n, and the canonical form

$$A(x, x) = \mu_1 \tau_1^2 + \mu_2 \tau_2^2 + \cdots + \mu_k \tau_k^2$$

in the subspace R_k. We now find the relation between the coefficients $\mu_1, \mu_2, \ldots, \mu_k$ and the coefficients $\lambda_1, \lambda_2, \ldots, \lambda_n$. For convenience, we assume that the canonical coefficients are arranged in decreasing order, i.e., that

$$\lambda_1 \geqslant \lambda_2 \geqslant \cdots \geqslant \lambda_n, \qquad \mu_1 \geqslant \mu_2 \geqslant \cdots \geqslant \mu_k.$$

As we know, the quantity λ_1 is the maximum value of the quadratic form $A(x, x)$ on the unit sphere of the space R_n; similarly, μ_1 is the maximum value of $A(x, x)$ on the unit sphere of the subspace R_k. This implies that $\mu_1 \leqslant \lambda_1$. Moreover, we also have $\mu_1 \geqslant \lambda_{n-k+1}$. To see this, we first note the following result, which is an easy consequence of the considerations of Sec. 15.3:

> **LEMMA.** *If in an n-dimensional space R_n, we choose two subspaces R_p and R_q, with dimensions p and q, and if the sum $p + q$ exceeds n, then the dimension of the intersection of R_p and R_q is not less than $p + q - n$.*

We now prove the inequality $\mu_1 \geqslant \lambda_{n-k+1}$. Let $e_1, e_2, \ldots, e_n$ be the canonical basis of $A(x, x)$ relative to which it takes the form (4). Consider the $(n - k + 1)$-dimensional subspace R' spanned by the vectors $e_1, e_2, \ldots, e_{n-k+1}$. Since $k + (n - k + 1) > n$, then by our lemma, the subspaces R' and R_k have at least one nonzero vector in common. Let this vector be

$$x_0 = (\gamma_1, \ldots, \gamma_{n-k+1}, 0, \ldots, 0),$$

and assume that x_0 is normalized, i.e., that $|x_0| = 1$.   According to (4), we have

$$A(x_0, x_0) = \lambda_1 \gamma_1^2 + \cdots + \lambda_{n-k+1} \gamma_{n-k+1}^2$$
$$\geq \lambda_{n-k+1}(\gamma_1^2 + \cdots + \gamma_{n-k+1}^2) = \lambda_{n-k+1}.$$

This implies that μ_1, the *maximum* value of the quadratic form $A(x, x)$ on the unit sphere of the space R_k, cannot be less than λ_{n-k+1}, Q.E.D.   Thus the quantity μ_1 satisfies the inequalities

$$\lambda_1 \geq \mu_1 \geq \lambda_{n-k+1}. \tag{5}$$

Naturally, the quantity μ_1 takes different values for different k-dimensional subspaces.   We now show that *there exist k-dimensional subspaces for which the equality signs hold in* (5).   Consider the subspace R' spanned by the first k vectors $e_1, e_2, \ldots, e_k$ of the canonical basis of the form $A(x, x)$.   In the subspace R', $A(x, x)$ has the form

$$A(x, x) = \lambda_1 \xi_1^2 + \lambda_2 \xi_2^2 + \cdots + \lambda_k \xi_k^2$$

relative to the basis $e_1, e_2, \ldots, e_k$.   Relative to any other canonical basis of the subspace R', $A(x, x)$ has the form

$$A(x, x) = \mu_1 \tau_1^2 + \mu_2 \tau_2^2 + \cdots + \mu_k \tau_k^2.$$

The fact that $\mu_1 = \lambda_1$ follows from the uniqueness considerations of Sec. 69 and the assumed arrangement of the canonical coefficients.   Thus, the quantity μ_1 takes its maximum value λ_1 on the subspace R'.

We now consider the subspace R'' spanned by the last k vectors e_{n-k+1}, $e_{n-k+2}, \ldots, e_n$ of the canonical basis of the form $A(x, x)$.   In the subspace R'', $A(x, x)$ has the form

$$A(x, x) = \lambda_{n-k+1} \xi_{n-k+1}^2 + \cdots + \lambda_n \xi_n^2$$

relative to the basis $e_{n-k+1}, \ldots, e_n$.   Relative to an arbitrary canonical basis of the subspace R'', $A(x, x)$ takes the form

$$A(x, x) = \mu_1 \tau_1^2 + \mu_2 \tau_2^2 + \cdots + \mu_k \tau_k^2.$$

Just as before, we conclude that $\mu_1 = \lambda_{n-k+1}$.   Therefore, the quantity μ_1 takes its minimum value λ_{n-k+1} on the subspace R''.   Thus, we obtain a new definition of the coefficient λ_{n-k+1}:

The coefficient λ_{n-k+1} in the canonical representation of the quadratic form $A(x, x)$ equals the smallest value of the maximum of $A(x, x)$ on the unit spheres of all possible k-dimensional subspaces of the space R_n.

Using this property, we can estimate the other canonical coefficients of the quadratic form $A(x, x)$ on the subspace R_k.   For example, if we fix the subspace R_k, μ_2 is the smallest value of the maximum of $A(x, x)$ on the unit spheres of all the $(k - 1)$-dimensional subspaces of R_k, whereas λ_{n-k+2} is the smallest value of the maximum of $A(x, x)$ on the unit spheres of all the $(k - 1)$-dimensional subspaces of the *whole* space R_n. Therefore, we have $\mu_2 \geqslant \lambda_{n-k+2}$, and similarly $\mu_3 \geqslant \lambda_{n-k+3}, \ldots, \mu_k \geqslant \lambda_n$. On the other hand, λ_2 is the smallest value of the maximum of the quadratic form $A(x, x)$ on the unit spheres of all the $(n - 1)$-dimensional subspaces of the whole space R_n. Moreover, according to our lemma, the intersection of every $(n - 1)$-dimensional subspace with the subspace R_k is a subspace with no less than $(n - 1) + k - n = k - 1$ dimensions, so that λ_2 is not less than the smallest value of the maximum of $A(x, x)$ on the unit spheres of all such subspaces; in particular, λ_2 is not less than μ_2, the smallest value of the maximum of $A(x, x)$ on the unit spheres of all the $(k - 1)$-dimensional subspaces of R_k. Therefore, we have $\lambda_2 \geqslant \mu_2$ and similarly $\lambda_3 \geqslant \mu_3, \ldots, \lambda_k \geqslant \mu_k$. Thus, the canonical coefficients $\mu_1, \mu_2, \ldots, \mu_k$ obey the inequalities

$$
\begin{aligned}
\lambda_1 &\geqslant \mu_1 \geqslant \lambda_{n-k+1}, \\
\lambda_2 &\geqslant \mu_2 \geqslant \lambda_{n-k+2}, \\
&\cdots\cdots \\
\lambda_k &\geqslant \mu_k \geqslant \lambda_n.
\end{aligned}
\tag{6}
$$

For $k = n - 1$, the inequalities (6) become

$$
\begin{aligned}
\lambda_1 &\geqslant \mu_1 \geqslant \lambda_2, \\
\lambda_2 &\geqslant \mu_2 \geqslant \lambda_3, \\
&\cdots\cdots \\
\lambda_{n-1} &\geqslant \mu_{n-1} \geqslant \lambda_n.
\end{aligned}
\tag{6a}
$$

Problem. Show that each of the quantities $\mu_1, \mu_2, \ldots, \mu_k$ can actually attain the upper and lower bounds indicated in (6).

71.2. Consider the behavior of the quadratic form

$$
A(x, x) = \sum_{k=1}^{n} \lambda_k \xi_k^2
$$

on the $(n - 1)$-dimensional subspace R_{n-1} specified by the equation

$$
\alpha_1 \xi_1 + \alpha_2 \xi_2 + \cdots + \alpha_n \xi_n = 0 \qquad (\alpha_1^2 + \alpha_2^2 + \cdots + \alpha_n^2 = 1). \tag{7}
$$

Assuming that all the coefficients $\lambda_1, \lambda_2, \ldots, \lambda_n$ are different, we can calculate the coefficients $\mu_1, \mu_2, \ldots, \mu_{n-1}$ (Sec. 71.1) by using a method due to M. G. Krein. At least one of the coefficients $\alpha_1, \alpha_2, \ldots, \alpha_n$ is nonzero; suppose, for example that $\alpha_n \neq 0$. Then (7) gives the relation

$$
\xi_n = -\frac{1}{\alpha_n} \sum_{j=1}^{n-1} \alpha_j \xi_j.
$$

Substituting this expression for ξ_n into $A(x, x)$, we find that $A(x, x)$ has the form

$$A(x, x) = \lambda_1 \xi_1^2 + \lambda_2 \xi_2^2 + \cdots + \lambda_{n-1} \xi_{n-1}^2 + \frac{\lambda_n}{\alpha_n^2} \left(\sum_{j=1}^{n-1} \alpha_j \xi_j \right)^2$$

on the subspace R_{n-1}, in terms of the variables $\xi_1, \xi_2, \ldots, \xi_{n-1}$. The canonical coefficients of this quadratic form are the same as its stationary values on the unit sphere in the space R_{n-1} (Sec. 70). In the variables $\xi_1, \xi_2, \ldots, \xi_{n-1}$, this sphere has the equation

$$E(x, x) = \xi_1^2 + \xi_2^2 + \cdots + \xi_{n-1}^2 + \frac{1}{\alpha_n^2} \left(\sum_{j=1}^{n-1} \alpha_j \xi_j \right)^2 = 1.$$

Just as before, we determine these stationary values by using Lagrange's method. Thus, we form the function

$$A(x, x) - \lambda E(x, x) = \sum_{j=1}^{n-1} (\lambda_j - \lambda)\xi_j^2 + \frac{\lambda_n - \lambda}{\alpha_n^2} \left(\sum_{j=1}^{n-1} \alpha_j \xi_j \right)^2,$$

and equate to zero its partial derivatives with respect to ξ_k ($k = 1, 2, \ldots, n - 1$), obtaining

$$\xi_k (\lambda_k - \lambda) + \frac{\lambda_n - \lambda}{\alpha_n^2} \left(\sum_{j=1}^{n-1} \alpha_j \xi_j \right) \alpha_k = 0. \tag{8}$$

The required coefficients $\mu_1, \mu_2, \ldots, \mu_{n-1}$ are the roots of the equation obtained by setting equal to zero the determinant $D(\lambda)$ of the system of linear equations (8). The coefficient matrix of this system is clearly the sum of two matrices; the first matrix is diagonal, with the numbers $\lambda_k - \lambda$ ($k = 1, 2, \ldots, n - 1$) along the main diagonal, while the second matrix has the form

$$\frac{\lambda_n - \lambda}{\alpha_n^2} \begin{Vmatrix} \alpha_1 \alpha_1 & \alpha_2 \alpha_1 & \cdots & \alpha_{n-1} \alpha_1 \\ \alpha_1 \alpha_2 & \alpha_2 \alpha_2 & \cdots & \alpha_{n-1} \alpha_2 \\ \cdot & \cdot & \cdots & \cdot \\ \alpha_1 \alpha_{n-1} & \alpha_2 \alpha_{n-1} & \cdots & \alpha_{n-1} \alpha_{n-1} \end{Vmatrix}.$$

By the linear property of determinants (Property 3 of Sec. 3), the determinant $D(\lambda)$ is the sum of the determinant of the first matrix and all the determinants obtained by replacing one or more columns of the determinant of the first matrix by the corresponding columns of the second matrix, and taking account of the factor $(\lambda_n - \lambda)/\alpha_n^2$. Since any two columns of the second matrix are proportional, we need only consider the case where *one* of the columns of the determinant of the first matrix is replaced by the corresponding column of the second matrix.

Thus, for example, if the kth column of the first matrix is replaced by the kth column of the second matrix, the resulting determinant has the form

$$\frac{\lambda_n - \lambda}{\alpha_n^2} \begin{vmatrix} \lambda_1 - \lambda & 0 & \dots & 0 & \alpha_k\alpha_1 & 0 & \dots & 0 \\ 0 & \lambda_2 - \lambda & \dots & 0 & \alpha_k\alpha_2 & 0 & \dots & 0 \\ \cdot & \cdot & \dots & \cdot & \cdot & \cdot & \dots & \cdot \\ 0 & 0 & \dots & \lambda_{k-1} - \lambda & \alpha_k\alpha_{k-1} & 0 & \dots & 0 \\ 0 & 0 & \dots & 0 & \alpha_k\alpha_k & 0 & \dots & 0 \\ 0 & 0 & \dots & 0 & \alpha_k\alpha_{k+1} & \lambda_{k+1} - \lambda & \dots & 0 \\ \cdot & \cdot & \dots & \cdot & \cdot & \cdot & \dots & \cdot \\ 0 & 0 & \dots & 0 & \alpha_k\alpha_{n-1} & 0 & \dots & \lambda_{n-1} - \lambda \end{vmatrix}$$

$$= \frac{\alpha_k^2}{\alpha_n^2} \frac{\prod\limits_{j=1}^{n} (\lambda_j - \lambda)}{\lambda_k - \lambda}.$$

Denote the determinant of the first matrix by

$$E(\lambda) = \prod_{k=1}^{n-1} (\lambda_k - \lambda),$$

and write

$$G(\lambda) = \prod_{k=1}^{n} (\lambda_k - \lambda).$$

Then, the required determinant $D(\lambda)$ becomes

$$D(\lambda) = E(\lambda) + \frac{1}{\alpha_n^2} G(\lambda) \sum_{k=1}^{n-1} \frac{\alpha_k^2}{\lambda_k - \lambda}. \tag{9}$$

Solving the equation $D(\lambda) = 0$, we find the quantities $\mu_1, \mu_2, \dots, \mu_{n-1}$ in which we are interested. We note that these quantities depend on the squares of the numbers α_j rather than on the numbers α_j themselves. Thus, changing the sign of one or more coefficients in (7) does not change the canonical coefficients of the form $A(x, x)$ on the subspace R_{n-1}.

Equation (9) is of particular interest since *it allows us to construct from given numbers* $\mu_1, \mu_2, \dots, \mu_{n-1}$ *satisfying the inequalities* (6a) *a subspace* R_{n-1} *on which the form* $A(x, x)$ *has the canonical coefficients* $\mu_1, \mu_2, \dots, \mu_{n-1}$. (Again, it is assumed that the numbers $\lambda_1, \lambda_2, \dots, \lambda_n$ are all different.) We now show how this is done.

First, we note that we can write (9) in the form

$$\alpha_n^2 \frac{D(\lambda)}{G(\lambda)} = \alpha_n^2 \frac{E(\lambda)}{G(\lambda)} + \sum_{k=1}^{n-1} \frac{\alpha_k^2}{\lambda_k - \lambda} = \sum_{k=1}^{n} \frac{\alpha_k^2}{\lambda_k - \lambda}. \tag{10}$$

Thus, the numbers $\alpha_1^2, \alpha_2^2, \ldots, \alpha_n^2$ are proportional to the coefficients obtained when we expand the rational function $D(\lambda)/G(\lambda)$ in partial fractions. Now, suppose that we are given numbers $\mu_1, \mu_2, \ldots, \mu_{n-1}$ satisfying the inequalities (6a). Write

$$D_1(\lambda) = \prod_{k=1}^{n-1} (\mu_k - \lambda),$$

and expand the rational function $D_1(\lambda)/G(\lambda)$ in partial fractions

$$\frac{D_1(\lambda)}{G(\lambda)} = \frac{c_1}{\lambda_1 - \lambda} + \frac{c_2}{\lambda_2 - \lambda} + \cdots + \frac{c_n}{\lambda_n - \lambda}. \tag{11}$$

The coefficients $c_1, c_2, \ldots, c_n$ are given by the familiar formula

$$c_k = \frac{D_1(\lambda_k)}{(\lambda_1 - \lambda_k) \ldots (\lambda_{k-1} - \lambda_k)(\lambda_{k+1} - \lambda_k) \ldots (\lambda_n - \lambda_k)} = -\frac{D_1(\lambda_k)}{G'(\lambda_k)},$$

and all have the same sign. To see this, note that the numbers $D_1(\lambda_1)$, $D_1(\lambda_2), \ldots, D_1(\lambda_n)$ alternate in sign, since by hypothesis, the roots of the polynomial $D_1(\lambda)$ alternate with the roots of the polynomial $G(\lambda)$. The numbers $G'(\lambda_1), G'(\lambda_2), \ldots, G'(\lambda_n)$ also alternate in sign, since the numbers $\lambda_1, \lambda_2, \ldots, \lambda_n$ are simple roots of the polynomial $G(\lambda)$. Thus, the numbers $D_1(\lambda_k)/G'(\lambda_k)$, and hence the coefficients c_k, all have the same sign. By supplying an appropriate factor, we can assume that the c_k are all positive and add up to 1. Then, we define the numbers $\alpha_1, \alpha_2, \ldots, \alpha_n$ by the formulas

$$\alpha_1^2 = c_1, \quad \alpha_2^2 = c_2, \quad \ldots, \quad \alpha_n^2 = c_n. \tag{12}$$

The numbers $\alpha_1, \alpha_2, \ldots, \alpha_n$ can have any sign.

Finally, we show that the subspace R_{n-1} defined by the equation

$$\alpha_1 \xi_1 + \alpha_2 \xi_2 + \cdots + \alpha_n \xi_n = 0$$

is the subspace on which the quadratic form $A(x, x)$ has the canonical coefficients $\mu_1, \mu_2, \ldots, \mu_{n-1}$. In fact, by what we proved before, the polynomial $D(\lambda)$ whose roots are the canonical coefficients of $A(x, x)$ on the subspace R_{n-1}, is given by the formula (9) or the equivalent formula (10). Comparing (10) with (11) and using (12), we find that the polynomial $D(\lambda)$ differs only by a numerical factor from the polynomial $D_1(\lambda)$ just constructed. But this implies that the roots of $D(\lambda)$ coincide with the numbers $\mu_1, \mu_2, \ldots, \mu_{n-1}$, Q.E.D.

Remark. It can be shown that the problem of constructing a subspace R_k on which the quadratic form $A(x, x)$ has the canonical coefficients $\mu_1, \mu_2, \ldots, \mu_k$, which satisfy the inequalities (6), can also be solved in the general case, i.e., it does not have to be assumed that $k = n - 1$ or that all the numbers $\lambda_1, \lambda_2, \ldots, \lambda_n$ are different.

Problem. The quadratic forms $A(x, x)$ and $B(x, x)$ are said to be *comparable* if the inequality $A(x, x) \leqslant B(x, x)$ holds for any $x \in R$. Let $\lambda_1 \geqslant \lambda_2 \geqslant \cdots \geqslant \lambda_n$ be the canonical coefficients of the form $A(x, x)$, and let $\mu_1 \geqslant \mu_2 \geqslant \cdots \geqslant \mu_n$ be the canonical coefficients of the form $B(x, x)$. Show that the inequality

$$\lambda_k \leqslant \mu_k$$

holds for any k $(1 \leqslant k \leqslant n)$. (This is obvious if the forms $A(x, x)$ and $B(x, x)$ have a canonical basis in common.)

Hint. The coefficient λ_k equals the smallest of the maxima of the form $A(x, x)$ on a system of subspaces, and the coefficient μ_k equals the smallest of the maxima of the form $B(x, x)$ on the same system of subspaces.

72. Simultaneous Reduction of Two Quadratic Forms

The following question plays an important role in certain problems of mathematics and physics: *Given two quadratic forms $A(x, x)$ and $B(x, x)$ defined on an n-dimensional affine space R, how does one find a basis in which both $A(x, x)$ and $B(x, x)$ are reduced to canonical form (i.e., to sums of squares of the components of x with certain coefficients)?* The following example in the plane $(n = 2)$ shows that this problem does not always have a solution. Consider the two forms

$$A(x, x) = \xi_1^2 - \xi_2^2,$$
$$B(x, x) = \xi_1 \xi_2.$$

Finding a common canonical basis for these two forms is the same as finding a common pair of conjugate vectors for the hyperbolas $A(x, x) = 1$ and $B(x, x) = 1$ (see Sec. 44.4). Since these are equilateral hyperbolas, we know from analytic geometry that the conjugate directions of the hyperbolas are symmetric with respect to their asymptotes. Therefore, the polar angles φ_1 and φ_2 corresponding to the pair of conjugate directions satisfy the relation

$$\varphi_1 + \varphi_2 = \frac{\pi}{2}$$

for the first hyperbola, and the relation

$$\varphi_1 + \varphi_2 = 0$$

for the second hyperbola, where $-\pi \leqslant \varphi_1, \varphi_2 \leqslant \pi$. Since these two relations are mutually exclusive, there does not exist a common pair of conjugate vectors in this case.

It turns out that the problem of simultaneous reduction of two quadratic forms does have a solution if we make the supplementary assumption that one of the forms, say $B(x, x)$, is positive definite, i.e., that $B(x, x) > 0$ for $x \neq 0$. In this case, the existence of a solution is easily proved as follows:

Let $B(x, y)$ be the symmetric bilinear form corresponding to the quadratic form $B(x, x)$ (see Sec. 41). Introduce a Euclidean metric in the affine space R by writing

$$(x, y) = B(x, y).$$

The fact that $B(x, y)$ is symmetric and positive definite guarantees that (x, y) satisfies the axioms for a scalar product (Sec. 49). By Sec. 68, there exists an orthonormal basis (with respect to this metric) in which $A(x, x)$ takes the canonical form

$$A(x, x) = \lambda_1 \eta_1^2 + \lambda_2 \eta_2^2 + \cdots + \lambda_n \eta_n^2, \tag{13}$$

where $\eta_1, \eta_2, \ldots, \eta_n$ denote the components of the vector x in the basis just found. In the same basis, the second quadratic form $B(x, x)$ becomes

$$B(x, x) = (x, x) = \eta_1^2 + \eta_2^2 + \cdots + \eta_n^2$$

[see equation (17) of Sec. 51]. Thus, there exists a basis in which both $A(x, x)$ and $B(x, x)$ have canonical form.

73. Explicit Construction of the Required Basis

To construct the components of the vectors of the basis which is simultaneously canonical for two quadratic forms, we use the extremal properties of quadratic forms. As was shown in Sec. 70, the vectors of the required basis are the vectors which obey the condition

$$(x, x) = B(x, x) = 1,$$

and for which the form $A(x, x)$ has stationary values. Suppose that $A(x, x)$ and $B(x, x)$ are given by

$$A(x, x) = \sum_{i, k = 1}^{n} a_{ik} \xi_i \xi_k,$$

$$B(x, x) = \sum_{i, k = 1}^{n} b_{ik} \xi_i \xi_k$$

in the original basis. Using Lagrange's method, we form the function

$$F(\xi_1, \xi_2, \ldots, \xi_n) = \sum_{i, k = 1}^{n} a_{ik} \xi_i \xi_k - \mu \sum_{i, k = 1}^{n} b_{ik} \xi_i \xi_k,$$

and then equate to zero its partial derivatives with respect to all the ξ_i:

$$\sum_{k = 1}^{n} a_{ik} \xi_k - \mu \sum_{k = 1}^{n} b_{ik} \xi_k = 0 \qquad (i = 1, 2, \ldots, n). \tag{14}$$

The resulting system of homogeneous equations

$$(a_{11} - \mu b_{11})\xi_1 + (a_{12} - \mu b_{12})\xi_2 + \cdots + (a_{1n} - \mu b_{1n})\xi_n = 0,$$
$$(a_{21} - \mu b_{21})\xi_1 + (a_{22} - \mu b_{22})\xi_2 + \cdots + (a_{2n} - \mu b_{2n})\xi_n = 0, \quad (15)$$
$$\cdots$$
$$(a_{n1} - \mu b_{n1})\xi_1 + (a_{n2} - \mu b_{n2})\xi_2 + \cdots + (a_{nn} - \mu b_{nn})\xi_n = 0$$

has a nontrivial solution if and only if its determinant vanishes, i.e.,

$$\begin{vmatrix} a_{11} - \mu b_{11} & a_{12} - \mu b_{12} & \ldots & a_{1n} - \mu b_{1n} \\ a_{21} - \mu b_{21} & a_{22} - \mu b_{22} & \ldots & a_{2n} - \mu b_{2n} \\ \cdots & \cdots & \cdots & \cdots \\ a_{n1} - \mu b_{n1} & a_{n2} - \mu b_{n2} & \ldots & a_{nn} - \mu b_{nn} \end{vmatrix} = 0. \quad (16)$$

Solving (16), we find n solutions $\mu = \mu_k$ ($k = 1, 2, \ldots, n$); substituting μ_k into the system (15), we find the components $\xi_1^{(k)}, \xi_2^{(k)}, \ldots, \xi_n^{(k)}$ of the corresponding basis vector e_k. The results of Sec. 72 guarantee that (16) has n real roots and that every root of multiplicity r corresponds to r linearly independent solutions of the system (15).

Turning to the calculation of the canonical coefficients, we now show that the coefficients $\lambda_1, \lambda_2, \ldots, \lambda_n$ in the canonical representation (13) of the form $A(x, x)$ coincide with the corresponding roots $\mu_1, \mu_2, \ldots, \mu_n$ of the determinant (16). We could use an argument like that given in Sec. 70, but we prefer to carry out a direct calculation. Given the root μ_m, we multiply the ith equation of the system (15) by $\xi_i^{(m)}$ (the ith component of the solution corresponding to μ_m) for $i = 1, 2, \ldots, n$, and then add all the resulting equations. The result is

$$A(e_m, e_m) = \sum_{i, k = 1}^{n} a_{ik}\xi_i^{(m)}\xi_k^{(m)}$$
$$= \mu_m \sum_{i, k = 1}^{n} b_{ik}\xi_i^{(m)}\xi_k^{(m)} \quad (17)$$
$$= \mu_m B(e_m, e_m) = \mu_m,$$

since $B(e_m, e_m) = 1$. On the other hand, in the canonical coordinate system, the vector e_m has components $\eta_1^{(m)}, \eta_2^{(m)}, \ldots, \eta_n^{(m)}$ where $\eta_i^{(m)} = 0$ if $i \neq m$ and $\eta_m^{(m)} = 1$. Therefore we have

$$A(e_m, e_m) = \sum_{i = 1}^{n} \lambda_i(\eta_i^{(m)})^2 = \lambda_m.$$

Comparing this with (17), we see that $\mu_m = \lambda_m$, Q.E.D. This result allows us to write $A(x, x)$ in canonical form without calculating the canonical basis.

Problem 1. Find a common pair of conjugate directions for the curves

$$\frac{x^2}{4} + \frac{y^2}{1} = 1, \qquad 2xy = 1.$$

$$\textit{Ans.} \quad \frac{y}{x} = \pm\tfrac{1}{2}.$$

Problem 2. Construct the linear transformation which reduces the two quadratic forms

$$A(x, x) = \xi_1^2 + 2\xi_1\xi_2 + 2\xi_2^2 - 2\xi_1\xi_3 + 3\xi_3^2,$$
$$B(x, x) = \xi_1^2 + 2\xi_1\xi_2 + 3\xi_2^2 + 2\xi_2\xi_3 - 2\xi_1\xi_3 + 6\xi_3^2$$

to canonical form. What are the corresponding canonical forms?

$$\textit{Ans.} \quad \xi_1 = \eta_1 - \eta_2 + 2\eta_3, \quad \xi_2 = \eta_2 - \eta_3, \quad \xi_3 = \eta_3;$$
$$A(x, x) = \eta_1^2 + \eta_2^2 + \eta_3^2, \quad B(x, x) = \eta_1^2 + 2\eta_2^2 + 3\eta_3^2.$$

74. Uniqueness Questions for a Pair of Quadratic Forms

The problem (posed in Sec. 72) of simultaneously reducing two quadratic forms $A(x, x)$ and $B(x, x)$ to canonical form, where one of the forms, e.g. $B(x, x)$, is positive definite, was solved in a rather strong form, i.e., we reduced $B(x, x)$ to a sum of squares with coefficients equal to 1. In general, this is not required, and hence the coefficients of the canonical forms are not uniquely determined. Nevertheless, as we now show, *the ratios of the corresponding canonical coefficients still do not depend on the method used to simultaneously reduce $A(x, x)$ and $B(x, x)$ to canonical form.*

Suppose that $A(x, x)$ and $B(x, x)$ have been simultaneously reduced to canonical form in two different ways, i.e., suppose that in the variables $\xi_1, \xi_2, \ldots, \xi_n$, we have

$$A(x, x) = \sum_{i=1}^{n} \lambda_i \xi_i^2, \qquad B(x, x) = \sum_{i=1}^{n} \nu_i \xi_i^2,$$

while in the variables $\eta_1, \eta_2, \ldots, \eta_n$, we have

$$A(x, x) = \sum_{i=1}^{n} \rho_i \eta_i^2, \qquad B(x, x) = \sum_{i=1}^{n} \tau_i \eta_i^2.$$

Since the form $B(x, x)$ is positive definite, the numbers ν_i and τ_i ($i = 1, \ldots, n$) are all positive. Consider the new coordinate transformation

$$\sqrt{\nu_i}\,\xi_i = \bar{\xi}_i, \qquad \sqrt{\tau_i}\,\eta_i = \bar{\eta}_i.$$

Then the forms $A(x, x)$ and $B(x, x)$ become

$$A(x, x) = \sum_{i=1}^{n} \frac{\lambda_i}{\nu_i} \bar{\xi}_i^2, \qquad B(x, x) = \sum_{i=1}^{n} \bar{\xi}_i^2$$

in the variables $\bar{\bar{\xi}}_i$, and

$$A(x, x) = \sum_{i=1}^{n} \frac{\rho_i}{\tau_i} \bar{\eta}_i^2, \qquad B(x, x) = \sum_{i=1}^{n} \bar{\eta}_i^2$$

in the variables $\bar{\eta}_i$. Let $e_1, e_2, \ldots, e_n$ be the basis corresponding to the variables $\bar{\bar{\xi}}_i$, and let $f_1, f_2, \ldots, f_n$ be the basis corresponding to the variables $\bar{\eta}_i$. Both these bases are orthonormal in the metric determined by the form $B(x, x)$. Moreover, according to the uniqueness considerations of Sec. 69, the set of canonical coefficients of the quadratic form $A(x, x)$ is uniquely determined. Therefore, the sequence of numbers $\lambda_1/\nu_1, \lambda_2/\nu_2, \ldots, \lambda_n/\nu_n$ must coincide with the sequence of numbers $\rho_1/\tau_1, \rho_2/\tau_2, \ldots, \rho_n/\tau_n$, except possibly for order, Q.E.D.

Problem. Show that if no two of the ratios $\lambda_1/\nu_1, \lambda_2/\nu_2, \ldots, \lambda_n/\nu_n$ are equal, then the basis in which $A(x, x)$ and $B(x, x)$ both take canonical form is uniquely determined to within numerical factors.

*75. Distribution of the Curvature of Normal Sections of a Smooth Surface

We assume that the reader is familiar with the following basic facts from differential geometry:

a) The vector equation $\mathbf{r} = \mathbf{r}(t)$, where $\mathbf{r}(t)$ is a twice differentiable vector function of the scalar variable t, describes a smooth space curve (i.e., a curve with a continuously turning tangent), provided that the derivative $d\mathbf{r}/dt$, henceforth abbreviated to $\mathbf{r}_t$, does not vanish; we shall denote this curve by Γ. For a given value $t = t_0$, the vector $\mathbf{r}_t$ is directed along the tangent to the curve Γ, at the point corresponding to the parameter value $t = t_0$. The length of this vector depends on how the parameter t varies along the curve; in particular, if we take the parameter to be the length s of the curve Γ, as measured from some fixed point, then the length of the vector $\mathbf{r}_s = \boldsymbol{\tau}$ is 1. The equation $|d\mathbf{r}/ds| = 1$ can be written in the form $|d\mathbf{r}| = ds$, which no longer depends on the choice of the parameter.

b) The vector $\mathbf{r}_{tt}$ is in general not collinear with the vector $\mathbf{r}_t$; the two vectors $\mathbf{r}_t$ and $\mathbf{r}_{tt}$ determine the *osculating plane* of the curve Γ at $t = t_0$; the position of the osculating plane does not depend on the choice of the parameter t. If we again choose this parameter to be the arc length, then $\mathbf{r}_{ss} = \boldsymbol{\tau}_s = k\boldsymbol{\nu}$, where $\boldsymbol{\nu}$ is a unit vector directed along the normal to the curve Γ which lies in the osculating plane (the so-called *principal normal*) and is directed toward the concave side of

Γ; k is a positive coefficient called the *curvature* of Γ at the point $t = t_0$, and is equal to the limiting ratio of the angle of rotation of the tangent to the corresponding increment of arc length, as the latter goes to zero.

c) The vector equation $\mathbf{r} = \mathbf{r}(u, v)$, where $\mathbf{r}(u, v)$ is a twice differentiable function of the two scalar parameters u and v, describes a surface Σ; it is assumed that $\mathbf{r}_u$ is not collinear with $\mathbf{r}_v$ (where we abbreviate $\partial \mathbf{r} / \partial u$ to $\mathbf{r}_u$ and $\partial \mathbf{r} / \partial v$ to $\mathbf{r}_v$). The lines $u = C_1$, $v = C_2$ form a *net of coordinate curves* on the surface Σ. The equations $u = u(t)$, $v = v(t)$, where $u(t)$ and $v(t)$ are twice differentiable functions and $u_t^2 + v_t^2 \neq 0$, describe a smooth curve Γ on the surface Σ. Differentiating the vector equation $\mathbf{r} = \mathbf{r}[u(t), v(t)]$ of this curve, we obtain

$$d\mathbf{r} = \mathbf{r}_u \, du + \mathbf{r}_v \, dv. \tag{18}$$

Thus, the tangent vector to every smooth curve on the surface Σ at a given point of Σ is a linear combination of the vectors $\mathbf{r}_u$ and $\mathbf{r}_v$, which are clearly tangent to the coordinate curves going through the given point. Therefore, all the tangent vectors at the given point of the surface generate a plane S called the *tangent plane to the surface Σ*.

Taking the scalar product of (18) with itself, we obtain the following expression for the square of the element of arc length of the curve Γ:

$$ds^2 = |d\mathbf{r}|^2 = (\mathbf{r}_u, \mathbf{r}_u) \, du^2 + 2(\mathbf{r}_u, \mathbf{r}_v) \, du \, dv + (\mathbf{r}_v, \mathbf{r}_v) \, dv^2.$$

Denoting the scalar products which are the coefficients of this quadratic form in the variables du and dv by the symbols

$$E = (\mathbf{r}_u, \mathbf{r}_u), \qquad F = (\mathbf{r}_u, \mathbf{r}_v), \qquad G = (\mathbf{r}_v, \mathbf{r}_v),$$

we have

$$ds^2 = E \, du^2 + 2F \, du \, dv + G \, dv^2. \tag{19}$$

We now calculate the curvature of the curve Γ at a given point of Γ. To do this, we take the second derivative of the function $\mathbf{r}[u(t), v(t)]$ with respect to the arc length s, obtaining

$$k\mathbf{v} = \frac{d^2\mathbf{r}}{ds^2} = \frac{d}{ds}(\mathbf{r}_u u_s + \mathbf{r}_v v_s) = \mathbf{r}_{uu} u_s^2 + 2\mathbf{r}_{uv} u_s v_s + \mathbf{r}_{vv} v_s^2 + \mathbf{r}_u u_{ss} + \mathbf{r}_v v_{ss}.$$

We then take the scalar product of this equation with the unit vector $\mathbf{m}$ which is normal to the plane S; either of the two possible directions of $\mathbf{m}$ is chosen and then held fixed. Writing

$$L = (\mathbf{r}_{uu}, \mathbf{m}), \qquad M = (\mathbf{r}_{uv}, \mathbf{m}), \qquad N = (\mathbf{r}_{vv}, \mathbf{m}),$$

we obtain

$$k(\mathbf{v}, \mathbf{m}) = L\left(\frac{du}{ds}\right)^2 + 2M \frac{du}{ds}\frac{dv}{ds} + N\left(\frac{dv}{ds}\right)^2, \tag{20}$$

since the scalar products $(\mathbf{r}_u, \mathbf{m})$ and $(\mathbf{r}_v, \mathbf{m})$ obviously vanish. The scalar product $(\mathbf{v}, \mathbf{m})$ is the cosine of the angle θ between the vector $\mathbf{v}$ which defines the principal normal to the smooth curve Γ and the vector $\mathbf{m}$ which defines the normal to the surface.

We now set $k \cos \theta = k_n$. The number k_n is called the *normal curvature* of the surface at the given point in the given direction, i.e., the direction in which the curve Γ is drawn; clearly, k_n does not depend on which curve is actually drawn in this direction. In particular, if the curve Γ is the intersection Γ_n of the surface with the plane going through the surface normal $\mathbf{m}$ (such curves are called *normal sections* of the surface), then $\theta = 0$ or π, and hence $k_n = \pm k$. Thus, the absolute value of k_n equals the curvature of the normal section drawn in the given direction. The sign of k_n shows which side of the curve Γ_n is concave; $k_n > 0$ if $\theta = 0$, i.e., if the principal normal $\mathbf{v}$ to the curve Γ_n coincides with the vector $\mathbf{m}$, whereas $k_n < 0$ if the vector $\mathbf{v}$ is directed in the opposite direction to $\mathbf{m}$. Equation (20) now takes the form

$$k_n = L\left(\frac{du}{ds}\right)^2 + 2M \frac{du}{ds}\frac{dv}{ds} + N\left(\frac{dv}{ds}\right)^2. \tag{21}$$

To see the way in which the normal curvature k_n at a given point depends on the direction determining k_n, we note that this direction is given by the vector

$$d\mathbf{r} = \mathbf{r}_u\, du + \mathbf{r}_v\, dv$$

which is tangent to the curve (or simply by the ratio dv/du), and (21) does in fact give k_n as a function of dv/du. We shall return to this formula later.

It should be noted that the quadratic form (19) is *positive definite*, since for any values of du and dv it equals the square of the corresponding element of arc. Thus, using the general result of Sec. 72, we can find a new basis $\mathbf{e}_1, \mathbf{e}_2$ in the tangent plane S, relative to which the *first* quadratic form

$$(d\mathbf{r}, d\mathbf{r})_1 \equiv ds^2 = E\, du^2 + 2F\, du\, dv + G\, dv^2 \tag{22}$$

and the *second* quadratic form

$$(d\mathbf{r}, d\mathbf{r})_2 \equiv k_n\, ds^2 = L\, du^2 + 2M\, du\, dv + N\, dv^2 \tag{23}$$

both take canonical form. To construct this basis, we define the scalar product of two vectors $d\mathbf{r}_1$ and $d\mathbf{r}_2$ in the tangent plane S as $(d\mathbf{r}_1, d\mathbf{r}_2)_1$, i.e., as the value of the symmetric bilinear form corresponding to the first quadratic form. If we denote the components of the vectors $d\mathbf{r}_1$ and $d\mathbf{r}_2$ relative to the basis $\mathbf{r}_u, \mathbf{r}_v$ by du_1, dv_1 and du_2, dv_2, respectively, then the bilinear form in question is

$$(d\mathbf{r}_1, d\mathbf{r}_2)_1 = E\, du_1\, du_2 + F\, du_1\, dv_2 + F\, du_2\, dv_1 + G\, dv_1\, dv_2.$$

Clearly, this expression is just the ordinary scalar product of the vectors

$$d\mathbf{r}_1 = \mathbf{r}_u\,du_1 + \mathbf{r}_v\,dv_1, \qquad d\mathbf{r}_2 = \mathbf{r}_u\,du_2 + \mathbf{r}_v\,dv_2.$$

Thus, the scalar product defined by the bilinear form $(d\mathbf{r}_1, d\mathbf{r}_2)_1$ coincides with the ordinary scalar product $(d\mathbf{r}_1, d\mathbf{r}_2)$, so that the required canonical basis $\mathbf{e}_1, \mathbf{e}_2$, which according to the construction of Sec. 72 is orthonormal in terms of the scalar product $(d\mathbf{r}_1, d\mathbf{r}_2)_1$, is also orthonormal in the usual sense. Therefore, in the tangent plane S there exists an orthonormal basis $\mathbf{e}_1, \mathbf{e}_2$ in which both $(d\mathbf{r}, d\mathbf{r})_1$ and $(d\mathbf{r}, d\mathbf{r})_2$ take canonical form, i.e., writing $d\mathbf{r} = \xi_1\mathbf{e}_1 + \xi_2\mathbf{e}_2$, we have

$$(d\mathbf{r}, d\mathbf{r})_1 = \xi_1^2 + \xi_2^2, \qquad (d\mathbf{r}, d\mathbf{r})_2 = \lambda_1\xi_1^2 + \lambda_2\xi_2^2.$$

Therefore, in the basis $\mathbf{e}_1, \mathbf{e}_2$, the formula (21) for the normal curvature becomes

$$k_n = \frac{\lambda_1\xi_1^2 + \lambda_2\xi_2^2}{\xi_1^2 + \xi_2^2}.$$

The ratio $\xi_1^2/(\xi_1^2 + \xi_2^2)$ is obviously the square of the cosine of the angle φ between the vectors $d\mathbf{r}$ and $\mathbf{e}_1$; similarly, $\xi_2^2/(\xi_1^2 + \xi_2^2)$ is $\sin^2\varphi$. Therefore, we have

$$k_n = k_n(\varphi) = \lambda_1\cos^2\varphi + \lambda_2\sin^2\varphi.$$

The coefficients λ_1 and λ_2 are easily determined by substituting the values $\varphi = 0$ and $\varphi = \pi/2$ into this formula. For $\varphi = 0$, we obtain

$$\lambda_1 = k_n(0),$$

i.e., λ_1 is the normal curvature corresponding to the vector $\mathbf{e}_1$. Similarly, λ_2 is the normal curvature corresponding to the vector $\mathbf{e}_2$. The numbers $k_1 = k_n(0)$ and $k_2 = k_n(\pi/2)$ are called the *principal curvatures* of the surface Σ at the given point. The directions of the vectors $\mathbf{e}_1$ and $\mathbf{e}_2$ are called the *principal directions* in the tangent plane S. We have just derived *Euler's formula*, which expresses the normal curvature for any direction in terms of the principal curvatures and the angle φ, i.e.,

$$k_n(\varphi) = k_1\cos^2\varphi + k_2\sin^2\varphi.$$

In particular, it is clear that the principal curvatures k_1 and k_2 are the largest and smallest values of the normal curvature of the surface Σ at the given point. It is also easy to deduce this fact from the general theory of extremal values of quadratic forms. In fact, according to Sec. 73, the canonical coefficients of a quadratic form are the same as the stationary values of the form with the constraint $(d\mathbf{r}, d\mathbf{r})_1 = 1$, while by (22) and (23), we have $(d\mathbf{r}, d\mathbf{r})_2 = k_n$ when $(d\mathbf{r}, d\mathbf{r})_1 = 1$. Therefore, the stationary values of the form $(d\mathbf{r}, d\mathbf{r})_2$, with the constraint $(d\mathbf{r}, d\mathbf{r})_1 = 1$, are the stationary values of the normal curvature. Since in the present case, there are just two stationary values, these values must be the maximum and the minimum of the normal curvature.

To represent graphically the dependence of the normal curvature $k_n(\varphi)$ on the angle φ, we draw in the plane S the curve with polar equation

$$\rho = \frac{1}{\sqrt{|k_n(\varphi)|}},$$

or equivalently

$$\rho^2 |k_n(\varphi)| = |k_1 \rho^2 \cos^2 \varphi + k_2 \rho^2 \sin^2 \varphi| = |k_1 \xi_1^2 + k_2 \xi_2^2| = 1.$$

This curve is the union of the two second-degree curves

$$k_1 \xi_1^2 + k_2 \xi_2^2 = \pm 1,$$

with principal axes along the vectors $\mathbf{e}_1$ and $\mathbf{e}_2$ (the *Dupin indicatrix*); it gives an intuitive geometric picture of the distribution of normal curvature at a given point of the surface.

We now derive formulas for calculating the principal curvatures and the principal directions at a given point of the surface Σ. Let (du, dv) be a vector (in the plane S) which determines a principal direction. In our case, the equations (15) of Sec. 73 take the form

$$\begin{aligned}
(L - \mu E)\, du + (M - \mu F)\, dv &= 0, \\
(M - \mu F)\, du + (N - \mu G)\, dv &= 0,
\end{aligned} \tag{24}$$

and equation (16) becomes

$$\begin{vmatrix} L - \mu E & M - \mu F \\ M - \mu F & N - \mu G \end{vmatrix} = 0. \tag{25}$$

By Sec. 73, the roots of (25) are the same as the coefficients λ_1 and λ_2 of the reduced form $(d\mathbf{r}, d\mathbf{r})_2$, i.e., by what was just proved, the roots are the same as the principal curvatures. Substituting each of these roots into the system (24), we find the components du and dv of the corresponding principal direction. In fact, in this case, we can write down directly an equation which determines the ratio dv/du. The system (24) can be rewritten in the form

$$\begin{aligned}
(L\, du + M\, dv) - \mu(E\, du + F\, dv) &= 0, \\
(M\, du + N\, dv) - \mu(F\, du + G\, dv) &= 0,
\end{aligned}$$

from which it follows that

$$\begin{vmatrix} L\, du + M\, dv & E\, du + F\, dv \\ M\, du + N\, dv & F\, du + G\, dv \end{vmatrix} = 0.$$

Dividing each row of this determinant by du, we obtain

$$\begin{vmatrix} L + M\dfrac{dv}{du} & E + F\dfrac{dv}{du} \\ M + N\dfrac{dv}{du} & F + G\dfrac{dv}{du} \end{vmatrix} = 0. \tag{26}$$

The two roots of the quadratic equation in dv/du obtained from (26) give the values of dv/du which determine the two principal directions. Taken over a whole region of the surface Σ, rather than at a single point, (26) represents the differential equation of a family of curves, which are tangent at every point of Σ to the principal normal sections; these lines are called the *lines of curvature* of the surface Σ.

Problem 1. Let $\mathbf{m}_u$ and $\mathbf{m}_v$ denote the derivatives of the unit vector $\mathbf{m}$ (the surface normal) with respect to the parameters u and v. Show that the linear operator which carries $\mathbf{r}_u$ into $\mathbf{m}_u$ and $\mathbf{r}_v$ into $\mathbf{m}_v$ is a symmetric operator in the tangent plane; this operator is called *Rodrigues' operator*.
Hint. Differentiate the relations $(\mathbf{m}, \mathbf{r}_u) = 0$ and $(\mathbf{m}, \mathbf{r}_v) = 0$ with respect to u and v, respectively, and then compare the results.

Problem 2. Show that the bilinear form $(A\mathbf{f}, \mathbf{g})$, where A is Rodrigues' operator and $\mathbf{f}$ and $\mathbf{g}$ are vectors in the tangent plane, equals the negative of the bilinear form corresponding to the second quadratic form (23).
Hint. Verify this by using $\mathbf{r}_u$ and $\mathbf{r}_v$ as basis vectors.

Problem 3. What are the eigenvectors and eigenvalues of Rodrigues' operator?
Ans. The eigenvectors are the vectors of the orthonormal basis of the two quadratic forms (22) and (23), i.e., the vectors lying in the principal directions. The eigenvalues are the canonical coefficients of the second quadratic form, i.e., the corresponding principal curvatures (taken with opposite sign).

Problem 4. Show that if every direction at every point of a surface is a principal direction, then the surface is a sphere or a plane.
Ans. Every vector is an eigenvector of Rodrigues' operator; therefore, the curvature λ is constant in all directions at the given point Q, i.e., $\lambda = \lambda(Q)$. Differentiating the relations $\mathbf{m}_u = \lambda\mathbf{r}_u$, $\mathbf{m}_v = \lambda\mathbf{r}_v$ with respect to u and v, we obtain $\lambda_u\mathbf{r}_v + \lambda_v\mathbf{r}_u = 0$, which implies that $\lambda_u = \lambda_v = 0$, since $\mathbf{r}_u$ and $\mathbf{r}_v$ are linearly independent. It follows that $\lambda(Q) = \text{const}$, $\mathbf{m} = \lambda(\mathbf{r} - \mathbf{r}_0)$. If $\lambda = 0$, we obtain a plane. If $\lambda \neq 0$, then $\mathbf{m}$ is collinear with $\mathbf{r} - \mathbf{r}_0$; taking the scalar product of $\mathbf{m} = \lambda(\mathbf{r} - \mathbf{r}_0)$ with $\mathbf{r} - \mathbf{r}_0$, we obtain $|\mathbf{r} - \mathbf{r}_0| = 1/\lambda$, i.e., the surface is a sphere of radius $1/\lambda$.

Problem 5. The three families of surfaces defined by the equations

$$F_1(x, y, z) = C_1, \quad F_2(x, y, z) = C_2, \quad F_3(x, y, z) = C_3$$

are said to form a *triply orthogonal system* in a region G, if no two surfaces of the same family intersect and if the surfaces of different families intersect each other at right angles (in the region G). Prove the following theorem due to Dupin: *The surfaces of a triply orthogonal system intersect each other in the lines of curvature.*

Hint. The quantities

$$u = F_1(x, y, z), \quad v = F_2(x, y, z), \quad w = F_3(x, y, z)$$

can be regarded as parameters specifying a point in space, and u and v can be regarded as parameters specifying a point on a surface of the first family, say. Because of the orthogonality, we have

$$(\mathbf{r}_u, \mathbf{r}_v) = (\mathbf{r}_u, \mathbf{r}_w) = (\mathbf{r}_v, \mathbf{r}_w) = 0.$$

Differentiating these relations with respect to w, v, and u, respectively, we obtain $(\mathbf{r}_w, \mathbf{r}_{uv}) = 0$, say. Since $\mathbf{r}_w$ is parallel to $\mathbf{m}$, we have $(\mathbf{m}, \mathbf{r}_{uv}) = 0$, and both quadratic forms take canonical form.

Problem 6. Prove the following theorem due to Liouville: *Every conformal (i.e., angle-preserving) transformation of space maps the family of all spheres and planes into itself* (i.e., every sphere goes into a sphere or a plane, and every plane goes into a sphere or a plane).

Hint. It is sufficient to prove that every direction at every point of the image of a sphere is a principal direction. To do this, include the sphere in a suitably chosen triply orthogonal system of surfaces, and then apply the result of Problem 5.

All the results of this section can be extended to the case of an $(n - 1)$-dimensional surface Σ in an n-dimensional space. In this case, a surface Σ is specified by $n - 1$ parameters $u_1, u_2, \ldots, u_{n-1}$, so that its equation has the form

$$\mathbf{r} = \mathbf{r}(u_1, u_2, \ldots, u_{n-1}).$$

Moreover, we have

$$d\mathbf{r} = \sum_{i=1}^{n-1} \mathbf{r}_i \, du_i, \qquad ds^2 = (d\mathbf{r}, d\mathbf{r}) = \sum_{i,j=1}^{n-1} g_{ij} \, du_i \, du_j,$$

where

$$\mathbf{r}_i = \frac{\partial \mathbf{r}}{\partial u_i}, \qquad g_{ij} = (\mathbf{r}_i, \mathbf{r}_j).$$

The second quadratic form can be written in the form

$$\sum_{i,j=1}^{n-1} \beta_{ij} \, du_i \, du_j, \quad \text{where } \beta_{ij} = (\mathbf{r}_{ij}, \mathbf{m}).$$

The simultaneous reduction of the first and second quadratic forms to canonical form corresponds to the problem of determining $n - 1$ principal directions and principal curvatures. We leave the details of these derivations to the reader.

*76. Small Oscillations of Mechanical Systems

We assume that the reader is familiar with the following facts from theoretical mechanics:

a) The configuration of a mechanical system with n degrees of freedom is specified by the values of n *generalized coordinates* $q_1, q_2, \ldots, q_n$. The *kinetic energy* of the system is given by a quadratic form

$$T = \sum_{i,j=1}^{n} a_{ij} \dot{q}_i \dot{q}_j \tag{27}$$

in the *generalized velocities* $\dot{q}_1, \dot{q}_2, \ldots, \dot{q}_n$ where the overdot denotes differentiation with respect to time; the coefficients in (27) are functions of the coordinates q_i ($i = 1, 2, \ldots, n$) and time. The *potential energy* of the system is a function

$$U = U(q_1, q_2, \ldots, q_n, t)$$

of the coordinates and time.

b) The equations of motion of the system (*Lagrange's equations*) have the following form

$$\frac{d}{dt}\left(\frac{\partial T}{\partial \dot{q}_i}\right) - \frac{\partial}{\partial q_i}(T - U) = 0 \qquad (i = 1, 2, \ldots, n). \tag{28}$$

When the external constraints are stationary, T and U do not depend explicitly on the time; it will henceforth be tacitly assumed that this is the case.

Now let $q_1 = q_1^0$, $q_2 = q_2^0, \ldots, q_n = q_n^0$ be an equilibrium position of the system. In an equilibrium position, the kinetic energy T of the system vanishes identically (in time), since $\dot{q}_i = 0$ ($i = 1, 2, \ldots, n$). Moreover, the derivatives of T with respect to $\dot{q}_i$, which are linear forms in $\dot{q}_1, \dot{q}_2, \ldots, \dot{q}_n$, also vanish identically in t. Therefore, the quantities $q_1^0, q_2^0, \ldots, q_n^0$ satisfy the equations

$$\frac{\partial U}{\partial q_i} = 0 \qquad (i = 1, 2, \ldots, n),$$

i.e., equilibrium positions are possible only at stationary points of the potential energy. It can be shown that a point where the potential energy is a minimum corresponds to a *stable* equilibrium position. Consider such a minimum point. Without loss of generality, we can assume that both the generalized coordinates $q_1, q_2, \ldots, q_n$ and the potential energy vanish at this point. If we study the motion only in a small neighborhood of this zero point, we can assume that the coefficients of the quadratic form T are constant, and are in fact equal to their values at the zero point. Moreover, neglecting terms of order higher than the second in the Taylor's series expansion of the potential energy U, we can assume that U is a quadratic form in the coordinates $q_1, q_2, \ldots, q_n$ with *constant* coefficients, i.e.,

$$U = \sum_{i,j=1}^{n} b_{ij}q_iq_j.$$

Since both forms U and T are positive definite, there exists a linear transformation

$$q_i = \sum_{j=1}^{n} c_{ij}\eta_j,$$

$$\dot{q}_i = \sum_{j=1}^{n} c_{ij}\dot{\eta}_j \quad (i = 1, 2, \ldots, n), \tag{29}$$

carrying the variables $q_i, \dot{q}_i$ into the variables $\eta_i, \dot{\eta}_i$, which reduces T and U to the form

$$T = \sum_{i=1}^{n} \dot{\eta}_i^2, \quad U = \sum_{i=1}^{n} \omega_i^2\eta_i^2 \quad (\omega_1 \geqslant \omega_2 \geqslant \cdots \geqslant \omega_n > 0).$$

In the generalized coordinates $\eta_1, \eta_2, \ldots, \eta_n$, Lagrange's equations (28) become

$$\ddot{\eta}_i + \omega_i^2\eta_i = 0 \quad (i = 1, 2, \ldots, n).$$

These equations are easily solved and have the solutions

$$\eta_i = A_i \cos \omega_i(t - t_i) \quad (i = 1, 2, \ldots, n),$$

where the constants t_i and A_i are determined from the initial conditions (i.e., the initial values of the coordinates η_i and $\dot{\eta}_i$). The frequencies ω_i are called the *natural frequencies* (or *resonant frequencies*) of the system. Therefore, *each of the coordinates η_i undergoes simple harmonic motion with a fixed natural frequency ω_i.* By the uniqueness considerations of Sec. 74, the natural frequencies of the system are uniquely determined by giving the potential energy and kinetic energy of the system, and do not depend on the choice of the transformation (29).

To find explicit expressions for the coordinates q_i in terms of the coordinates η_i (and *vice versa*), we use the method of Sec. 73.

Problem 1. How do the natural frequencies of the system behave when its potential energy is increased (as a function of the coordinates), i.e., when the "stiffness" of the system is increased?

Hint. Use the problem in Sec. 71.2.

Ans. They do not decrease.

Problem 2. How do the natural frequencies of the system behave when the kinetic energy is increased (as a function of the generalized velocities $\dot{q}_i$), i.e., when the "inertia" of the system is increased?

Ans.   They do not increase.

Problem 3. How do the natural frequencies of the system behave if an additional constraint of the form

$$\sum_{j=1}^{n} a_j q_j = 0$$

is imposed on the system?

Hint.   Use the inequalities (6) of Sec. 71.

11

QUADRIC SURFACES

77. Reduction of the General Equation of a Quadric Surface to Canonical Form

By a quadric (or second-degree) surface in an n-dimensional space R, we mean the locus of the points $x = (\xi_1, \xi_2, \ldots, \xi_n)$ which satisfy an equation of the form

$$\sum_{i,\,k=1}^{n} a_{ik}\xi_i\xi_k + 2\sum_{i=1}^{n} b_i\xi_i + c = 0 \tag{1}$$

or

$$A(x, x) + 2l(x) + c = 0,$$

where

$$A(x, x) = \sum_{i,\,k=1}^{n} a_{ik}\xi_i\xi_k$$

is a quadratic form in the components of the radius vector of the point x,

$$l(x) = \sum_{i=1}^{n} b_i\xi_i$$

is a linear form, and c is a constant.[1] We assume that the space R is Euclidean and that the numbers $\xi_1, \xi_2, \ldots, \xi_n$ are the coordinates of the point x with

---

[1] In the case $n = 2$, the geometric object defined by (1) is called a second-degree curve. However, we shall henceforth always use the word "surface," despite the fact that, strictly speaking, the word "curve" should be used whenever $n = 2$.

respect to an orthonormal basis. In this section, we study the problem of choosing a new orthonormal basis in the space R such that our quadric surface is defined by a particularly simple equation, called the *canonical equation* of the surface. Subsequently, we shall use the canonical equation to study the properties of the surface.

First of all, we perform the orthogonal coordinate transformation

$$\xi_i = \sum_{j=1}^{n} q_{ij}\eta_j \qquad (i = 1, 2, \ldots, n) \tag{2}$$

in R, which, according to Sec. 68, reduces the quadratic form $A(x, x)$ to the canonical form

$$A(x, x) = \sum_{i=1}^{n} \lambda_i \eta_i^2.$$

After substituting from (2), equation (1) becomes

$$\sum_{i=1}^{n} \lambda_i \eta_i^2 + 2 \sum_{i=1}^{n} l_i \eta_i + c = 0, \tag{1'}$$

where the l_i $(i = 1, 2, \ldots, n)$ are the new coefficients of the linear form $l(x)$.

If, for some i, $\lambda_i \neq 0$ in (1'), then we can eliminate the corresponding linear term by appropriately shifting the origin of coordinates. For example, if $\lambda_1 \neq 0$, we have

$$\lambda_1 \eta_1^2 + 2l_1\eta_1 = \lambda_1\left(\eta_1 + \frac{l_1}{\lambda_1}\right)^2 - \frac{l_1^2}{\lambda_1}.$$

We set $\eta_1' = \eta_1 + (l_1/\lambda_1)$, which is equivalent to shifting the origin to the point $(-l_1/\lambda_1, 0, 0, \ldots, 0)$. As a result of this substitution, the two terms $\lambda_1 \eta_1^2 + 2l_1\eta_1$ are changed to $\lambda_1 \eta_1'^2 - (l_1^2/\lambda_1^2)$, i.e., the quadratic term still has the same coefficient λ_1 as before, but the linear term disappears and l_1^2/λ_1^2 is subtracted from the constant term. After making a series of such substitutions, the equation of the surface becomes

$$\lambda_1 \eta_1^2 + \lambda_2 \eta_2^2 + \cdots + \lambda_r \eta_r^2 + 2l_{r+1}\eta_{r+1} + \cdots + 2l_n\eta_n + c = 0.$$

Here, for simplicity, we have dropped the primes on the variables η_i', and we have renumbered the variables in such a way that the variables appearing in the quadratic form appear first, i.e., $\lambda_1, \lambda_2, \ldots, \lambda_r$ are nonzero and $\lambda_k = 0$ for $k > r$. If all the numbers $l_{r+1}, l_{r+2}, \ldots, l_n$ turn out to be zero, then we obtain the *canonical equation of a central surface*

$$\lambda_1 \eta_1^2 + \lambda_2 \eta_2^2 + \cdots + \lambda_r \eta_r^2 + c = 0. \tag{3}$$

Next, assume that at least one of the numbers $l_{r+1}, l_{r+2}, \ldots, l_n$ is nonzero, and carry out a new orthogonal coordinate transformation using the formulas

$$
\begin{aligned}
\tau_1 &= \eta_1, \\
\tau_2 &= \eta_2, \\
&\cdots\cdots \\
\tau_r &= \eta_r, \\
\tau_{r+1} &= -\frac{1}{M}(l_{r+1}\eta_{r+1} + \cdots + l_n\eta_n), \\
&\cdots\cdots
\end{aligned}
\tag{4}
$$

where M is a positive factor which guarantees the orthogonality of the transformation matrix. Since the sum of the squares of the elements of every row of an orthogonal matrix must equal 1, we have

$$M^2 = l_{r+1}^2 + l_{r+2}^2 + \cdots + l_n^2.$$

The remaining rows (i.e., rows $r + 2, r + 3, \ldots, n$) can be arbitrary, provided only that the resulting matrix is orthogonal. This last requirement can be formulated as follows: The elements of rows $r + 2, r + 3, \ldots, n$ of the transformation matrix must be the components of vectors forming an orthonormal basis for the orthogonal complement of a certain $(r + 1)$-dimensional subspace of the Euclidean space T_n, i.e., the subspace spanned by the $r + 1$ vectors

$$(1, 0, 0, \ldots, 0), (0, 1, 0, \ldots, 0), \ldots, (0, 0, \ldots, 1, 0, \ldots, 0),$$

$$-\frac{1}{M}(0, 0, \ldots, 0, l_{r+1}, l_{r+2}, \ldots, l_n).$$

This shows that it is indeed possible to construct the required matrix.

As a result of the transformation (4), the equation of the surface takes the form

$$\lambda_1\tau_1^2 + \cdots + \lambda_r\tau_r^2 = 2M\tau_{r+1} - c.$$

If $c \neq 0$, another shift of the origin given by the formula

$$\tau_{r+1}' = \tau_{r+1} - \frac{c}{2M} \quad \text{or} \quad 2M\tau_{r+1}' = 2M\tau_{r+1} - c$$

allows us to eliminate the constant term; then, dropping the prime on τ_{r+1}', we obtain the equation

$$\lambda_1\tau_1^2 + \cdots + \lambda_r\tau_r^2 = 2M\tau_{r+1};
\tag{5}$$

this is the *canonical equation of a noncentral surface.*

A central surface will be called a *degenerate conical surface* if $c = 0$ in its canonical equation (3). Any surface will be called a *degenerate cylindrical surface* if there are less than n variables in its canonical equation. The meaning of this terminology will be explained later.

78. Central Surfaces

By a *center* of a surface is meant a point

$$x_0 = (\xi_1^0, \xi_2^0, \ldots, \xi_n^0)$$

with the following property: If the point

$$(\xi_1^0 + \xi_1, \xi_2^0 + \xi_2, \ldots, \xi_n^0 + \xi_n)$$

lies on the surface, then the point

$$(\xi_1^0 - \xi_1, \xi_2^0 - \xi_2, \ldots, \xi_n^0 - \xi_n),$$

which is symmetric with respect to x_0, also lies on the surface. A surface with the canonical equation (3) has at least one center, since every point for which

$$\eta_1 = \eta_2 = \cdots = \eta_r = 0$$

is obviously a center. This explains why surfaces with the canonical equation (3) are called *central* surfaces.

We now show that a surface with the canonical equation (3) has no centers other than the point

$$\eta_1 = \eta_2 = \cdots = \eta_r = 0;$$

this fact will be used later. To see this, let $(\xi_1^0, \xi_1^0, \ldots, \xi_r^0)$ be a center of the surface. Then the relation

$$\lambda_1(\xi_1^0 + \xi_1)^2 + \lambda_2(\xi_2^0 + \xi_2)^2 + \cdots + \lambda_r(\xi_r^0 + \xi_r)^2 + c = 0$$

implies

$$\lambda_1(\xi_1^0 - \xi_1)^2 + \lambda_2(\xi_2^0 - \xi_2)^2 + \cdots + \lambda_r(\xi_r^0 - \xi_r)^2 + c = 0.$$

Subtracting the second equation from the first, we obtain

$$\lambda_1\xi_1^0\xi_1 + \lambda_2\xi_2^0\xi_2 + \cdots + \lambda_r\xi_r^0\xi_r = 0.$$

Now take a point on the surface for which $\xi_2 = \cdots = \xi_r = 0$, $\xi_1 \neq 0$. [It is obvious that (3) can be satisfied for such values of the ξ_i.] Then we obtain

$$\lambda_1\xi_1^0\xi_1 = 0,$$

so that $\xi_1^0 = 0$. Similarly, we show that

$$\xi_2^0 = \cdots = \xi_r^0 = 0, \quad \text{Q.E.D.}$$

We begin our study of central surfaces by considering *nondegenerate* central surfaces, i.e., we assume that $r = n$ and $c \neq 0$. Then (3) is easily transformed into the form

$$\pm \frac{\eta_1^2}{a_1^2} \pm \frac{\eta_2^2}{a_2^2} \pm \cdots \pm \frac{\eta_n^2}{a_n^2} = 1,$$

where the numbers a_k are defined by

$$a_k = + \sqrt{\left|\frac{c}{\lambda_k}\right|} \qquad (k = 1, 2, \ldots, n),$$

and are called the *semiaxes* of the surface. We renumber the coordinates in such a way that the positive terms appear first, i.e.,

$$\frac{\eta_1^2}{a_1^2} + \frac{\eta_2^2}{a_2^2} + \cdots + \frac{\eta_k^2}{a_k^2} - \frac{\eta_{k+1}^2}{a_{k+1}^2} - \cdots - \frac{\eta_n^2}{a_n^2} = 1. \tag{6}$$

It is natural to exclude from consideration the case $k = 0$, since for $k = 0$, no real values $\eta_1, \eta_2, \ldots, \eta_n$ can satisfy (6); in this case, one sometimes says that (6) defines an "imaginary" surface. Thus, there remain n different types of nondegenerate central surfaces, corresponding to the values $k = 1, 2, \ldots, n$. In the two-dimensional case ($n = 2$), when $k = 1$ and $k = 2$, equation (6) defines the two curves

$$(k = 1) \qquad \frac{\eta_1^2}{a_1^2} - \frac{\eta_2^2}{a_2^2} = 1 \qquad \text{(a hyperbola)},$$

$$(k = 2) \qquad \frac{\eta_1^2}{a_1^2} + \frac{\eta_2^2}{a_2^2} = 1 \qquad \text{(an ellipse)},$$

which are familiar from analytic geometry. When $n = 3$, we have $k = 1$, $k = 2$, $k = 3$, and the corresponding nondegenerate central surfaces in three-dimensional space are given by the equations

$$(k = 1) \qquad \frac{\eta_1^2}{a_1^2} - \frac{\eta_2^2}{a_2^2} - \frac{\eta_3^2}{a_3^2} = 1,$$

$$(k = 2) \qquad \frac{\eta_1^2}{a_1^2} + \frac{\eta_2^2}{a_2^2} - \frac{\eta_3^2}{a_3^2} = 1,$$

$$(k = 3) \qquad \frac{\eta_1^2}{a_1^2} + \frac{\eta_2^2}{a_2^2} + \frac{\eta_3^2}{a_3^2} = 1.$$

We now remind the reader of the construction of these surfaces.

Consider the sections of each of the surfaces made by the horizontal planes $\eta_3 = Ca_3$ ($-\infty < C < \infty$). These sections are respectively hyperbolas with real axis η_1:

$$\frac{\eta_1^2}{a_1^2} - \frac{\eta_2^2}{a_2^2} = 1 + C^2,$$

ellipses defined for all values of C:

$$\frac{\eta_1^2}{a_1^2} + \frac{\eta_2^2}{a_2^2} = 1 + C^2,$$

and ellipses defined only for $|C| \leqslant 1$:

$$\frac{\eta_1^2}{a_1^2} + \frac{\eta_2^2}{a_2^2} = 1 - C^2.$$

To locate the vertices of these sections, we consider the sections made by the coordinate planes $\eta_1 = 0$, $\eta_2 = 0$. In the case $k = 1$, only the coordinate plane $\eta_2 = 0$ gives a real section, i.e., the hyperbola

$$\frac{\eta_1^2}{a_1^2} - \frac{\eta_3^2}{a_3^2} = 1.$$

The vertices of the hyperbolas formed by the horizontal sections lie on this

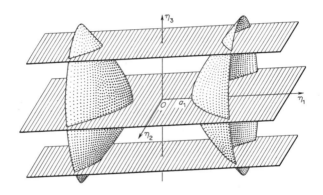

FIGURE 2

curve; as a result of this construction, we obtain a surface called a *hyperboloid of two sheets* (see Figure 2). In the case $k = 2$, the sections made by both planes $\eta_1 = 0$ and $\eta_2 = 0$ are hyperbolas with imaginary axis η_3:

$$\frac{\eta_2^2}{a_2^2} - \frac{\eta_3^2}{a_3^2} = 1, \qquad \frac{\eta_1^2}{a_1^2} - \frac{\eta_3^2}{a_3^2} = 1.$$

The set of ellipses formed by the horizontal sections have vertices lying on these hyperbolas, and form a surface called a *hyperboloid of one sheet* (see Figure 3). Finally, in the case $k = 3$, the sections made by the coordinate planes $\eta_1 = 0$, $\eta_2 = 0$ are ellipses; drawing the ellipses made by the horizontal sections, we obtain an *ellipsoid* (see Figure 4).

Quadric surfaces in spaces with more than three dimensions are not easily visualized. Nevertheless, even in the multidimensional case, we can show essential differences between the types of central surfaces corresponding to

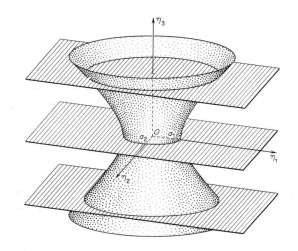

FIGURE 3

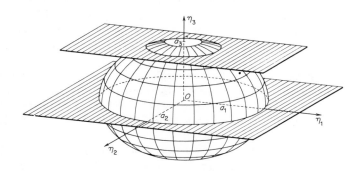

FIGURE 4

the different values $k = 1, 2, \ldots, n$. We begin by pointing out differences which are geometrically obvious in three dimensions. On the hyperboloid of two sheets ($k = 1$), there exists a pair of points which cannot be made to coincide by a continuous displacement of the points along the surface; to obtain such a pair of points, it is sufficient to take the first point on one sheet and the second point on the other sheet. On the hyperboloid of one sheet ($k = 2$), every two points can be made to coincide by means of a continuous displacement along the surface; however, there exists a closed curve, e.g., a curve going around the "throat" of the hyperboloid, which cannot be continuously deformed into a point. On the ellipsoid ($k = 3$), any closed curve can be deformed into a point. These facts can serve as the starting point for classifying the geometric differences between central surfaces in an n-dimensional space, as we now show.

We introduce the following definitions: A geometric figure A is said to be *homeomorphic* to a figure B if there exists a one-to-one, bicontinuous (i.e., continuous in both directions) mapping of the points of the figure A into the points of the figure B. A figure A lying on a surface S is said to be *homotopic* to a figure B lying on the same surface, if the figure A can be mapped into the figure B by means of a continuous deformation, during which the figure A always remains on the surface S.

Using these definitions, we can formulate the geometric differences between the central surfaces corresponding to different values of k as follows: For $k = 1$, we can find a pair of points on the surface which are not homotopic to each other; for $k = 2$, every point on the surface is homotopic to every other point, but there exists a curve which is homeomorphic to a circle and which is not homotopic to a point; for $k = 3$, every curve which is homeomorphic to a circle is homotopic to a point, but there exists a part of the surface which is homeomorphic to a sphere (more precisely, to a two-dimensional sphere, i.e., a sphere in three-dimensional space) and which is not homotopic to a point. Continuing in this way, we can formulate the following distinguishing property of the central surface corresponding to a given value of k: every part of such a surface which is homeomorphic to a $(k - 1)$-dimensional sphere is homotopic to a point, but there exists a part of the surface which is homeomorphic to a k-dimensional sphere and which is not homotopic to a point. In particular, this implies that the central surfaces in n-dimensional space which correspond to different values of k are not homeomorphic to each other. However, the proof of these interesting facts will not be given here.

Next, we consider the case $c = 0$, i.e., the case of *degenerate* central surfaces. In this case, equation (3) becomes homogeneous, i.e., if the point $(\eta_1, \eta_2, \ldots, \eta_n)$ satisfies (3), so does the point $(t\eta_1, t\eta_2, \ldots, t\eta_n)$, for any t. This means that the surface is formed of straight lines going through the origin of coordinates [except when all the terms in the sum (3) have the same sign, in which case (3) defines a single point, i.e., the origin]; such surfaces are said to be *conical*. In the case $r = n$, the canonical equation of a conical surface can be written in the form

$$\frac{\eta_1^2}{a_1^2} + \cdots + \frac{\eta_k^2}{a_k^2} - \frac{\eta_{k+1}^2}{a_{k+1}^2} - \cdots - \frac{\eta_n^2}{a_n^2} = 0.$$

We now find the number of different types of conical surfaces corresponding to a given value of n. If the number of negative coefficients

$m = n - k$ in the canonical equation (3) is greater than $n/2$, then, multiplying the equation by -1, we obtain an equation describing the same surface, but which now has a number of negative coefficients less than $n/2$. Therefore, it is sufficient to consider the cases corresponding to the values $m \leqslant n/2$. If n is even, then, excluding the case of a point $(m = 0)$, we obtain $n/2$ different types of conical surfaces, corresponding to the values $m = 1, 2, \ldots, n/2$. If n is odd, there are $(n - 1)/2$ different types of conical surfaces, i.e., those corresponding to

$$m = 1, 2, \ldots, \frac{n - 1}{2}.$$

In the plane $(n = 2)$, in addition to a point, there is only one other type of conical surface $(m = 1)$, with the canonical equation

$$\frac{\eta_1^2}{a_1^2} - \frac{\eta_2^2}{a_2^2} = 0.$$

The corresponding geometric figure is a pair of intersecting straight lines with the equations

$$\frac{\eta_1}{a_1} = \pm \frac{\eta_2}{a_2}.$$

In three-dimensional space $(n = 3)$, in addition to a point, there is also only one other type of conical surface, corresponding to

$$m = \frac{n - 1}{2} = \frac{3 - 1}{2} = 1,$$

with canonical equation

$$\frac{\eta_1^2}{a_1^2} + \frac{\eta_2^2}{a_2^2} - \frac{\eta_3^2}{a_3^2} = 0.$$

The corresponding geometric object is a cone; in the particular case where $a_1 = a_2$, this is a right circular cone (see Figure 5).

To visualize the form of the conical surface in the general case, we consider its intersection with the hyperplane $\eta_n = Ca_n$ $(-\infty < C < \infty)$, i.e.,

$$\frac{\eta_1^2}{a_1^2} + \cdots + \frac{\eta_k^2}{a_k^2} - \frac{\eta_{k+1}^2}{a_{k+1}^2} - \cdots - \frac{\eta_{n-1}^2}{a_{n-1}^2} = C^2.$$

This is the equation of a *nondegenerate* central surface in an $(n - 1)$-dimensional space. The surfaces corresponding to different values of C are all similar to each other, with semiaxes proportional to the value of C. Thus, *every conical surface in the n-dimensional space R_n can be obtained from a central surface in the $(n - 1)$-dimensional space R_{n-1} by displacing the* *central surface along a perpendicular to R_{n-1}, and at the same time proportionately stretching the surface in all directions.* Moreover, to obtain in this way all possible types of conical surfaces, it is sufficient to use only the central surfaces in R_{n-1} for which the number of negative terms in the canonical equation does not exceed $(n - 1)/2$.

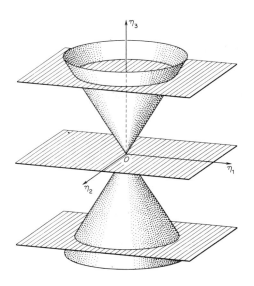

FIGURE 5

79. Nondegenerate Noncentral Surfaces (Paraboloids)

Just as in Sec. 78, we can reduce the canonical equation of a nondegenerate noncentral surface to the form

$$\frac{\eta_1^2}{a_1^2} + \cdots + \frac{\eta_k^2}{a_k^2} - \frac{\eta_{k+1}^2}{a_{k+1}^2} - \cdots - \frac{\eta_{n-1}^2}{a_{n-1}^2} = 2\eta_n. \tag{7}$$

We now find the different types of nondegenerate noncentral surfaces. If the number of negative terms in the left-hand side of (7) is greater than $(n - 1)/2$, then, multiplying (7) by -1, we obtain the equation of the same surface, but with a number of negative terms in the left-hand side which is less than $(n - 1)/2$ and with a change of sign of the right-hand side. The sign of the right-hand side is restored by the mirror reflection $\eta_n' = -\eta_n$.

Thus, if we do not count surfaces obtained from each other by mirror reflections as being of different types, the number of different types of non-degenerate noncentral surfaces is equal to the number of integers satisfying the inequality $0 \leqslant m \leqslant (n - 1)/2$; this number equals $n/2$ if n is even and $(n + 1)/2$ if n is odd.

In the plane ($n = 2$), there is only one nondegenerate noncentral curve, i.e., the parabola with canonical equation

$$\eta_1^2 = 2a_1^2\eta_2 \qquad (m = 0).$$

In three dimensions, there are two nondegenerate noncentral surfaces

$$\left(n = 3, \; \frac{n + 1}{2} = \frac{3 + 1}{2} = 2\right):$$

$$\frac{\eta_1^2}{a_1^2} + \frac{\eta_2^2}{a_2^2} = 2\eta_3 \qquad (m = 0),$$

$$\frac{\eta_1^2}{a_1^2} - \frac{\eta_2^2}{a_2^2} = 2\eta_3 \qquad (m = 1).$$

In the first case ($m = 0$), the sections of the surface made by the plane $\eta_3 = C > 0$ is an ellipse.   To find the position of the vertices of this ellipse, we construct the sections of the surface made by the coordinate planes $\eta_1 = 0$ and $\eta_2 = 0$.   Each of these sections is a parabola, and the intersections of these parabolas with the plane $\eta_3 = C$ locate the vertices of the ellipse.   The resulting surface is called an *elliptic paraboloid*; in the special case where $a_1 = a_2$, the surface is called a *circular paraboloid* (see Figure 6).

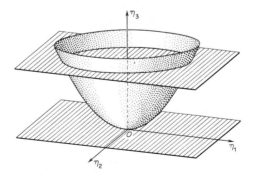

FIGURE 6

In the second case ($m = 1$), the section of the surface made by the plane $\eta_3 = C > 0$ is a hyperbola with real axis η_1.   To find the position of the vertices, we note that the section of the surface made by the coordinate plane $\eta_2 = 0$ is the parabola

$$\eta_1^2 = 2a_1^2\eta_3,$$

whose intersection with the plane $\eta_3 = C$ gives the position of the vertices of the hyperbola.   The section made by the plane $\eta_3 = C < 0$ is a hyperbola with real axis η_2.   The vertices of this hyperbola lie on the parabola

$$\eta_2^2 = -2a_2^2\eta_3$$

in the plane $\eta_1 = 0$.   The section made by $\eta_3 = 0$ is a pair of straight lines, which serve as asymptotes for the projections on the plane $\eta_3 = 0$ of all the hyperbolas lying in horizontal sections of the surface.   The surface itself is called a *hyperbolic paraboloid* (see Figure 7).

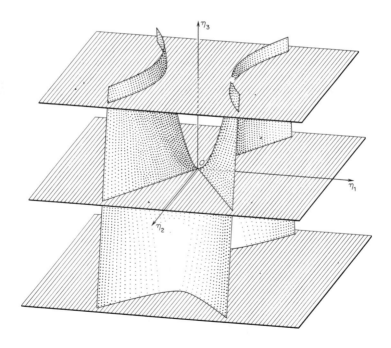

FIGURE 7

To visualize the form of the surface (7) in the general case, we investigate the way the sections made by the hyperplanes $\eta_n = C$ change when C varies from 0 to $+\infty$. Every such section is a central surface in $n - 1$ dimensions. All these surfaces are similar to each other, and their semiaxes (unlike the case of conical surfaces) vary according to a parabolic law, i.e., are proportional to the square root of C. For $C = 0$, the central surface becomes conical. For $C < 0$, the central surface goes into the *conjugate surface*, i.e., the positive and negative coefficients in the canonical equation exchange their roles. In the special case where all the terms of (7) have the same sign, which, to be explicit, we take to be positive, the surface exists only in the half-space $\eta_n \geqslant 0$.

The reason for calling this class of nondegenerate surfaces *noncentral* is that such surfaces *actually have no centers*. To prove this, assume the contrary, i.e., suppose that the surface (7) has a center$(\eta_1^0, \eta_2^0, \ldots, \eta_n^0)$. Since in particular, this center must be a center of symmetry for the section $\eta_n = \eta_n^0$, which represents a nondegenerate central surface in $n - 1$ dimensions, we must have (see Sec. 78)

$$\eta_1^0 = \eta_2^0 = \cdots = \eta_{n-1}^0 = 0.$$

Thus, the center must lie on the η_n axis. Now, if we go from an arbitrary point$(\eta_1, \eta_2, \ldots, \eta_{n-1}, \eta_n^0 + \delta)$ lying on the surface to the symmetric point $(-\eta_1, -\eta_2, \ldots, \eta_n^0 - \delta)$, the equation (7) must still be satisfied. But the left-hand side of (7) remains the same when we make this transition, and hence its right-hand side cannot change, which implies that $\delta = 0$. Thus, we find that there are no points on the surface for which $\eta_n \neq \eta_n^0$. But obviously (7) has solutions $\eta_1, \eta_2, \ldots, \eta_n$ with $\eta_n \neq \eta_n^0$. This contradiction shows that our surface *cannot have a center*.

80. Degenerate Cylindrical Surfaces

By a *degenerate cylindrical surface* we mean a surface whose canonical equation contains less than n coordinates. For example, suppose that the coordinate η_n is absent in the canonical equation. Then, all the sections of the surface made by the $(n - 1)$-dimensional hyperplanes

$$\eta_n = C \qquad (-\infty < C < \infty)$$

give the same surface in $n - 1$ dimensions. Therefore, *every degenerate cylindrical surface in the n-dimensional space R_n is generated by translating a quadric surface in the $(n - 1)$-dimensional space R_{n-1} along a perpendicular to R_{n-1}.*

We now examine degenerate cylindrical surfaces in two and three dimensions. In the case of a plane ($n = 2$), the canonical equation contains only one coordinate and is just

$$\frac{\eta_1^2}{a_1^2} = C.$$

For $C > 0$, we obtain a pair of parallel lines, for $C = 0$ a pair of coincident lines, and for $C < 0$ an "imaginary curve." To construct cylindrical surfaces in three-dimensional space ($n = 3$), we must translate all the second-degree curves in the (η_1, η_2) plane along the η_3 axis. When this is done, ellipses, hyperbolas and parabolas give elliptic, hyperbolic and parabolic cylinders, respectively (see Figure 8), while pairs of intersecting, parallel and coincident lines lead to intersecting, parallel and coincident planes (see Figure 9).

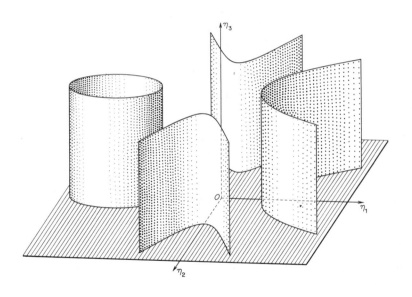

FIGURE 8

Problem 1. What quadric surfaces in three-dimensional space (with coordinates x, y, z) do the following equations represent?

a) $\dfrac{x^2}{4} - \dfrac{y^2}{9} + \dfrac{z^2}{1} = 1$;   b) $\dfrac{x^2}{4} - \dfrac{y^2}{9} - \dfrac{z^2}{1} = -1$;   c) $x = y^2 + z^2$;

d) $y = x^2 + z^2 + 1$;   e) $y = xz$.

Ans. a) A hyperboloid of one sheet with its axis along the y-axis;

b) a hyperboloid of one sheet with its axis along the x-axis; c) a circular paraboloid with its axis along the x-axis; d) a circular paraboloid with its axis along the y-axis, displaced one unit along this axis; e) a hyperbolic paraboloid.

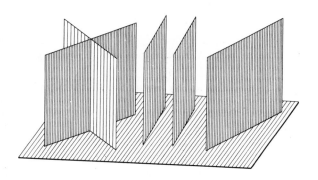

FIGURE 9

Problem 2. Prove that the midpoints of the chords of a quadric surface parallel to the vector $y = (\eta_1, \eta_2, \ldots, \eta_n)$ lie on an $(n-1)$-dimensional hyperplane (the diametral plane conjugate to the vector y).

Problem 3. Simplify the following equations of quadric surfaces in three-dimensional space, and give the corresponding transformations of coordinates:

a) $5x^2 + 6y^2 + 7z^2 - 4xy + 4yz - 10x + 8y + 14z - 6 = 0$,
b) $x^2 + 2y^2 - z^2 + 12xy - 4xz - 8yz + 14x + 16y - 12z - 3 = 0$,
c) $4x^2 + y^2 + 4z^2 - 4xy + 8xz - 4yz - 12x - 12y + 6z = 0$.

Ans. a) $x_1^2 + 2y_1^2 + 3z_1^2 = 6$; $3(x - 1) = -x_1 + 2y_1 + 2z_1,$
$$3y = 2x_1 - y_1 + 2z_1,$$
$$3(z + 1) = 2x_1 + 2y_1 - z_1;$$

b) $x_1^2 + 2y_1^2 - 3z_1^2 = 6$; $3(x + 1) = -x_1 + 2y_1 + 2z_1,$
$$3(y + 1) = 2x_1 - y_1 + 2z_1,$$
$$3z = 2x_1 + 2y_1 - z_1;$$

c) $y_1^2 = 2x_1$; $3(x - m) = 2x_1 + 2y_1 + z_1,$
$$3(y + 2m) = 2x_1 - y_1 - 2z_1,$$
$$3(z + 2m) = -x_1 + 2y_1 - 2z_1,$$
$$\text{where } m \text{ is arbitrary.}$$

Problem 4. Show that the intersection of an ellipsoid with semiaxes $a_1 \geqslant a_2 \geqslant \cdots \geqslant a_n$ with a k-dimensional hyperplane going through the center of the ellipsoid is another ellipsoid with semiaxes $b_1 \geqslant b_2 \geqslant \cdots \geqslant b_k$, where

$$a_1 \geqslant b_1 \geqslant a_{n-k+1},$$
$$a_2 \geqslant b_2 \geqslant a_{n-k+2},$$
$$\cdots \cdots$$
$$a_k \geqslant b_k \geqslant a_n.$$

Hint. The semiaxes of the ellipsoid are determined from the canonical coefficients of the corresponding quadratic form. Use the results of Sec. 71.

*81. Analysis of a Quadric Surface from its General Equation

We have just described all possible types of quadric surfaces in an n-dimensional Euclidean space; the type of the surface was determined from its canonical equation. However, the surface is often specified by its general equation (1) rather than by its canonical equation, and it is sometimes important to determine the type of the surface, i.e., to construct its canonical equation, without carrying out all the transformations described in Sec. 77. It turns out that to write down the canonical equation of the surface specified by equation (1), it is sufficient to know the following two quantities:

1. The roots of the nth degree polynomial

$$\Delta(\lambda) = \begin{vmatrix} a_{11} - \lambda & a_{12} & \cdots & a_{1n} \\ a_{21} & a_{22} - \lambda & \cdots & a_{2n} \\ \cdot & \cdot & \cdots & \cdot \\ a_{n1} & a_{n2} & \cdots & a_{nn} - \lambda \end{vmatrix};$$

2. The coefficients of the nth degree polynomial

$$\Delta_1(\lambda) = \begin{vmatrix} a_{11} - \lambda & a_{12} & \cdots & a_{1n} & b_1 \\ a_{21} & a_{22} - \lambda & \cdots & a_{2n} & b_2 \\ \cdot & \cdot & \cdots & \cdot & \cdot \\ a_{n1} & a_{n2} & \cdots & a_{nn} - \lambda & b_n \\ b_1 & b_2 & \cdots & b_n & c \end{vmatrix}.$$

To obtain explicit expressions for the coefficients of $\Delta_1(\lambda)$, we use the linear property of determinants (Sec. 3.3). Every column of the determinant $\Delta_1(\lambda)$, except the last one, can be written as a sum of two columns, the first consisting of the numbers a_{ij} ($i = 1, 2, \ldots, n$; j fixed) and the number b_j, and the second consisting of n zeros and the number $-\lambda$. As a result, the determinant $\Delta_1(\lambda)$ can be written as a sum of determinants, each of which is obtained by replacing certain columns (except the last one) in the matrix

$$A_1 = \begin{vmatrix} a_{11} & a_{12} & \ldots & a_{1n} & b_1 \\ a_{21} & a_{22} & \ldots & a_{2n} & b_2 \\ \cdot & \cdot & \ldots & \cdot & \cdot \\ a_{n1} & a_{n2} & \ldots & a_{nn} & b_n \\ b_1 & b_2 & \ldots & b_n & c \end{vmatrix} \tag{8}$$

by columns consisting of n zeros and the single element $-\lambda$, with the number $-\lambda$ appearing on the principal diagonal of the matrix. After expansion with respect to the columns containing the number $-\lambda$, each of these determinants becomes

$$(-\lambda)^k M_{n+1-k},$$

where k is the number of columns containing the element $-\lambda$, and M_{n+1-k} is a minor of order $n + 1 - k$ of the matrix A_1. This minor is characterized by the fact that if it uses the ith row ($i = 1, 2, \ldots, n$) of A_1, it also uses the ith column, and moreover, it has to use the last row and column of A_1. Minors with this property will be called *bordered minors*. It is obvious that every bordered minor of the matrix A_1 appears in the expansion of the determinant $\Delta_1(\lambda)$. From this, we immediately conclude that *the coefficient of $(-\lambda)^k$ in the expansion of the determinant $\Delta_1(\lambda)$ in powers of $-\lambda$ equals the sum of all the bordered minors of order $n + 1 - k$*. It is convenient to write the expansion of $\Delta_1(\lambda)$ in the form

$$\Delta_1(\lambda) = \alpha_{n+1} - \alpha_n \lambda + \alpha_{n-1} \lambda^2 - \cdots + \alpha_1 (-\lambda)^n,$$

where the coefficient α_k is the sum of all the kth order bordered minors of the matrix A_1.

As we already know, the roots of the characteristic polynomial $\Delta(\lambda)$ give us the coefficients of the squared variables in the canonical equation. To find the remaining term, which is of degree 0 if the canonical equation has the form (3) and of degree 1 if it has the form (5), we have to examine the behavior of the polynomial $\Delta_1(\lambda)$ under coordinate transformations. Thus, consider the quadratic form

$$A_1(x, x) = \sum_{i, k=1}^{n} a_{ik} \xi_i \xi_k + 2 \sum_{i=1}^{n} b_i \xi_i \xi_{n+1} + c \xi_{n+1}^2 \tag{9}$$

in the $(n + 1)$-dimensional Euclidean space R_{n+1}; here $\xi_1, \xi_2, \ldots, \xi_n, \xi_{n+1}$ are the components of the vector $x \in R_{n+1}$ with respect to some orthonormal basis $e_1, e_2, \ldots, e_n, e_{n+1}$. The operator corresponding to this quadratic form is the symmetric operator A_1 which has the matrix (8) relative to the basis $e_1, e_2, \ldots, e_n, e_{n+1}$ in the space R_{n+1}; we shall also denote this operator by $A_{(e)}$. In addition to this operator, consider the operator E_1

defined by the relations $E_1 e_k = e_k$ for $k \leqslant n$ and $E_1 e_{n+1} = 0$. This operator has the following matrix relative to the basis $e_1, e_2, \ldots, e_n, e_{n+1}$

$$E_1 = \begin{Vmatrix} 1 & & & & & & \\ & 1 & & & & & \\ & & 1 & & & & \\ & & & \cdot & & & \\ & & & & \cdot & & \\ & & & & & 1 & \\ & & & & & & 0 \end{Vmatrix}, \tag{10}$$

where all unwritten elements are zero. We denote by R_n the subspace with a basis consisting of the vectors $e_1, e_2, \ldots, e_n$; obviously, the operator E_1 is the identity operator in this subspace.

Now, suppose that we are given an isometric transformation Q in the space R_n. Then Q carries the orthonormal basis $e_1, e_2, \ldots, e_n$ into another orthonormal basis $f_1, f_2, \ldots, f_n$. We construct an isometric transformation Q_1 in the space R_{n+1} by setting

$$Q_1 e_1 = f_1, \quad Q_1 e_2 = f_2, \ldots, \quad Q_1 e_n = f_n, \quad Q_1 e_{n+1} = e_{n+1} = f_{n+1}.$$

If the matrix of the operator Q has the form

$$\begin{Vmatrix} q_{11} & q_{12} & \cdots & q_{1n} \\ q_{21} & q_{22} & \cdots & q_{2n} \\ \cdot & \cdot & \cdots & \\ q_{n1} & q_{n2} & \cdots & q_{nn} \end{Vmatrix}$$

in the space R_n, then the matrix of the operator Q_1 just constructed has the form

$$\begin{Vmatrix} q_{11} & q_{12} & \cdots & q_{1n} & 0 \\ q_{21} & q_{22} & \cdots & q_{2n} & 0 \\ \cdot & \cdot & \cdots & \cdot & \cdot \\ q_{n1} & q_{n2} & \cdots & q_{nn} & 0 \\ 0 & 0 & \cdots & 0 & 1 \end{Vmatrix}.$$

This matrix corresponds to the following coordinate transformation (Sec. 54):

$$\begin{aligned} \xi_1 &= q_{11}\eta_1 + q_{21}\eta_2 + \cdots + q_{n1}\eta_n, \\ \xi_2 &= q_{12}\eta_1 + q_{22}\eta_2 + \cdots + q_{n2}\eta_n, \\ \cdot \quad & \cdots \qquad \cdots \qquad \cdots \qquad \cdots \\ \xi_n &= q_{1n}\eta_1 + q_{2n}\eta_2 + \cdots + q_{nn}\eta_n, \\ \xi_{n+1} &= \eta_{n+1}. \end{aligned} \tag{11}$$

In the new basis $f_1, f_2, \ldots, f_n, f_{n+1}$, the operator A has the matrix

$$A_{(f)} = Q^{-1} A_{(e)} Q$$

(Sec. 38), while the operator E_1 has the same matrix (10) as before, and according to Sec. 38.2, we have the relation

$$\det(A_{(f)} - \lambda E_1) = \det(A_{(e)} - \lambda E_1).$$

We now assume that the transformation Q is the transformation (cf. Sec. 77) which reduces the quadratic form

$$A(x, x) = \sum_{i,k=1}^{n} a_{ik} \xi_i \xi_k$$

to the canonical form

$$A(x, x) = \sum_{i=1}^{n} \lambda_i \eta_i^2.$$

It follows from (11) that Q_1 transforms the quadratic form $A_1(x, x)$ in $n + 1$ variables, given by (9), into

$$\sum_{i=1}^{n} \lambda_i \eta_i^2 + 2 \sum_{i=1}^{n} l_i \eta_i \eta_{n+1} + c \eta_{n+1}^2.$$

After this transformation, the matrix of the operator A_1, which, as we know, transforms in the same way as the matrix of the quadratic form (Sec. 55.1), becomes

$$A_{(f)} = \begin{Vmatrix} \lambda_1 & & & & & & & & & & l_1 \\ & \lambda_2 & & & & & & & & & l_2 \\ & & \cdot & & & & & & & & \cdot \\ & & & \cdot & & & & & & & \cdot \\ & & & & \cdot & & & & & & \cdot \\ & & & & & \lambda_r & & & & & l_r \\ & & & & & & 0 & & & & l_{r+1} \\ & & & & & & & \cdot & & & \cdot \\ & & & & & & & & \cdot & & \cdot \\ & & & & & & & & & 0 & l_n \\ l_1 & l_2 & \ldots & l_r & l_{r+1} & \ldots & & & & l_n & c \end{Vmatrix},$$

and the polynomial $\Delta_1(\lambda) = \det(A_{(f)} - \lambda E_1)$ equals the determinant

$$\begin{vmatrix} \lambda_1 - \lambda & & & & & & & & l_1 \\ & \lambda_2 - \lambda & & & & & & & l_2 \\ & & \cdot & & & & & & \cdot \\ & & & \cdot & & & & & \cdot \\ & & & & \cdot & & & & \cdot \\ & & & & & \lambda_r - \lambda & & & l_r \\ & & & & & & -\lambda & & l_{r+1} \\ & & & & & & & \cdot & \cdot \\ & & & & & & & \cdot & \cdot \\ & & & & & & & -\lambda & l_n \\ l_1 & l_2 & \ldots & l_r & l_{r+1} & \ldots & & l_n & c \end{vmatrix}.$$

The coefficients of this polynomial can be calculated by using the bordered minors of the matrix $A_{(f)}$, just as they were calculated before by using the bordered minors of the matrix $A_{(e)} = A_1$.

We note that for $r < n$, all the bordered minors of the matrix $A_{(f)}$ which are of order higher than $r + 2$ must vanish, since they contain two proportional columns. Thus, for $r < n$, the coefficients $\alpha_{r+3}, \alpha_{r+4}, \ldots, \alpha_{n+1}$ vanish. Moreover, for $r < n$, the nonvanishing minors of order $r + 2$ must use the first r rows and columns of the matrix $A_{(f)}$. In general, the bordered minors of order $r + 1$ need not use these r rows and columns. However, we note the following two cases where a bordered minor of order $r + 1$ must in fact use the first r rows and columns:

1) $r = n$; in this case, it is obvious that the matrix $A_{(f)}$ has only one minor of order $r + 1$ (i.e., of order $n + 1$), namely, its determinant; in other words, this minor contains *all* the rows and columns of $A_{(f)}$.

2) $r < n, l_{r+1} = l_{r+2} = \cdots = l_n = 0$; in this case, there is only one non-vanishing bordered minor of order $r + 1$. It is formed from the rows and columns with numbers $1, 2, \ldots, r, n + 1$.

We now show how the next step in the transformation of equation (1) of Sec. 77, i.e., the elimination of the quantities $l_1, l_2, \ldots, l_r$, affects the matrix of the operator A_1. First consider the transformation

$$\eta_1' = \eta_1 + \frac{l_1}{\lambda_1}\eta_{n+1}, \quad \eta_k' = \eta_k \qquad (k = 2, 3, \ldots, n + 1).$$

As a result of this transformation, the matrix $A_{(f)}$ goes into the matrix

$$A_{(f)}^{(1)} = \left\|\begin{array}{ccccccccc} \lambda_1 & & & & & & & & 0 \\ & \lambda_2 & & & & & & & l_2 \\ & & \cdot & & & & & & \cdot \\ & & & \cdot & & & & & \cdot \\ & & & & \cdot & & & & \cdot \\ & & & & & \lambda_r & & & l_r \\ & & & & & & 0 & & l_{r+1} \\ & & & & & & & \cdot & \cdot \\ & & & & & & & \cdot & \cdot \\ & & & & & & & 0 & l_n \\ 0 & l_2 & \ldots & l_r & l_{r+1} & \ldots & l_n & & c - \dfrac{l_1^2}{\lambda_1} \end{array}\right\|.$$

This operation on $A_{(f)}$ can be described as follows: The first column is multiplied by l_1/λ_1 and subtracted from the last column, and then the first row is also multiplied by l_1/λ_1 and subtracted from the last row. The subsequent transformations required to eliminate the quantities $l_2, l_3, \ldots, l_r$ can be described similarly. As a result of all these transformations, the matrix $A_{(f)}$ goes into the matrix

$$
A_{(f)}^{(r)} = \left\|
\begin{array}{cccccccc}
\lambda_1 & & & & & & & 0 \\
& \lambda_2 & & & & & & 0 \\
& & \cdot & & & & & \cdot \\
& & & \cdot & & & & \cdot \\
& & & & \cdot & & & \cdot \\
& & & & & \lambda_r & & 0 \\
& & & & & 0 & & l_{r+1} \\
& & & & & & & \cdot \\
& & & & & & & \cdot \\
& & & & & & 0 & l_n \\
0 & 0 & \ldots & 0 & l_{r+1} & \ldots & l_n & c'
\end{array}
\right\|.
$$

As a result of these transformations, the bordered minors of the matrix $A_{(f)}$ which use the first r rows and columns of $A_{(f)}$ do not change their values.

Next, we consider the polynomial

$$\det(A_{(f)}^{(r)} - \lambda E_1) \equiv \Delta_1^{(r)}(\lambda)$$

$$
= \left|
\begin{array}{cccccccc}
\lambda_1 - \lambda & & & & & & & 0 \\
& \lambda_2 - \lambda & & & & & & 0 \\
& & \cdot & & & & & \cdot \\
& & & \cdot & & & & \cdot \\
& & & & \lambda_r - \lambda & & & 0 \\
& & & & & -\lambda & & l_{r+1} \\
& & & & & & \cdot & \cdot \\
& & & & & & \cdot & \cdot \\
& & & & & & -\lambda & l_n \\
0 & 0 & \ldots & 0 & l_{r+1} & \ldots & l_n & c
\end{array}
\right|
$$

$$= \alpha'_{n+1} - \alpha'_n \lambda + \alpha'_{n-1} \lambda^2 - \cdots + \alpha'_1 (-\lambda)^n,$$

where we have dropped the prime on c'. The coefficients of this polynomial are calculated by using the bordered minors of the matrix $A_{(f)}^{(r)}$ in just the same way as the coefficients of the polynomial $\Delta_1(\lambda)$ are calculated by using the bordered minors of the matrix $A_{(f)}$. In view of the property proved above, to the effect that the bordered minors of order $r + 2$ (where $r < n$) are invariant under the transformation leading from $A_{(f)}$ to $A_{(f)}^{(r)}$, we find that $\alpha'_{r+2} = \alpha_{r+2}$; moreover, in the two special cases noted above, we have $\alpha'_{r+1} = \alpha_{r+1}$. When $r = n$, the coefficient α'_{n+1} of the polynomial $\Delta_1^{(r)}(\lambda)$ is obviously equal to the product $\lambda_1\lambda_2\ldots\lambda_n c$, so that the quantity c in the canonical equation (3) is just

$$c = \frac{\alpha'_{n+1}}{\lambda_1\lambda_2\ldots\lambda_n} = \frac{\alpha_{n+1}}{\lambda_1\lambda_2\ldots\lambda_n}.$$

Suppose now that $r < n$. Then, we must determine the coefficient α_{r+2} of the polynomial $\Delta_1^{(r)}(\lambda)$; this coefficient will be needed below. (It is easily verified that in this case, all the coefficients α_m of the polynomial $\Delta_1^{(r)}(\lambda)$ with $m > r + 2$ vanish.) The nonvanishing bordered minors of $A_{(f)}^{(r)}$ of order $r + 2$ have the form

$$\begin{vmatrix} \lambda_1 & & & & & 0 \\ & \lambda_2 & & & & 0 \\ & & \cdot & & & \cdot \\ & & & \cdot & & \cdot \\ & & & & \cdot & \cdot \\ & & & \lambda_r & & 0 \\ & & & & 0 & l_m \\ 0 & 0 & \ldots & 0 & l_m & c \end{vmatrix} = -\lambda_1\lambda_2\ldots\lambda_r l_m^2 \quad (m = r+1, \ldots, n),$$

and their sum, which equals the coefficient $\alpha'_{r+2} = \alpha_{r+2}$, is given by

$$-\lambda_1\lambda_2\ldots\lambda_r(l_{r+1}^2 + l_{r+2}^2 + \cdots + l_n^2).$$

We recall that the condition for reducing equation (1) to the canonical form (5) is that at least one of the coefficients $l_{r+1}, l_{r+2}, \ldots, l_n$ be nonvanishing. We can now formulate this condition equivalently in the form of the inequality

$$\alpha_{r+2} \neq 0,$$

and we have the following formula for calculating the coefficient M of the canonical form (5):

$$M^2 = l_{r+1}^2 + l_{r+2}^2 + \cdots + l_n^2 = -\frac{\alpha_{r+2}}{\lambda_1\lambda_2\ldots\lambda_r}.$$

However, if $\alpha_{r+2} = 0$, then $l_{r+1} = l_{r+2} = \cdots = l_n = 0$, and (1) reduces to the canonical form (3). Thus, we have arrived at another special case In this case, the coefficient $\alpha'_{r+1} = \alpha_{r+1}$ is obviously equal to the product $\lambda_1\lambda_2\ldots\lambda_r c$, so that the coefficient c of the canonical form (3) is given by

$$\frac{\alpha'_{r+1}}{\lambda_1\lambda_2\ldots\lambda_r} = \frac{\alpha_{r+1}}{\lambda_1\lambda_2\ldots\lambda_r}.$$

We now summarize these results in the form of a table. As before, we agree to arrange the roots $\lambda_1, \lambda_2, \ldots, \lambda_n$ of the characteristic polynomial

$\Delta(\lambda)$ in such a way that the nonzero roots come first. We denote the product $\lambda_1\lambda_2 \ldots \lambda_r$ by Λ_r.

Data	Canonical Equation
$\lambda_n \neq 0$	$\lambda_1\eta_1^2 + \lambda_2\eta_2^2 + \cdots + \lambda_n\eta_n^2 + \dfrac{\alpha_{n+1}}{\Lambda_n} = 0$
$\begin{array}{l}\lambda_n = 0 \\ \lambda_{n-1} \neq 0\end{array}\Big\}\quad \begin{array}{l}\alpha_{n+1} \neq 0 \\ \alpha_{n+1} = 0\end{array}$	$\lambda_1\eta_1^2 + \lambda_2\eta_2^2 + \cdots + \lambda_{n-1}\eta_{n-1}^2 + 2\sqrt{-\dfrac{\alpha_{n+1}}{\Lambda_{n-1}}}\,\eta_n = 0$ $\lambda_1\eta_1^2 + \lambda_2\eta_2^2 + \cdots + \lambda_{n-1}\eta_{n-1}^2 + \dfrac{\alpha_n}{\Lambda_{n-1}} = 0$
$\begin{array}{l}\lambda_{n-1} = 0 \\ \lambda_{n-2} \neq 0\end{array}\Big\}\quad \begin{array}{l}\alpha_n \neq 0 \\ \alpha_n = 0\end{array}$	$\lambda_1\eta_1^2 + \lambda_2\eta_2^2 + \cdots + \lambda_{n-2}\eta_{n-2}^2 + 2\sqrt{-\dfrac{\alpha_n}{\Lambda_{n-2}}}\,\eta_{n-1} = 0$ $\lambda_1\eta_1^2 + \lambda_2\eta_2^2 + \cdots + \lambda_{n-2}\eta_{n-2}^2 + \dfrac{\alpha_{n-1}}{\Lambda_{n-2}} = 0$
$\cdots\cdots$ $\cdots\cdots$	$\cdots\cdots\cdots\cdots$ $\cdots\cdots\cdots\cdots$
$\begin{array}{l}\lambda_2 = 0 \\ \lambda_1 \neq 0\end{array}\Big\}\quad \begin{array}{l}\alpha_3 \neq 0 \\ \alpha_3 = 0\end{array}$	$\lambda_1\eta_1^2 + 2\sqrt{-\dfrac{\alpha_3}{\lambda_1}}\,\eta_2 = 0$ $\lambda_1\eta_1^2 + \dfrac{\alpha_2}{\lambda_1} = 0$

The scheme given in this section for analyzing the general equation of a quadric surface is due to P. S. Modenov.

12

INFINITE-DIMENSIONAL EUCLIDEAN SPACES

We have already encountered infinite-dimensional Euclidean spaces earlier in the book; an example of such a space is the set $C_2(a, b)$ of continuous functions on the interval $a \leqslant t \leqslant b$, with the scalar product defined by the formula

$$(x, y) = \int_a^b x(t)y(t) \, dt.$$

The study of the geometry of infinite-dimensional Euclidean spaces allows us to give a geometric treatment of many facts of analysis; such a geometric treatment, by fostering new points of view, subsequently leads to the anticipation of new facts and the discovery of new properties of analytical objects. Thus, in this chapter, we shall use the geometry of function space to construct a theory of Fourier series and a theory of integral equations.

82. Limits in a Euclidean Space

Henceforth, we shall denote the norm of a vector x in a Euclidean space R by $\|x\|$. By the *distance* $\rho(x, y)$ between two vectors x and y, we mean the norm of their difference, i.e.,

$$\rho(x, y) = \|x - y\|.$$

The quantity $\rho(x, y)$ obeys the triangle inequality

$$\rho(x, z) \leqslant \rho(x, y) + \rho(y, z),$$

which is obtained from the inequality (12) of Sec. 50 by replacing x by $x - y$ and y by $y - z$.

82.1. Definition of convergence. We say that the sequence of points $x_1, x_2, \ldots, x_m, \ldots$ of a Euclidean space R *converges to the point x of R if*[1]

$$\lim_{m \to \infty} \rho(x, x_m) = 0.$$

The point x is called the *limit* of the sequence $x_1, x_2, \ldots, x_m, \ldots$. It is easy to see that the element x is uniquely defined. In fact, if we have two relations

$$\lim_{m \to \infty} \rho(x, x_m) = 0, \qquad \lim_{m \to \infty} \rho(y, x_m) = 0,$$

then, given any $\varepsilon > 0$, we can find a number N such that the inequalities

$$\rho(x, x_m) < \frac{\varepsilon}{2}, \qquad \rho(y, x_m) < \frac{\varepsilon}{2}$$

hold if $m \geqslant N$. Therefore, by the triangle inequality, we have

$$\rho(x, y) \leqslant \rho(x, x_m) + \rho(x_m, y) < \frac{\varepsilon}{2} + \frac{\varepsilon}{2} = \varepsilon;$$

since ε is arbitrarily small, we have $\rho(x, y) = 0$, so that $x = y$, Q.E.D.

Example 1. Let R_n be an *n*-dimensional Euclidean space, and let $e_1, e_2, \ldots, e_n$ be an orthonormal basis in R_n. If we write

$$x = \xi_1 e_1 + \xi_2 e_2 + \cdots + \xi_n e_n,$$
$$x_m = \xi_1^{(m)} e_1 + \xi_2^{(m)} e_2 + \cdots + \xi_n^{(m)} e_n \qquad (m = 1, 2, \ldots),$$

then

$$x - x_m = (\xi_1 - \xi_1^{(m)})e_1 + \cdots + (\xi_n - \xi_n^{(m)})e_n.$$

By equation (17) of Sec. 50, we have

$$\|x - x_m\|^2 = (\xi_1 - \xi_1^{(m)})^2 + \cdots + (\xi_n - \xi_n^{(m)})^2.$$

It is obvious that the distance $\|x - x_m\|$ from the point x to the point x_m converges to zero as $m \to \infty$ if and only if *all the numerical sequences*

$$\xi_1^{(m)}, \xi_2^{(m)}, \ldots, \xi_n^{(m)} \qquad (m = 1, 2, \ldots)$$

converge to the limits $\xi_1, \xi_2, \ldots, \xi_n$, *respectively*, as $m \to \infty$. In short, *convergence of a vector x in R_n is equivalent to convergence of all the components of x.*

Example 2. Convergence of the sequence of functions $x_m(t) \in C_2(a, b)$ to the function $x(t)$ means that

$$\rho^2(x, x_m) = \|x - x_m\|^2 = \int_a^b |x(t) - x_m(t)|^2 \, dt \to 0 \tag{1}$$

---

[1] The elements of R will be called either *vectors* or *points*, synonymously.

as $m \to \infty$. In analysis, this kind of convergence is called *convergence in the mean*. Other kinds of convergence of a sequence of functions are also studied in analysis. For example, the sequence $x_m(t)$ is said to converge to the limit $x(t)$ *uniformly* if the quantity

$$\max_{a \leqslant t \leqslant b} |x(t) - x_m(t)| \to 0$$

as $m \to \infty$. It follows from an estimate of the integral (1) that every uniformly convergent sequence of functions also converges in the mean. However, it is easy to construct a sequence of functions which converges in the mean but does not converge uniformly. For example, let the function $x_m(t)$ be a function which takes values between 0 and 1, and let $x_m(t)$ be different from zero only on an interval Δ_m (contained in $a \leqslant t \leqslant b$) of length less than $1/m$; moreover, let $x_m(t)$ take the value 1 somewhere in Δ_m. Then, it is obvious that

$$\int_a^b x_m^2(t)\, dt < \frac{1}{m},$$

so that the sequence $x_m(t)$ *converges in the mean to zero* as $m \to \infty$. However,

$$\max_{a \leqslant t \leqslant b} x_m(t) = 1$$

for any m, so that the sequence *does not converge uniformly to zero*. It can also be verified that this sequence does not converge uniformly to any function. In fact, the intervals Δ_m can be chosen in such a way that the sequence does not even converge for any value of t.

LEMMA 1. *In a Euclidean space R, the scalar product (x, y) is a continuous function of both variables x and y, i.e., if $x_n \to x$, $y_n \to y$, then $(x_n, y_n) \to (x, y)$.*

Proof. Write $x - x_n = h_n$, $y - y_n = k_n$; by hypothesis, $\|h_n\| \to 0$, $\|k_n\| \to 0$. By the Schwarz inequality (Sec. 50.2), we have

$$|(x, y) - (x_n, y_n)| = |(x, y) - (x - h_n, y - k_n)|$$
$$= |(x, k_n) + (y, h_n) - (h_n, k_n)| \leqslant \|x\|\,\|k_n\| + \|y\|\,\|h_n\| + \|h_n\|\,\|k_n\|;$$

as n increases, this quantity approaches zero. It follows that

$$(x, y) = \lim_{n \to \infty} (x_n, y_n), \quad \text{Q.E.D.}$$

A consequence of Lemma 1 is the fact that the norm of a vector x, i.e., the quantity $+\sqrt{(x, x)}$, is a continuous function of x.

82.2. Limit points and closed sets. A point x of the Euclidean space R is called a *limit point* of the set $F \subset R$, if there exists a sequence $x_1, x_2, \ldots$ of points of F which converges to the point x. Another definition of a limit point, which is obviously equivalent to the one just given, is the following: A point x is a limit point of the set F if for any $\varepsilon > 0$, there exists a point $y \in F$ for which $\rho(x, y) < \varepsilon$, where $y \neq x$. A subset $F \subset R$ is said to be *closed* if it contains all its limit points.

Example. If R'' denotes the orthogonal complement of a subspace R' of a Euclidean space R (Sec. 50.3), then R'' is closed for any R' whatsoever. To see this, let x be a limit point of the subspace R'', and form a sequence $x_1, x_2, \ldots, x_m, \ldots$ of points of R'' which converges to the point x. If x' is any point of the subspace R', then $(x', x_m) = 0$ for any m. Using this fact and the Schwarz inequality, we have

$$|(x', x)| = |(x', x) - (x', x_m)| = |(x', x - x_m)| \leqslant \|x'\| \, \|x - x_m\| \to 0,$$

i.e., $(x', x) = 0$. Thus, the element x is orthogonal to any $x' \in R'$, and hence, by the very definition of an orthogonal complement, $x \in R''$, Q.E.D.

82.3. Closure of a set. Let A be an arbitrary set of elements of a Euclidean space, and let $\overline{A}$ denote the set consisting of all points of the set A and also all limit points of the set A (which are not elements of A). This set $\overline{A}$ is called the *closure* of the set A.

We now show that *the closure of an arbitrary set A is always a closed set.* Let a be any limit point of the set $\overline{A}$; this means that for any preassigned $\varepsilon > 0$, we can find a point $\bar{x} \in \overline{A}$ for which $\rho(a, \bar{x}) < \varepsilon/2$. Since $\bar{x} \in \overline{A}$, the point $\bar{x}$ is either a point of the set A or a limit point of the set A. In both cases, we can find a point $x \in A$ for which $\rho(x, \bar{x}) < \varepsilon/2$; in particular, in the first case we can set $x = \bar{x}$. Using the triangle inequality, we find that

$$\rho(a, x) \leqslant \rho(a, \bar{x}) + \rho(x, \bar{x}) < \frac{\varepsilon}{2} + \frac{\varepsilon}{2} = \varepsilon.$$

Therefore, for any $\varepsilon > 0$, we can find a point $x \in A$ for which $\rho(a, x) < \varepsilon$. By giving ε the values $1, 1/2, 1/3, \ldots, 1/n, \ldots$ in succession, we obtain a sequence of points $x_1, x_2, x_3, \ldots, x_n, \ldots$ in A which converges to the point a. Thus, the point a is a limit point of the set A, and hence is an element of $\overline{A}$. It follows that *every limit point of the set $\overline{A}$ is itself an element of $\overline{A}$*, i.e., the set $\overline{A}$ is closed, Q.E.D. Moreover, we note that every closed set F containing the set A must also contain all the limit points of A, and hence must contain the whole set $\overline{A}$. Since, as we have just proved, the set $\overline{A}$ is closed, it can now be characterized as *the smallest closed set containing the set A.*

Example 1. The closure of the set A of all rational numbers on the real line is the set of all points of the line, rational and irrational.

Example 2. Every finite-dimensional subspace $F \subset R$ is closed. In fact, if f is not in F, then there exists an expansion $f = g + h$, where $g \in F$ and h is the perpendicular dropped from the end of the vector f onto the subspace F (Sec. 56). For any $q \in F$, we have $\|f - q\| \geqslant \|h\| > 0$; therefore, f cannot be a limit point of the set F.

Example 3. Let R be a Euclidean space, and let A be a subspace of R. Then the set $\bar{A}$ is also a subspace. To prove this, we must verify the following facts:

a) If $x \in \bar{A}$, $y \in \bar{A}$, then $x + y \in \bar{A}$.

b) If $x \in \bar{A}$, and λ is a real number, then $\lambda x \in \bar{A}$.

To prove a), consider sequences $x_n \in A$ and $y_n \in A$ which converge to the limits x and y, respectively. Then

$$\|(x + y) - (x_n + y_n)\| = \|(x - x_n) + (y - y_n)\| \leqslant \|x - x_n\| + \|y - y_n\| \to 0,$$

and therefore, the sequence $x_n + y_n$ converges to $x + y$. Since $x_n + y_n \in A$, then $x + y \in \bar{A}$, as required. To prove b), consider a sequence $x_n \in A$ which converges to the element x. Then

$$\|\lambda x - \lambda x_n\| = \|\lambda(x - x_n)\| = |\lambda| \, \|x - x_n\| \to 0,$$

and therefore, the sequence λx_n converges to the element λx; since $\lambda x_n \in A$, then $\lambda x \in \bar{A}$, as asserted.

82.4. Dense sets. A set B of elements of a Euclidean space is said to be *dense* in the set $A \subset R$, if the closure of the set B contains the set A. Thus, every element $f \in A$ can be represented as the limit of a sequence of elements f_n $(n = 1, 2, \ldots)$, where the f_n are elements of the set B.

Example 1. The set of rational points on the line R_1 forms a dense set in the line R_1.

Example 2. The set of polynomials $P(t)$ in the space $C_2(a, b)$ is a dense set in this space. In fact, by Weierstrass' theorem,[2] every continuous function $f(t)$ on the interval $a \leqslant t \leqslant b$ is the limit as $n \to \infty$ of a sequence of polynomials $P_n(t)$, which converge to $f(t)$ both uniformly and in the metric of the space $C_2(a, b)$.

LEMMA 2. *If the set $B \subset R$ is dense in the set $A \subset R$, and if the set A is dense in R, then the set B is also dense in R.*

Proof. By hypothesis, the closure of the set B contains the set A. But the smallest closed set containing A is the closure of the set A, and by hypothesis, this last set is just the whole space R. Therefore B is dense in R, Q.E.D.

---

[2] See, e.g.: L. Brand, *Advanced Calculus*, John Wiley & Sons, Inc., New York (1955), p. 529.

Example. The set B of polynomials in t with rational coefficients is obviously dense in the set A of all polynomials in t $(a \leqslant t \leqslant b)$. The set A is dense in the space $C_2(a, b)$; see Example 2 above. Therefore, the set of polynomials with rational coefficients is a dense (and countable!) set in $C_2(a, b)$.

83. Complete Spaces

83.1. Fundamental sequences. A sequence of points $x_1, x_2, \ldots, x_m, \ldots$ of points of a Euclidean space R is said to be *fundamental*, if for any $\varepsilon > 0$, there exists a number N such that the inequality

$$\rho(x_n, x_m) \leqslant \varepsilon$$

holds for $n, m > N$. For brevity, we write this as

$$\lim_{n, m \to \infty} \rho(x_n, x_m) = 0.$$

If R is the real line with the usual metric, the concept of a fundamental sequence of points is just the classical concept of a fundamental sequence of numbers. In the theory of real numbers, one proves the Cauchy criterion, according to which every fundamental sequence of numbers is convergent. In a general Euclidean space, the Cauchy criterion is no longer valid.

Example. We now show that the Cauchy criterion is not valid in the Euclidean space $C_2(a, b)$. Let $x_m(t)$ be a sequence of continuous functions which take values between 0 and 1, and as $m \to \infty$, let $x_m(t)$ converge uniformly to 0 on every interval $[a, c - \varepsilon]$ and to 1 on every interval $[c + \varepsilon, b]$, where c is a fixed number between a and b. This sequence satisfies the Cauchy criterion, since

$$\int_a^b |x_n(t) - x_m(t)|^2 \, dt = \int_a^{c-\varepsilon} + \int_{c-\varepsilon}^{c+\varepsilon} + \int_{c+\varepsilon}^b \leqslant \varepsilon + 2\varepsilon + \varepsilon = 4\varepsilon,$$

for sufficiently large n and m. *However, at the same time, the sequence $x_m(t)$ does not converge in the mean to any continuous function.* To prove this assertion, we first note the following fact: If the sequence of functions $f_m(t)$ $(m = 1, 2, \ldots)$ converges in the mean on the interval $\Delta = \{a \leqslant t \leqslant b\}$ to the continuous function $f(t)$, and if $f_m(t)$ converges uniformly on an interval $\delta = \{c \leqslant t \leqslant d\}$ within Δ to the function $\varphi(t)$, then the identity $\varphi(t) \equiv f(t)$ holds in the interval δ. In fact, in the space $C_2(c, d)$, we have the relations

$$\rho^2(f_n, f) = \int_c^d |f_n(t) - f(t)|^2 \, dt \leqslant \int_a^b |f_n(t) - f(t)|^2 \, dt \to 0,$$

$$\rho^2(f_n, \varphi) = \int_c^d |f_n(t) - \varphi(t)|^2 \, dt \leqslant \max_{t \in \delta} |f_n(t) - \varphi(t)|^2 \, (d - c) \to 0.$$

By the uniqueness of the limit (Sec. 82.1), we have $f(t) \equiv \varphi(t)$, as required. Thus, returning to the sequence $x_1(t), x_2(t), \ldots, x_m(t), \ldots$ constructed above, if we require that this sequence converge in the mean to a continuous function $f(t)$, then by what has just been proved, we must have $f(t) = 0$ for $a \leqslant t < c$ and $f(t) = 1$ for $c < t \leqslant b$. But in this case, it is obvious that regardless of the value of $f(c)$, the function $f(t)$ cannot be a continuous function on the interval $a \leqslant t \leqslant b$.

83.2. We now establish some simple properties of fundamental sequences.

LEMMA 1. *Every convergent sequence is fundamental. In fact, by the triangle inequality*

$$\rho(x_n, x_m) \leqslant \rho(x_n, x) + \rho(x, x_m),$$

and if $x_n \to x$, the right-hand side becomes less than any preassigned number ε, for sufficiently large n and m.

LEMMA 2. *Every fundamental sequence is bounded.*

Proof. Let $x_1, x_2, \ldots, x_m, \ldots$ be the given fundamental sequence, and let x be any point of the space. For any given $\varepsilon > 0$, we find an N such that $\rho(x_N, x_m) \leqslant \varepsilon$ for any $m > N$. Moreover, let M denote the largest of the distances $\rho(x, x_1), \rho(x, x_2), \ldots, \rho(x, x_N)$. Then, by the triangle inequality

$$\rho(x, x_m) \leqslant \rho(x, x_N) + \rho(x_N, x_m) \leqslant M + \varepsilon$$

for any $m > N$, and by the definition of M

$$\rho(x, x_m) \leqslant M$$

for $m \leqslant N$. Thus we have

$$\rho(x, x_m) \leqslant M + \varepsilon,$$

for any m, Q.E.D.

83.3. Definition of a complete space. A Euclidean space R is said to be *complete*, if every fundamental sequence in R is a convergent sequence.

Example. First we verify that an n-dimensional Euclidean space R is a complete space. Let $e_1, e_2, \ldots, e_n$ be an orthonormal basis in R and let

$$x_m = \{\xi_1^{(m)}, \xi_2^{(m)}, \ldots, \xi_n^{(m)}\}, \quad \text{where } m = 1, 2, \ldots,$$

be a fundamental sequence. Since

$$|\xi_j^{(p)} - \xi_j^{(q)}|^2 \leqslant \sum_{i=1}^{n} |\xi_i^{(p)} - \xi_i^{(q)}|^2 = \|x_p - x_q\|^2,$$

the numerical sequence $\xi_j^{(p)}$ $(p = 1, 2, \ldots)$ is fundamental for every fixed $j = 1, 2, \ldots, n$, and as such has a limit ξ_j. The numbers $\xi_1, \xi_2, \ldots, \xi_n$ define a vector $x \in R$. Since

$$\|x - x_m\|^2 = \sum_{i=1}^{n} |\xi_i - \xi_i^{(m)}|^2 \to 0 \quad \text{as } m \to \infty,$$

the vector x is the limit of the given fundamental sequence. Thus, every fundamental sequence in the space R has a limit in R, Q.E.D.

83.4. The space l_2. An element of this space is any sequence of numbers $x = (\xi_1, \xi_2, \ldots, \xi_n, \ldots)$ the sum of whose squares converges, i.e., such that

$$\sum_{n=1}^{\infty} \xi_n^2 < \infty.$$

Linear operations are defined in the natural way:

$$(\xi_1, \xi_2, \ldots, \xi_n, \ldots) + (\eta_1, \eta_2, \ldots, \eta_n, \ldots)$$
$$= (\xi_1 + \eta_1, \xi_2 + \eta_2, \ldots, \xi_n + \eta_n, \ldots),$$
$$\alpha(\xi_1, \xi_2, \ldots, \xi_n, \ldots) = (\alpha\xi_1, \alpha\xi_2, \ldots, \alpha\xi_n, \ldots).$$

The scalar product of the vectors $x = (\xi_1, \xi_2, \ldots, \xi_n, \ldots)$ and $y = (\eta_1, \eta_2, \ldots, \eta_n, \ldots)$ is given by the formula

$$(x, y) = \sum_{n=1}^{\infty} \xi_n \eta_n. \tag{2}$$

We must verify the legitimacy of these definitions. First of all, the Schwarz inequality (Sec. 50.2)

$$\left| \sum_{k=n}^{n+m} \xi_k \eta_k \right| \leqslant \sqrt{\sum_{k=n}^{n+m} \xi_k^2} \sqrt{\sum_{k=n}^{n+m} \eta_k^2},$$

together with the classical Cauchy convergence criterion, shows that the series (2) is always convergent for $x \in l_2$, $y \in l_2$. Moreover, the equalities

$$\sum_{k=n}^{n+m} (\alpha\xi_k)^2 = \alpha^2 \sum_{k=n}^{n+m} \xi_k^2,$$

$$\sum_{k=n}^{n+m} (\xi_k + \eta_k)^2 = \sum_{k=n}^{n+m} \xi_k^2 + 2\sum_{k=n}^{n+m} \xi_k \eta_k + \sum_{k=n}^{n+m} \eta_k^2$$

show that the series

$$\sum_{k=1}^{\infty} (\alpha\xi_k)^2, \qquad \sum_{k=1}^{\infty} (\xi_k + \eta_k)^2$$

converge for $x \in l_2$, $y \in l_2$, so that the operations of addition of vectors and multiplication of vectors by numbers, as just defined, can be carried out within the space l_2. All the axioms for a scalar product (Sec. 49) are obviously satisfied. Thus, the space l_2 is Euclidean. We now verify that *it is complete.*

Let $x_m = (\xi_1^{(m)}, \xi_2^{(m)}, \ldots, \xi_n^{(m)}, \ldots)$, where $m = 1, 2, \ldots$, be a fundamental sequence of vectors in the space l_2. Since when m and p approach infinity, we have

$$\|x_m - x_p\|^2 = \sum_{n=1}^{\infty} |\xi_n^{(m)} - \xi_n^{(p)}|^2 \to 0$$

by hypothesis, then in particular, every term $|\xi_n^{(m)} - \xi_n^{(p)}|^2$ converges to zero (for fixed n), when m and p increase without limit. Therefore, for any fixed n, the sequence of components $\xi_n^{(m)}$ $(m = 1, 2, \ldots)$ converges by the classical Cauchy criterion. We write

$$\xi_n = \lim_{m \to \infty} \xi_n^{(m)}$$

and show that the vector $x = (\xi_1, \xi_2, \ldots, \xi_n, \ldots)$ belongs to the space l_2. The proof is as follows: By Lemma 2 of Sec. 83.2, we have

$$\|x_m\|^2 = \sum_{n=1}^{\infty} |\xi_n^{(m)}|^2 \leqslant K,$$

where K does not depend on m. Therefore, for any fixed N

$$\sum_{n=1}^{N} \xi_n^2 = \lim_{m \to \infty} \sum_{n=1}^{N} |\xi_n^{(m)}|^2 \leqslant K,$$

which implies at once that the series $\sum_{n=1}^{\infty} \xi_n^2$ is convergent.

It remains to show that $\|x - x_m\| \to 0$ as $m \to \infty$. To show this, we pass to the limit $p \to \infty$ in the inequality

$$\sum_{n=1}^{N} |\xi_n^{(m)} - \xi_n^{(p)}|^2 \leqslant \varepsilon,$$

which is valid for any given $\varepsilon > 0$ and any N, for sufficiently large m and p. As a result, we obtain the inequality

$$\sum_{n=1}^{N} |\xi_n^{(m)} - \xi_n|^2 \leqslant \varepsilon,$$

from which, by passing to the limit $N \to \infty$, we obtain the inequality

$$\|x_m - x\|^2 = \sum_{n=1}^{\infty} |\xi_n^{(m)} - \xi_n|^2 \leqslant \varepsilon,$$

which is valid for all sufficiently large m, Q.E.D.

84. Completion of a Euclidean Space

84.1. THEOREM 37. *Let R be a Euclidean space, which is in general not complete. Then there exists a Eucidean space E, called the completion of R, with the following properties:*

a) *There exists a subspace $E_1 \subset E$ which is isomorphic to R;*

b) *E_1 is dense in E.*

Proof. Two fundamental sequences $\{x_n\}$ and $\{y_n\}$ of the space R are said to be *coterminal* if

$$\lim_{n \to \infty} \|x_n - y_n\| = 0.$$

It is easily verified that two fundamental sequences which are coterminal with a third are also coterminal with each other. Therefore, all coterminal sequences which can be constructed from the elements of the space R can be divided into classes of coterminal sequences. All the sequences of a given class are coterminal with each other, and any sequence which is not a member of a given class is not coterminal with any sequence of the class. *The new space E is defined as the set of all these classes.*

Next, denoting the classes of coterminal sequences by the symbols X, Y, Z, etc., we define linear operations and a scalar product for these symbols. This is done as follows:

1. *Addition.* Given two classes X and Y, let $\{x_n\}$ be a sequence from the class X, and let $\{y_n\}$ be a sequence from the class Y. Form the sequence $\{x_n + y_n\}$. Since $\{x_n\}$ and $\{y_n\}$ are fundamental and since

$$\|(x_n + y_n) - (x_m + y_m)\| \leq \|x_n - x_m\| + \|y_n - y_m\|,$$

the sequence $\{x_n + y_n\}$ is also fundamental and hence defines a class Z. By definition, we set $Z = X + Y$. This definition is *unique*, since if $\{x'_n\}$ is coterminal with $\{x_n\}$, and if $\{y'_n\}$ is coterminal with $\{y_n\}$, then

$$\|(x'_n + y'_n) - (x_n + y_n)\| \leq \|x'_n - x_n\| + \|y'_n - y_n\| \to 0,$$

so that $\{x'_n + y'_n\}$ is coterminal with $\{x_n + y_n\}$. We leave it to the reader to verify that the addition axioms 1a to 1d of Sec. 11 are satisfied. We note that the zero element in the space E is the class 0 consisting of all sequences $\{x_n\}$ converging to zero.

2. *Multiplication by a number.* Given a class X and a number λ, we take any $\{x_n\} \in X$ and form the sequence $\{\lambda x_n\}$. Since $\{x_n\}$ is fundamental, so is $\{\lambda x_n\}$, since

$$\|\lambda x_n - \lambda x_m\| = |\lambda| \, \|x_n - x_m\|;$$

hence $\{\lambda x_n\}$ defines a class Y. By definition, we set $Y = \lambda X$. We leave it to the reader to verify that this definition is unique and that axioms 2e to 3h of Sec. 11 are satisfied.

3. *The scalar product.* Given two classes X and Y, we take any sequences $\{x_n\} \in X$ and $\{y_n\} \in Y$. We assert that the expression (x_n, y_n) has a limit as $n \to \infty$, which we define as the scalar product (X, Y) of X and Y. In fact, we have

$$|(x_n, y_n) - (x_m, y_m)| \leqslant |(x_n, y_n - y_m)| + |(x_n - x_m, y_m)|$$
$$\leqslant \|x_n\| \, \|y_n - y_m\| + \|x_n - x_m\| \, \|y_m\|.$$

The numbers $\|x_n\|$ and $\|y_m\|$ are bounded by Lemma 1 of Sec. 83. Therefore, the sequence (x_n, y_n) satisfies the ordinary Cauchy criterion and hence has a limit, as asserted. Moreover, this limit does not depend on the choice of the sequences $\{x_n\}$ and $\{y_n\}$ in the classes X and Y. In fact, if $\{x_n'\} \in X$, $\{y_n'\} \in Y$, we have

$$|(x_n', y_n') - (x_n, y_n)| = |(x_n' - x_n, y_n') + (x_n, y_n' - y_n)|$$
$$\leqslant \|x_n' - x_n\| \, \|y_n'\| + \|x_n\| \, \|y_n' - y_n\| \to 0,$$

since $\|y_n'\|$ and $\|x_n\|$ are bounded, while $\{x_n'\}$ is coterminal with $\{x_n\}$ and $\{y_n'\}$ is coterminal with $\{y_n\}$. We leave it to the reader to verify that all the axioms for a scalar product (Sec. 49) are satisfied.

Consider the Euclidean space E consisting of all classes of coterminal fundamental sequences, with the operations just defined. We now verify that E has the following properties:

a) *E contains a subspace E_1 which is isomorphic to the space R.* In fact, with every element $x \in R$ associate the class $X \in E$ containing the sequence $x, x, x, \ldots$, i.e., the class of *all* sequences converging to the point x. It is obvious that the subspace $E_1 \subset E$ of all such classes is isomorphic to the space R.

b) *E_1 is dense in E.* Let Y be an arbitrary class from E, and take any sequence $\{x_n\} \in Y$. For any $\varepsilon > 0$, the inequality $\|x_n - x_m\| \leqslant \varepsilon$ holds if $n, m \geqslant n_0$, starting from some n_0. We now show that the distance from Y to the class X_{n_0} containing the sequence $x_{n_0}, x_{n_0}, x_{n_0}, \ldots$, i.e., the class of sequences converging to x_{n_0}, does not exceed ε. In fact, according to the definition of the scalar product of classes, we have

$$\|Y - X_{n_0}\|^2 = (Y - X_{n_0}, Y - X_{n_0}) = \lim_{n \to \infty} (x_n - x_{n_0}, x_n - x_{n_0})$$
$$= \lim_{n \to \infty} \|x_n - x_{n_0}\|^2 \leqslant \varepsilon^2.$$

Thus, we can find an element $X_{n_0} \in E_1$ which is arbitrarily close to the element $Y \in E$. Therefore, E_1 is dense in E.

c) *E is a complete space.* Let $\{X_n\}$ be a fundamental sequence of elements of E. Construct a sequence $\{y_n\} \in R$ such that the class Y_n corresponding to the element y_n differs in norm from the element X_n of the space E by less than $1/n$. Then the sequence y_n is fundamental, since

$$\|y_n - y_m\| = \|Y_n - Y_m\| = \|X_n - X_m + (Y_n - X_n) - (Y_m - X_m)\|$$
$$\leqslant \|X_n - X_m\| + \frac{1}{n} + \frac{1}{m}.$$

Let $Y \in E$ be the class containing the sequence $\{y_n\}$. Then, for any $\varepsilon > 0$ and $n \geqslant n_0$, we have

$$\|Y - X_n\| \leqslant \|Y - Y_n\| + \|Y_n - X_n\|$$
$$\leqslant \lim_{p \to \infty} \|y_p - y_n\| + \frac{1}{n} \leqslant \varepsilon + \frac{1}{n} \leqslant 2\varepsilon,$$

which implies that

$$Y = \lim_{n \to \infty} X_n.$$

Thus, every fundamental sequence $X_n \in E$ has a limit in E, as required.

This completes the proof of Theorem 37. Next, we supplement this theorem by proving that the completion of a given space R is unique to within an isomorphism.

84.2. *Remark. Two spaces E' and E'' which are completions of a space R are isomorphic.*

Proof. Let E'_1 and E''_1 be the subspaces of E' and E'' which are isomorphic to the space R and which are therefore isomorphic to each other. This isomorphism can easily be extended to give an isomorphism between E' and E''. In fact, if X' is any element of the space E', then by property b) of a completion (Sec. 84.1), there exists a sequence $X'_n \in E'_1$ which converges to X'. The corresponding sequence $X''_n \in E''_2$ is also fundamental, since the distances between elements of E'_1 are the same as those between the corresponding elements of E''_1, because of the isomorphism between E'_1 and E''_1. Since E'' is complete, it contains an element

$$X'' = \lim_{n \to \infty} X''_n.$$

We associate this element $X'' \in E''$ with the element $X' \in E'$ originally chosen; X' is uniquely defined, since coterminal sequences in E'_1 correspond to coterminal sequences in E''_1, so that replacing the sequence

X'_n by a coterminal sequence corresponds to replacing the sequence X''_n by a coterminal sequence also. Thus, we have a one-to-one correspondence between all the elements of E' and those of E''. This correspondence is an isomorphism, since linear operations and scalar products are continuous operations and the correspondence $E' \leftrightarrow E''$ has been set up by using continuity.

85. The Space $L_2(a, b)$

In this section, we use certain theorems from the theory of functions of a real variable. The reader can find the proofs of these theorems in I. P. Natanson's book,[3] for example.

85.1. The Euclidean space $C_2(a, b)$, consisting of the continuous functions on the interval $a \leqslant x \leqslant b$, provided with the scalar product

$$(f, g) = \int_a^b f(x)g(x)\, dx,$$

is not complete (as we have seen in Sec. 83), i.e., not every fundamental sequence in $C_2(a, b)$ has a limit in $C_2(a, b)$. By the theorem proved in Sec. 84, the space $C_2(a, b)$ has a completion, which we denote by $L_2(a, b)$. The elements of the space $L_2(a, b)$ are defined abstractly by the procedure of Sec. 84; thus, it is natural to ask if they can be ascribed a concrete meaning and interpreted as some kind of functions. It turns out that this can in fact be done, but first, we must introduce some important concepts.

By definition, a set E of points of the interval $[a, b]$ *has measure less than the positive number* ε, if E can be contained in a finite or countably infinite set of intervals of total length less than ε. A function $f(x)$ defined on the interval $[a, b]$ is said to be *measurable*, if for any $\varepsilon > 0$ we can convert it into a continuous function by changing its values on a set of measure less than ε.

Example 1. The function $f_1(x)$, equal to 0 for $a \leqslant x < c$ and equal to 1 for $c \leqslant x \leqslant b$, is measurable; it is converted into a continuous function if we replace it on the interval $c - \varepsilon < x < c$ by a linear function ranging from 0 to 1.

Example 2. The function $f_2(x)$, equal to x^{-r} $(r > 0)$ for $0 < x \leqslant 1$ and equal to 0 (say) at $x = 0$, is measurable; it is converted into a continuous function if we replace it on the interval $0 < x < \varepsilon$ by a linear function ranging from 0 to ε^{-r}.

---

[3] Natanson, I. P., *The Theory of Functions of a Real Variable* (translated by L. F. Boron), Frederick Ungar Publishing Co., New York, vol. 1 (1955); vol. 2 (1960).

Next, we introduce the concept of the *Lebesgue integral* for nonnegative measurable functions $f(x)$. This is done as follows: Consider a sequence of positive numbers $\varepsilon_n \to 0$, and for each n construct a continuous function $f_n(x)$ which differs from $f(x)$ only on a set of measure less than ε_n. It is obvious that the function $f_n(x)$ can be taken to be nonnegative. Then, if the functions $f_n(x)$ can be constructed in such a way that their integrals (in the ordinary Riemann sense) have a common bound, then we say that $f(x)$ is *Lebesgue integrable*. In this case, the functions $f_n(x)$ can always be chosen in such a way that their integrals form a convergent sequence. In general, the limit of such a sequence is not uniquely defined by the function $f(x)$ itself, and can depend on the choice of the sequence $f_n(x)$. Now consider the smallest of the possible limits of integrals of the functions $f(x)$; we call this limit the *Lebesgue integral* of the function $f(x)$, and denote it by

$$(L) \int_a^b f(x) \, dx.$$

(This scheme for constructing the Lebesgue integral is due to Tonelli.) It can be shown that if the function $f(x)$ is integrable in the ordinary Riemann sense, e.g., if $f(x)$ is continuous, then $f(x)$ is also integrable in the Lebesgue sense, and its Lebesgue integral coincides with its Riemann integral. Therefore, the symbol (L) is not essential and will henceforth be omitted.

If the function $f(x)$ has an *improper* Riemann integral (e.g., $f(x) = x^{-r}$, $0 < x \leqslant 1, 0 < r < 1$), then $f(x)$ is also Lebesgue integrable, and once again its (improper) Riemann integral coincides with its Lebesgue integral. However, there are a large number of functions which have neither proper nor improper Riemann integrals but which are Lebesgue integrable. An example of such a function is the *Dirichlet function* $\chi(x)$, equal to 1 for rational x and 0 for irrational x; its Riemann integral does not exist, whereas its Lebesgue integral exists and equals 1. Moreover, if the function $f(x)$ is Lebesgue integrable, while the function $g(x)$ is measurable and does not exceed $f(x)$, then $g(x)$ is also Lebesgue integrable and

$$\int_a^b g(x) \, dx \leqslant \int_a^b f(x) \, dx.$$

We shall say that a function $f(x)$ which takes both positive and negative values is Lebesgue integrable, if its absolute value is an integrable function. In this case, the nonnegative functions

$$f^+(x) = \max \{0, f(x)\} \leqslant |f(x)|, \qquad f^-(x) = \max \{0, -f(x)\} \leqslant |f(x)|$$

are also integrable, by the inequality just given. Since

$$(x) = f^+(x) - f^-(x),$$

we define the integral of $f(x)$ as

$$\int_a^b f(x)\,dx = \int_a^b f^+(x)\,dx - \int_a^b f^-(x)\,dx.$$

It can be verified that the Lebesgue integral has the linear properties of the ordinary integral, i.e.,

$$\int_a^b \alpha f(x)\,dx = \alpha \int_a^b f(x)\,dx,$$

$$\int_a^b [f(x) + g(x)]\,dx = \int_a^b f(x)\,dx + \int_a^b g(x)\,dx.$$

85.2. We now let L_2 denote the set of all functions $f(x)$ on the interval $a \leqslant x \leqslant b$ whose squares are Lebesgue integrable, i.e., are such that

$$\int_a^b f^2(x)\,dx < \infty.$$

We now show that L_2 is a linear space.

If $f \in L_2$ and α is a real number, then $\alpha f \in L_2$. To see this, we note that if $f_n(x)$ is a continuous function which differs from $f(x)$ on a set of measure less than ε_n, then $\alpha f_n(x)$ is a continuous function which differs from $\alpha f(x)$ on the same set; therefore, if $f(x)$ is measurable, so is $\alpha f(x)$. If the integrals of $f_n^2(x)$, $n = 1, 2, \ldots$, are bounded, then the integrals of $\alpha^2 f_n^2(x)$ are also bounded, so that if $f^2(x)$ is integrable, so is $\alpha^2 f^2(x)$.

If $f \in L_2$, $g \in L_2$, then $f + g \in L_2$. To see this, we note that if $f_n(x)$ is a continuous function which differs from $f(x)$ on a set of measure less than ε_n, and if $g_n(x)$ is a continuous function which differs from $g(x)$ on a set of measure less than ε_n, then $f_n(x) + g_n(x)$ is a continuous function which differs from $f(x) + g(x)$ on a set of measure less than $2\varepsilon_n$. Thus, if f and g are measurable, so is $f + g$. If the integrals of $f_n^2(x)$ and $g_n^2(x)$, where $n = 1, 2, \ldots$, are bounded, then the integrals of

$$(f_n + g_n)^2 = f_n^2 + 2f_n g_n + g_n^2 \leqslant 2f_n^2 + 2g_n^2$$

are also bounded. Therefore, if f^2 and g^2 are integrable, so is $(f + g)^2$. We leave it to the reader to verify that the axioms for a linear space (Sec. 11) are satisfied.

85.3. Next, we introduce a scalar product in the space L_2 by the formula

$$(f, g) = \int_a^b f(x)g(x)\,dx.$$

In this regard, we note that the product fg is integrable, because of the inequality

$$|fg| \leqslant \tfrac{1}{2}(f^2 + g^2).$$

The validity of the first three axioms for a scalar product (Axioms I, II, III of Sec. 49) is obvious.   Thus, we devote our attention only to Axiom IV, which states that the scalar product of any vector with itself must be non-negative and vanishes only when the vector is the zero vector of the space. The first part of this assertion is obvious from the definition of the integral of a nonnegative function.   The second part of the assertion can give rise to complications, since we have nonnegative functions, like the Dirichlet function mentioned above, whose integrals are zero, even though the functions themselves can take positive values.   We avoid this difficulty as follows: It can be shown that the integral of a nonnegative function $u(x)$ equals zero if and only if the set of x for which $u(x) > 0$ has measure zero, i.e., can be covered by a finite or countably infinite set of intervals whose total length is less than any preassigned $\varepsilon > 0$.   Thus, we shall regard two functions f and g in L_2 as *equivalent* if they differ only on a set of measure zero.   In particular, every nonnegative function with a zero integral is different from zero on a set of measure zero, and hence is equivalent to zero. From now on, we shall take the elements of L_2 to be not just square integrable functions, but classes of equivalent square integrable functions.   This introduces no change in the definition of the scalar product, since the values of integrals do not change when the values of functions are changed on sets of measure zero.   Thus, we finally obtain a space which satisfies all the requirements of a Euclidean space.   As in Sec. 85.1, we denote this space by $L_2(a, b)$.   Moreover, $C_2(a, b)$ is a dense subset of $L_2(a, b)$, since the sequence of‑continuous functions $f_n(x)$, participating in the definition of the integral of $f^2(x)$, converges to the function $f(x)$ in the sense of the metric of $L_2(a, b)$.

We now cite a fundamental theorem concerning $L_2(a, b)$, proved by Riesz and Fischer, i.e., *the space $L_2(a, b)$ is a complete Euclidean space*.   In particular, this theorem implies that the space $L_2(a, b)$ is isomorphic with (or, one might say, coincides with) the completion of the space $C_2(a, b)$ [with respect to the metric of $C_2(a, b)$].   We shall not give the proof of this theorem, which requires some subtleties from the theory of functions of a real variable; instead, we refer the reader to the book by Natanson, cited above.

85.4. The construction given above is easily generalized to the case of functions of several independent variables.   It is only necessary to define what is meant by a set of measure less than ε.   To be explicit, we consider the case of two variables, where we have a function $f(x, y)$ defined on the rectangle

$$\square = \{a \leqslant x \leqslant b, \quad c \leqslant y \leqslant d\}.$$

Then, by definition, a set E has measure less than the positive number ε if E can be covered by a finite or countably infinite set of rectangles with total area less than ε.   The rest of the construction proceeds in just the same way as in the case of one variable.   The Lebesgue integral of the function $f(x, y)$ over the rectangle $\square$ is denoted by

$$(L) \int_a^b \int_c^d f(x, y)\, dx\, dy,$$

or simply by

$$\int_a^b \int_c^d f(x, y)\, dx\, dy.$$

The set of all functions $f(x, y)$ whose squares are Lebesgue integrable again forms a linear Euclidean space, which we denote by $L_{22}(\square)$.

We now cite an important theorem (*Fubini's theorem*), which allows us to calculate a double Lebesgue integral in terms of iterated integrals, just as in the case of ordinary Riemann integrals.

Let the function $f(x, y)$ be integrable over the rectangle

$$\square = \{a \leqslant x \leqslant b, \quad c \leqslant y \leqslant d\}.$$

If y is fixed, the function $f(x, y)$ becomes a function of x, which is integrable except possibly on a set of measure zero.   Moreover, its integral with respect to x is an integrable function of y, and

$$\int_c^d \left\{ \int_a^b f(x, y)\, dx \right\} dy = \int_a^b \int_c^d f(x, y)\, dx\, dy,$$

Similarly, $f(x, y)$ is an integrable function of y for all x, except possibly on a set of measure zero; its integral with respect to y is an integrable function of x, and

$$\int_a^b \left\{ \int_c^d f(x, y)\, dy \right\} dx = \int_a^b \int_c^d f(x, y)\, dx\, dy.$$

86. Orthogonal Complements

In Sec. 50.3, we defined the orthogonal complement of a subspace F of a Euclidean space R as the set of all vectors $x \in R$ which are orthogonal to every vector in the subspace F.   However, in general, we must first prove the existence of nonzero vectors x in the space R which satisfy this requirement.   First, we show that if x is a vector orthogonal to a subspace A of the Euclidean space R, then x is orthogonal to the closure $\overline{A}$ of the space A (Sec. 82.3).   To see this, we note that if $y \in \overline{A}$, there exists a sequence $\{y_n\} \in A$ which converges to A, and by hypothesis, $(x, y_n) = 0$.   It follows by Lemma 1 of Sec. 82 that

$$(x, y) = \lim_{n \to \infty} (x, y_n) = 0,$$

i.e., y is also orthogonal to x.   Since y is any vector of the subspace $\overline{A}$, then x is orthogonal to all of $\overline{A}$, Q.E.D.

Thus, for example, *there does not exist a vector $x \neq 0$ which is orthogonal to a dense subspace $A \subset R$* (Sec. 82.4). In fact, if such a vector x existed, then, by what has just been proved, x would also be orthogonal to the closure $\bar{A} = R$, i.e., x would be orthogonal to any vector in the space R. In particular, we would have $(x, x) = 0$, which contradicts the assumption that $x \neq 0$.

Since if x is orthogonal to $A \subset R$, then x is also orthogonal to $\bar{A}$, we can confine ourselves henceforth to the case of closed subspaces F. In particular, a finite-dimensional subspace F is always closed (Sec. 82.3). Moreover, we already know that in this case vectors orthogonal to F always exist (Sec. 56), provided only that $F \neq R$.

We now prove that *for any closed subspace F of a complete Euclidean space R, there exist nonzero vectors orthogonal to F.*

LEMMA 1. (*Parallelogram lemma*) *Given any two vectors x and y of a Euclidean space R, the relation*

$$\|x + y\|^2 + \|x - y\|^2 = 2\|x\|^2 + 2\|y\|^2$$

holds, i.e., the sum of the squares of the diagonals of a parallelogram equals the sum of the squares of its sides. The proof is a result of the following simple calculation:

$$\|x + y\|^2 + \|x - y\|^2 = (x + y, x + y) + (x - y, x - y)$$
$$= 2(x, x) + 2(y, y) = 2\|x\|^2 + 2\|y\|^2.$$

LEMMA 2. *Let f be a vector which does not belong to a closed subspace $F \subset R$. Then there exists an expansion*

$$f = g + h, \tag{3}$$

where $g \in F$ and h is orthogonal to F; moreover, the vectors g and h are uniquely defined.

Proof. We write

$$d = \inf_{q \in F} \rho(f, q),$$

where inf denotes the greatest lower bound (or infimum). There are two possibilities: either $d = 0$ or $d > 0$. The first possibility must be rejected, for if $d = 0$, we could find a sequence $q_n \in F$ such that $\rho(f, q_n) \to 0$, which would imply that f is a limit point of the subspace F and hence that f itself is an element of F (since F is closed); this contradicts the hypothesis. Thus, we have $d > 0$.

Now consider a sequence $g_n \in F$ for which $\rho(f, g_n) \to d$. Applying Lemma 1 to the parallelogram determined by the vectors $x = f - g_n$ and $y = f - g_m$, we obtain

$$2\|f - g_n\|^2 + 2\|f - g_m\|^2 = 4\|f - \frac{g_n + g_m}{2}\|^2 + \|g_n - g_m\|^2.$$

The left-hand side approaches $4d^2$ as $n, m \to \infty$. The first term of the right-hand side is not less than $4d^2$, since $(g_n + g_m)/2 \in F$, so that

$$\left\| f - \frac{g_n + g_m}{2} \right\| \geqslant d.$$

Therefore, the last term in the right-hand side converges to zero, which implies that the sequence g_n is fundamental. Since R is a complete space, the sequence g_n approaches a limit $g \in R$ as $n \to \infty$, and since the subspace F is closed, g belongs to F.

To show that the vector $h = f - g$ is orthogonal to F, we proceed as follows: For any $q \in F$ and any λ, we have

$$d^2 \leqslant \| f - (g - \lambda q)\|^2 = \| h - \lambda q \|^2$$
$$= (h - \lambda q, h - \lambda q) = d^2 - 2\lambda(h, q) + \lambda^2 \|q\|^2,$$

so that

$$-2\lambda(h, q) + \lambda^2 \|q\|^2 \geqslant 0,$$

which is possible for arbitrary λ only if $(h, q) = 0$; together with the fact that $g \in F$, this proves the expansion (3).

Finally, we show that the expansion (3) is unique. Suppose that

$$f = g + h,$$
$$f = g' + h',$$

where g, g' belong to F, and h, h' are orthogonal to F. Subtracting the second relation from the first, we obtain

$$0 = (g - g') + (h - h'),$$

where $g - g' \in F$ and $h - h'$ is orthogonal to F. By Lemma 1 of Sec. 50, $g - g' = h - h' = 0$, so that $g = g', h = h'$, Q.E.D.

Lemma 2 establishes the existence of orthogonal vectors for any closed subspace F of a complete Euclidean space R. As we already know (Sec. 50.3), the set of all such vectors forms the *orthogonal complement* G of the subspace F. This orthogonal complement G is a subspace, and is in fact closed (Sec. 82.2).

We can summarize all these considerations in the following theorem:

THEOREM 38. *For any closed subspace F of a complete Euclidean space R, there exists a closed orthogonal complement G. Every vector $z \in R$ can be represented in the form $z = u + v$, where $u \in G$, $v \in F$. The components u and v defined by the vector z and the subspace F are unique.*

87. Orthogonal Expansions

87.1. As we know (Sec. 51.2), every vector in a finite-dimensional Euclidean space can be expanded with respect to a given orthonormal basis.

In this section, we shall establish an analogous expansion theorem for a complete infinite-dimensional Euclidean space.

An orthonormal system of vectors $\{e_n\}$ in a Euclidean space R is said to be a *complete system* or a *basis*, if there are no nonzero vectors in R which are orthogonal to all the vectors e_n. In other words, a system of vectors $\{e_n\}$ is complete if the equations $(g, e_n) = 0$, where $n = 1, 2, \ldots$, imply that $g = 0$. Now let $\{e_n\}$ be a basis of the space R and let f be a vector in R. Suppose that we have an expansion

$$f = \sum_{n=1}^{\infty} a_n e_n, \tag{4}$$

where the series converges in the sense of the metric of the space R. To find the coefficients in this expansion, we denote the sum of the first m terms of (4) by s_m and take the scalar product of the relation $s_m \to f$ with the vector e_n (n fixed). By Lemma 1 of Sec. 82, the result is $(s_m, e_n) \to (f, e_n)$. For $m > n$, we have

$$(s_m, e_n) = (a_1 e_1 + \cdots + a_m e_m, e_n) = a_n,$$

since

$$(e_j, e_n) = \begin{cases} 0 & \text{for } j \neq n, \\ 1 & \text{for } j = n. \end{cases}$$

Therefore,

$$(f, e_n) = \lim_{m \to \infty} (s_m, e_n) = a_n,$$

and we obtain the formula

$$a_n = (f, e_n). \tag{5}$$

Moreover, at the same time, we have proved the uniqueness of the expansion (4), assuming that the expansion exists.

87.2. The numbers $a_n = (f, e_n)$ are called the *Fourier coefficients* of the vector f with respect to the basis $\{e_n\}$ (cf. Sec. 51); obviously, these numbers can be defined whether or not the expansion (4) actually exists. The geometric meaning of the Fourier coefficients is clear from formula (5), i.e.,

$$a_n = (f, e_n) = \|f\| \|e_n\| \cos (\widehat{f, e_n}) = \|f\| \cos (\widehat{f, e_n}),$$

where $(\widehat{f, e_n})$ is the angle between the vectors f and e_n. Thus, a_n is the projection of the vector f onto the direction of the vector e_n.

We have also encountered the numbers (5) in Sec. 56 when we solved the problem of representing the vector f as a sum of two orthogonal vectors g and h, where g lies in the finite-dimensional subspace spanned by the orthonormal basis $e_1, e_2, \ldots, e_n$. As was shown in Sec. 56, the numbers $a_j = (f, e_j)$ satisfy Bessel's inequality

$$\sum_{j=1}^{k} a_j^2 \leq \|f\|^2$$

for any finite k. Since this inequality holds for any k, by letting k go to infinity, we conclude that the infinite series with terms a_j^2 converges and that the inequality

$$\sum_{j=1}^{\infty} a_j^2 \leq \|f\|^2 \tag{6}$$

holds. As before, this inequality is called *Bessel's inequality*.

87.3. We now return to our original problem. Suppose that we have a complete orthonormal system $\{e_n\}$ and a vector f. Form the series

$$\sum_{n=1}^{\infty} a_n e_n, \tag{7}$$

where $a_n = (f, e_n)$. We now show that (7) *converges in the metric of the space R to the vector f*.

Proof. Let s_p be the sum of the first p terms of the series (7), and let $q > p$. Then we have

$$\|s_q - s_p\|^2 = \left\|\sum_{p+1}^{q} a_n e_n\right\|^2 = \sum_{p+1}^{q} a_n^2.$$

As $p, q \to \infty$, this quantity converges to zero, because of the convergence of the series formed from the numbers a_n^2. Therefore, the sums s_p form a fundamental sequence. Because of the assumption that the space R is complete, the sums s_p converge to a limit $s \in R$, as $p \to \infty$; moreover, $s = f$. To see this, we note that

$$(s, e_m) = \lim_{p \to \infty} (s_p, e_m) = \lim_{p \to \infty} \left(\sum_{j=1}^{p} a_j e_j, e_m\right) = \lim_{p \to \infty} a_m = a_m = (f, e_m)$$

for fixed m and $p > m$; therefore

$$(f - s, e_m) = (f, e_m) - (s, e_m) = 0, \tag{8}$$

for any m. Since, by hypothesis, the system $\{e_m\}$ is complete, it follows from the relation (8) that $f = s$. Thus we have

$$f = \lim_{p \to \infty} s_p = \sum_{n=1}^{\infty} a_n e_n,$$

Q.E.D.   Our result can be formulated as follows:

THEOREM 39. *Let f be a vector of a complete Euclidean space, and let f have the Fourier series*

$$\sum_{n=1}^{\infty} (f, e_n)e_n$$

with respect to a given complete orthonormal system $\{e_n\}$. Then this series converges in norm to the vector f.

Remark. Suppose that the conditions of Theorem 39 are met and we have two expansions

$$f = \sum_{k=1}^{\infty} a_k e_k, \qquad g = \sum_{k=1}^{\infty} b_k e_k.$$

Then, by Lemma 1 of Sec. 82

$$(f, g) = \left(\lim_{n \to \infty} \sum_{k=1}^{n} a_k e_k, \lim_{n \to \infty} \sum_{k=1}^{n} b_k e_k \right)$$

$$= \lim_{n \to \infty} \left(\sum_{k=1}^{n} a_k e_k, \sum_{k=1}^{n} b_k e_k \right) = \lim_{n \to \infty} \sum_{k=1}^{n} a_k b_k = \sum_{k=1}^{\infty} a_k b_k,$$

and in particular

$$(f, f) = \|f\|^2 = \sum_{k=1}^{\infty} a_k^2$$

for $g = f$. Thus, the existence of the expansion (4) guarantees that Bessel's inequality (6) reduces to an equality, i.e., *if an element f has an expansion of the form* (4), *then the square of its length equals the sum of the squares of its projections onto all the axes $e_1, e_2, \ldots, e_n, \ldots$* This statement constitutes a generalization of the Pythagorean theorem (Sec. 50) to the case of an infinite-dimensional space.

87.4. To apply Theorem 39, we must have available a complete orthonormal system of vectors $e_1, e_2, \ldots, e_n, \ldots$. As we know (Sec. 57), a general method of constructing orthogonal systems is to apply an orthogonalization process to any given system $g_1, g_2, \ldots, g_n, \ldots$. In the general case, however, the result may be an *incomplete* system. To obtain a complete system $e_1, e_2, \ldots, e_n, \ldots$ from a given system $g_1, g_2, \ldots, g_n, \ldots$, we have to impose further restrictions.

THEOREM 40. *A necessary and sufficient condition for the system $e_1, e_2, \ldots, e_n, \ldots$, obtained by orthogonalizing a given system $g_1, g_2, \ldots, g_n, \ldots$, to be complete is that the linear combination of the vectors $g_1, g_2, \ldots, g_n, \ldots$ be dense in the space R.*

Proof.

Necessity. By construction, the linear manifolds $L(e_1, e_2, \ldots, e_n, \ldots)$ and $L(g_1, g_2, \ldots, g_n, \ldots)$ coincide. If the system $e_1, e_2, \ldots, e_n, \ldots$ is complete, then by Theorem 39, every vector $f \in R$ is the sum of a Fourier series

$$\sum_{n=1}^{\infty} a_n e_n,$$

and hence is the limit of finite linear combinations of vectors of the system $\{e_n\}$. This means that the linear manifold $L(e_1, e_2, \ldots)$ is dense in the space R, and therefore that the linear manifold $L(g_1, g_2, \ldots)$ is also dense in R.

Sufficiency. Suppose that the linear manifold $L(g_1, g_2, \ldots) = L(e_1, e_2, \ldots)$ is dense in the space R. Then, as proved at the beginning of Sec. 86, there does not exist a vector $g \neq 0$ orthogonal to the subspace $L(e_1, e_2, \ldots)$. Therefore, there does not exist a vector $g \neq 0$ orthogonal to each of the vectors $e_1, e_2, \ldots, e_n, \ldots$, i.e., the system $\{e_n\}$ is complete, Q.E.D.

Example 1. The linear manifold spanned by the powers $1, t, \ldots, t^n, \ldots$, i.e., the set of all polynomials in t, is dense in the space $C_2(a, b)$ (see Sec. 82.4). Moreover, the space $C_2(a, b)$ is dense in its completion $L_2(a, b)$. Therefore, by Lemma 2 of Sec. 82, the set of all polynomials is dense in $L_2(a, b)$. It follows that the set of polynomials obtained by orthogonalizing the powers of t is dense in the space $L_2(a, b)$. In particular, the Legendre polynomials (Sec. 58) form a complete orthogonal system in the space $L_2(-1, 1)$.

Example 2. The system of trigonometric functions $1, \cos t, \sin t, \ldots$ is orthogonal in the space $C_2(-\pi, \pi)$. It follows from the well known theorem to the effect that a periodic function with a continuous derivative can be expanded in a uniformly convergent Fourier series[4] that the linear manifold spanned by the system of trigonometric functions is dense in the set B of functions $f(t)$ on the interval $-\pi \leqslant t \leqslant \pi$ which can be extended over the whole real line as continuous periodic functions with continuous first derivatives. An example of such a function is a function which has a continuous derivative on the interval $[-\pi, \pi]$ and which vanishes on the intervals $[-\pi, -\pi + \delta]$ and $[\pi - \delta, \pi]$ for some $\delta > 0$. We now show that the set B is in its turn dense in the set of all polynomials $P(t)$ on $[-\pi, \pi]$.

---

[4] See, e. g., L. Brand, *op. cit.*, p. 534.

Consider a sequence of functions $h_n(t)$, $n = 1, 2, \ldots$, where $0 \leqslant h_n(t) \leqslant 1$, which have continuous first derivatives and satisfy the conditions

$$
h_n(t) = \begin{cases} 0 & \text{for} \quad -\pi \leqslant t \leqslant -\pi + \dfrac{1}{2^n}, \quad \pi - \dfrac{1}{2^n} \leqslant t \leqslant \pi, \\[2mm] 1 & \text{for} \quad -\pi + \dfrac{1}{2^{n-1}} \leqslant t \leqslant \pi - \dfrac{1}{2^{n-1}}. \end{cases}
$$

We could specify the functions $h_n(t)$ by explicit formulas, but this is not necessary. The prcduct $P(t)h_n(t)$ belongs to the set B for any n and any polynomial $P(t)$. On the other hand, as $n \to \infty$, we have

$$
\begin{aligned}
\int_{-\pi}^{\pi} |P(t) - P(t)h_n(t)|\, dt &= \int_{-\pi}^{\pi} |P(t)[1 - h_n(t)]|\, dt \\
&\leqslant \int_{-\pi}^{-\pi+2^{1-n}} |P(t)|\, dt + \int_{\pi-2^{1-n}}^{\pi} |P(t)|\, dt \to 0,
\end{aligned}
$$

so that the function $P(t)h_n(t)$ converges in the mean to the function $P(t)$. This proves that the set B is dense in the set of all polynomials, as asserted.

Thus, by Lemma 2 of Sec. 82, the linear manifold spanned by the trigonometric functions is dense in the set of all polynomials. But, by Example 1, this latter set is dense in $L_2(-\pi, \pi)$, and hence, the linear manifold spanned by the trigonometric functions is dense in $L_2(-\pi, \pi)$. It follows by Theorem 40 that *the trigonometric functions form a complete system in the space* $L_2(-\pi, \pi)$. Normalizing this system, we obtain the functions

$$
\frac{1}{\sqrt{2\pi}}, \quad \frac{1}{\sqrt{\pi}} \sin t, \quad \frac{1}{\sqrt{\pi}} \cos t, \ldots,
$$

which form a *complete orthonormal system* in $L_2(-\pi, \pi)$. By Theorem 39, the expansion of a function $f(t) \in L_2(-\pi, \pi)$ with respect to this system [which is easily seen to be just the usual Fourier series expansion of $f(t)$] converges in the mean to $f(t)$.

Example 3. In the space $L_2(0, \pi)$, each of the two orthogonal systems

$$1, \cos t, \cos 2t, \ldots$$

and

$$\sin t, \sin 2t, \ldots$$

spans a linear manifold dense in $L_2(0, \pi)$. In fact, a function $f(t)$ defined on the interval $[0, \pi]$ can be extended into the interval $[-\pi, 0]$ either as an even function or as an odd function, as one prefers, and the Fourier series expansions of the functions so obtained will contain only cosines in the first case and only sines in the second case. The convergence in the mean of the expansion on $[-\pi, \pi]$ guarantees its convergence in the mean on $[0, \pi]$.

Example 4. The system of trigonometric functions $\sin nt$, $\sin ms$, $\sin nt \sin ms$ $(m, n = 1, 2, \ldots)$ is orthogonal in the space of functions $f(t, s)$ defined on the region $0 \leqslant t, s \leqslant \pi$ [a square in the (t, s) plane] with the scalar product

$$(f(t, s), g(t, s)) = \int_0^\pi \int_0^\pi f(t, s)g(t, s) \, dt \, ds.$$

If, instead of the theorem on simple Fourier series discussed in Example 2, we use the corresponding theorem on double Fourier series, then, by the same method as used in Examples 2 and 3, we can prove the completeness of our orthogonal system in the space of all square integrable functions defined on the region $0 \leqslant t, s \leqslant \pi$.

*87.5. Using the completeness of the systems of trigonometric functions and the system of Legendre polynomials, we can strengthen the results obtained in Sec. 67 concerning the corresponding Sturm-Liouville operators. Thus, we prove the following propositions:

1a) The equation $x''(t) = \lambda x(t)$ has solutions only for $\lambda = -n^2$ $(n = 0, 1, 2, \ldots)$ in the space of twice differentiable functions on the interval $-\pi \leqslant t \leqslant \pi$, satisfying the conditions (9) of Sec. 67.

1b) All solutions are given by the formula
$$x_n(t) = A_n \cos nt + B_n \sin nt \qquad (n = 0, 1, 2, \ldots),$$
where A_n and B_n are arbitrary constants.

2a) The equation $[(t^2 - 1)x'(t)]' = \lambda x(t)$ has solutions only for $\lambda = n(n + 1)$, $n = 0, 1, 2, \ldots$, in the class of twice differentiable functions on the interval $-1 \leqslant t \leqslant 1$.

2b) All solutions are given by the formula

$$x_n(t) = C_n P_n(t),$$

where $P_n(t)$ denotes the Legendre polynomial of order n (Sec. 58) and C_n is an arbitrary constant.

The assertions 1a and 2a follow from the fact that by Lemma 1 of Sec. 66, assuming the contrary leads to the existence of functions orthogonal to all the trigonometric functions $\sin nt$, $\cos nt$ $(n = 0, 1, 2, \ldots)$ or to all the Legendre polynomials $P_n(t)$ $(n = 0, 1, 2, \ldots)$; however, this is impossible, since both of these orthogonal systems are complete.

The assertions 1b and 2b mean that the eigenspaces $R^{(1)}, R^{(2)}, \ldots, R^{(n)}, \ldots$ of the Sturm-Liouville operators under consideration have dimensions 2 and 1, respectively. This can be seen as follows: It is clear that the dimension of the eigenspace $R^{(n)}$ of the operator $Lx = x''$, corresponding to the value $\lambda = -n^2$, is not less than 2, since the equation $x'' = -n^2x$ has two linearly independent solutions $x_1 = \cos nt$ and $x_2 = \sin nt$, satisfying the boundary conditions. It is also clear that the dimension of the eigenspace $R^{(n)}$ of the operator $Lx = [(t^2 - 1)x']'$, corresponding to the value $\lambda = n(n + 1)$, is not less than 1, since the equation $Lx = n(n + 1)x$ has the nonzero solution $x = P_n(t)$. Moreover, if the dimension of either eigenspace $R^{(n)}$ were greater than that just given, we could find a vector z in $R^{(n)}$ orthogonal to all the eigenvectors already known to be in $R^{(n)}$. Then, by Lemma 1 of Sec. 66 again, this vector would be orthogonal to all the other known eigenvectors for the other values of n. Therefore, because of the completeness of the system of eigenvectors, the vector z would have to be zero.

Comment. The results 1a and 1b can be obtained directly by using the fact that the general solution of the equation $x''(t) = \lambda x(t)$ is given by the formula

$$x(t) = A \cos \sqrt{-\lambda} \, t + B \sin \sqrt{-\lambda} \, t.$$

*87.6. Using Theorem 39, we can generalize the theorem on isomorphism of finite-dimensional Euclidean spaces (Sec. 52) to a certain class of infinite-dimensional spaces. First, we introduce the following definition:

A Euclidean space R is said to be *separable* if we can find a countable set $E \subset R$ which has any point $x \in R$ as a limit point (Sec. 82.2).

Example 1. The real line, considered as a one-dimensional Euclidean space R_1, is separable. For the countable set E, we can choose the set of all rational numbers.

Example 2. A finite-dimensional Euclidean space R_n is separable. If we choose an orthonormal basis $e_1, e_2, \ldots, e_n$ in R_n, then for the countable set E we can take the set of all vectors whose components with respect to the basis $e_1, e_2, \ldots, e_n$ are rational numbers.

Example 3. The space l_2 (Sec. 83.3) is separable. For the set E, we take the set of elements of l_2 of the form $(\xi_1, \xi_2, \ldots, \xi_n, 0, 0, \ldots)$, where the components $\xi_1, \xi_2, \ldots, \xi_n$ are rational numbers and n takes all integral values.

THEOREM 41. *Every complete separable infinite-dimensional Euclidean space R is isomorphic to the space l_2.*

Proof. First, we show that there exists a complete orthogonal system of vectors in the space R. To see this, we write down all the points of the countable set $E \subset R$ as a sequence $f_1, f_2, \ldots, f_n, \ldots$, and then subject this sequence to the orthogonalization process (Sec. 57). Let $\{e_k\}$, $(k = 1, 2, \ldots)$ be the orthogonal system of vectors which is obtained in this way. We now show that $\{e_k\}$ is complete. Let the vector $x \neq 0$ be orthogonal to every vector e_k $(k = 1, 2, \ldots)$. Since for any n, the vector f_n belongs to the linear manifold spanned by the vectors $e_1, e_2, \ldots, e_n$, the element

x is also orthogonal to every function f_n $(n = 1, 2, \ldots)$. By the Pythagorean theorem, we have

$$\|x - f_n\|^2 = \|x\|^2 + \|f_n\|^2 \geqslant \|x\|^2 > 0,$$

so that from the vectors $\{f_n\}$ we cannot form a sequence converging to the vector x, which contradicts the separability hypothesis. Thus, there cannot exist a vector $x \neq 0$ which is orthogonal to all the vectors $e_1, e_2, \ldots, e_n, \ldots$, and hence the system $\{e_k\}$ is complete, as asserted.

As always, we can assume that the system $\{e_k\}$ is normalized. By Theorem 39, every vector $x \in R$ has the expansion

$$x = \sum_{k=1}^{\infty} (x, e_k)e_k,$$

where the coefficients (x, e_k) form a sequence with a convergent sum of squares, i.e., a sequence in the space l_2. We now associate the vector

$$\bar{x} = \{(x, e_1), (x, e_2), \ldots\} \in l_2$$

with a given vector $x \in R$. Moreover, if $\bar{x} = (\xi_1, \xi_2, \ldots)$ is any vector in the space l_2, we associate with $\bar{x}$ the series $\sum_{k=1}^{\infty} \xi_k e_k$. Arguing as we did in the proof of Theorem 39, and using the completeness of the space R, we find that this series converges to an element $x \in R$. By the uniqueness of a Fourier series, the Fourier coefficients of the vector x coincide with the numbers ξ_n $(n = 1, 2, \ldots)$. Thus, the correspondence just established between R and l_2 is one-to-one. It is obvious that this correspondence preserves linear operations.

We now verify that scalar products are preserved by this correspondence. If

$$x = \sum_{k=1}^{\infty} \xi_k e_k, \quad y = \sum_{k=1}^{\infty} \eta_k e_k, \quad \bar{x} = \{\xi_k\} \in l_2, \quad \bar{y} = \{\eta_k\} \in l_2,$$

then, by the Remark to Theorem 39 of Sec. 37

$$(x, y) = \sum_{k=1}^{\infty} \xi_k \eta_k,$$

while

$$(\bar{x}, \bar{y}) = \sum_{k=1}^{\infty} \xi_k \eta_k$$

by the very definition of the scalar product in l_2. Thus

$$(x, y) = (\bar{x}, \bar{y})$$

as required. This completes the proof of Theorem 41.

The space $L_2(a, b)$ of square integrable functions on the interval $[a, b]$ is also a separable space. In fact, we saw in Sec. 82.4 that there is a countable set of polynomials S which is dense in the space $C_2(a, b)$. On the other hand, $C_2(a, b)$ is dense in its completion $L_2(a, b)$, and hence, by Lemma 2 of Sec. 82.4, the set S is dense in the space $L_2(a, b)$, i.e., the space $L_2(a, b)$ is separable. Since $L_2(a, b)$ is complete, applying Theorem 39, we find that the space $L_2(a, b)$ is isomorphic to the space l_2.

88. Bounded Operators. Completely Continuous Operators

We saw in Sec. 66 that a symmetric linear operator defined on a finite-dimensional Euclidean space always has eigenvectors. In an infinite-dimensional Euclidean space, this fact is in general no longer true. In fact, using the example of the operator corresponding to multiplication by the independent variable in the space $C_2(a, b)$ (Example 7 of Sec. 26, Example 5 of Sec. 64, Example 1 of Sec. 67), we convinced ourselves that there exist symmetric linear operators with no eigenvectors at all. Nevertheless, there is a large class of symmetric linear operators, of importance in mathematical physics, which do have eigenvectors (in fact, a "sufficiently large number" of them). This and subsequent sections will be devoted to a description of these operators.

88.1. Suppose we are given a linear operator A, defined on a Euclidean space R. Assume that the numerical function $\|Ax\|$ is bounded as x varies over the unit sphere in R. We considered this case in Sec. 53, and we called the quantity

$$\|A\| = \sup_{\|x\| \leqslant 1} \|Ax\|$$

the *norm* of the operator A. An operator with a finite norm is said to be *bounded*. If A is any bounded operator, we have the inequality

$$\|Ax\| \leqslant \|A\| \, \|x\|$$

for any $x \in R$. This inequality implies that *every bounded operator is continuous*, in the sense that if a sequence of vectors $x_n \in R$ converges to a vector x as $n \to \infty$, then $Ax_n \to Ax$. This follows from the relation

$$\|Ax - Ax_n\| = \|A(x - x_n)\| \leqslant \|A\| \, \|x - x_n\|.$$

An important example of a bounded linear operator defined on the Euclidean space $L_2(a, b)$ is the *Fredholm integral operator*

$$y(t) \equiv Ax(t) = \int_a^b K(t, s)x(s) \, ds. \tag{9}$$

The function $K(s, t)$, called the *kernel* of the integral operator A, is assumed to be square integrable in the region $a \leqslant s, t \leqslant b$, i.e.,

$$\int_a^b \int_a^b K^2(t, s) \, dt \, ds = K^2 < \infty.$$

By Fubini's theorem (Sec. 85), the function

$$k^2(t) = \int_a^b K^2(t, s) \, ds$$

exists and has the integral

$$\int_a^b k^2(t) \, dt = \int_a^b \int_a^b K^2(t, s) \, dt \, ds = K^2,$$

i.e., $k(t) \in L_2(a, b)$. Since $K(t, s)$, regarded as a function of s, belongs to $L_2(a, b)$, the integral (9), which represents the scalar product of the functions $K(t, s)$ and $x(s)$, exists for any $x(t) \in L_2(a, b)$. Estimating this scalar product by using the Schwarz inequality, we obtain

$$|y(t)|^2 \leqslant \int_a^b K^2(t, s) \, ds \int_a^b x^2(s) \, ds = k^2(t)\|x\|^2.$$

If $\|x\| \leqslant 1$, then

$$|y(t)|^2 \leqslant k^2(t),$$

so that

$$\|y\|^2 = \int_a^b |y(t)|^2 \, dt \leqslant \int_a^b k^2(t) \, dt \leqslant K^2.$$

Thus, the Fredholm operator A is bounded on the space $L_2(a, b)$, and its norm does not exceed K, i.e.,

$$\|A\|^2 \leqslant \int_a^b \int_a^b K^2(t, s) \, dt \, ds. \tag{10}$$

88.2. A set F of points in a Euclidean space R is called *compact* if every infinite set $F' \subset F$ contains at least one fundamental sequence.

Example 1. Every bounded infinite set of points F on the real line R is compact, according to the familiar Bolzano-Weierstrass theorem.[5] An unbounded set of points of R is not compact, since we can always choose a subset (which one?) of this set which does not contain a convergent sequence.

Example 2. We now verify the analogous assertion for a bounded infinite set F in an n-dimensional Euclidean space R_n. Let F' be any infinite subset of F. Choose an arbitrary orthonormal basis in R_n; in this basis, every vector x has components $\xi_1, \xi_2, \ldots, \xi_n$. The set of all first components $\{\xi_1\}$

---

[5] See, e.g.: L. Brand, *op. cit.*, p. 28.

of vectors in F' is bounded and therefore, by the Bolzano-Weierstrass theorem, contains a convergent sequence. Thus, from the set F', we can choose a sequence of vectors $x_1, x_2, \ldots, x_m, \ldots$, whose first components form a convergent sequence (of numbers). The set of second components $\{\xi_2\}$ of the vectors of this sequence is again bounded; therefore, by discarding certain superfluous vectors and renumbering the rest, we obtain a new sequence of vectors $x_1, x_2, \ldots, x_m, \ldots$, such that not only their first components but also their second components form a convergent sequence. Continuing this process, we ultimately arrive at a sequence of vectors $x_1, x_2, \ldots, x_m, \ldots$, which is such that each of their n components forms a convergent sequence as $m \to \infty$. But then the sequence $x_1, x_2, \ldots, x_m, \ldots$ itself is convergent in the metric of the space R_n. Thus, *every bounded infinite set in R_n is compact.* In particular, *the unit sphere in R_n is compact.*

Remark. Finite subsets of a metric space can also be regarded as compact, and all theorems which we shall prove for compact sets are trivially true for finite sets.

LEMMA 1. *A compact set in any Euclidean space is bounded.*

Proof. Suppose that a compact set $F \subset R$ is unbounded. Then, for any integer n, we can find a point $x_n \in F$ for which $\|x_n\| \geqslant n$. The subset $F' \subset F$ consisting of the points $x_1, x_2, \ldots, x_n, \ldots$ cannot contain a fundamental sequence, since by Lemma 2 of Sec. 83, every fundamental sequence is bounded. Thus, F is not compact, which proves the lemma.

In particular, for infinite subsets of an n-dimensional Euclidean space, compactness and boundedness are equivalent. However, in infinite-dimensional Euclidean spaces, bounded infinite sets need not be compact.

Example. The unit sphere of an infinite-dimensional Euclidean space is not *compact.* On the one hand, it contains an infinite system of orthonormal vectors $e_1, e_2, \ldots, e_n, \ldots$; on the other hand, since

$$\|e_n - e_m\|^2 = (e_n - e_m, e_n - e_m) = 2,$$

we cannot select a convergent sequence from this system.

Problem. Show that in the space l_2, the set of points $x = (\xi_1, \xi_2, \ldots, \xi_n, \ldots)$, obeying the inequalities $|\xi_n| \leqslant 1/n$ $(n = 1, 2, \ldots)$, is compact.

88.3. A linear operator A, defined on a Euclidean space R, is said to be *completely continuous* if it maps the unit sphere of the space R into a compact set. Any linear operator defined on a finite-dimensional Euclidean space is completely continuous, since in this case any linear operator maps the unit sphere into a bounded set (see Sec. 53), and every bounded set is compact in a finite-dimensional space. However, in an infinite-dimensional

Euclidean space, even the unit operator E is not completely continuous, since it maps the unit sphere into itself, i.e., into a noncompact set. If a bounded linear operator A maps the space R into a finite-dimensional space R_0, then A is completely continuous, for in this case the image of the unit sphere of the space R is a bounded set in the space R_0, a set which is compact, since R_0 is finite-dimensional.

LEMMA 2. *Let* $A_1, A_2, \ldots, A_n, \ldots$ *be a sequence of linear operators in a Euclidean space* R *which converges to the operator* A *in the sense that* $\|A - A_n\| \to 0$ *as* $n \to \infty$. *If the operators* A_n $(n = 1, 2, \ldots)$ *are completely continuous, then the linear operator* A *is also completely continuous.*

Proof. Let $f_1, f_2, \ldots, f_n, \ldots$ be any sequence of vectors of the space R whose norms do not exceed 1. Since the operator A_1 is completely continuous, the sequence $A_1 f_1, A_1 f_2, \ldots, A_1 f_n, \ldots$ contains a fundamental sequence $A_1 f_{m1}$. Similarly, using the complete continuity of the operator A_2, we can choose from the sequence $A_2 f_{m1}$ a fundamental sequence $A_2 f_{m2}$. Continuing in this way, we select from each sequence $A_k f_{m,k-1}$ a fundamental sequence $A_{km} f_{mk}$ $(m \to \infty)$. Thus, the diagonal sequence $f_{11}, f_{22}, \ldots, f_{mm}, \ldots$ is mapped into a fundamental sequence by every operator A_k.

We now show that the operator A also maps the sequence f_{mm} into a fundamental sequence. Thus, for any given $\varepsilon > 0$, we first find a number k such that

$$\|A - A_k\| < \frac{\varepsilon}{4},$$

and then find another number m such that

$$\|A_k f_{pp} - A_k f_{qq}\| < \frac{\varepsilon}{2}$$

for any $p, q \geqslant m$. Then we have

$$
\begin{aligned}
\|A f_{pp} - A f_{qq}\| &= \|A(f_{pp} - f_{qq})\| \\
&\leqslant \|A_k(f_{pp} - f_{qq})\| + \|(A - A_k)(f_{pp} - f_{qq})\| \\
&\leqslant \frac{\varepsilon}{2} + \|A - A_k\| \, \|f_{pp} - f_{qq}\| \leqslant \frac{\varepsilon}{2} + \frac{\varepsilon}{4} \cdot 2 = \varepsilon.
\end{aligned}
$$

Thus, the sequence $A f_{mm}$ satisfies the Cauchy criterion, i.e., is fundamental, Q.E.D.

Lemma 2 allows us to prove easily that *the Fredholm integral operator* (9) *is completely continuous.* First we note that if the function $K(t, s)$ has the form

$$K(t, s) = \sum_{k=1}^{n} \varphi_k(t)\psi_k(s),$$

where $\varphi_k(s)$, $\psi_k(t)$ $(k = 1, 2, \ldots, n)$ are square integrable functions [such a kernel $K(t, s)$ is called *degenerate*], then the operator (9) corresponding to $K(t, s)$ is bounded and

$$Ax(t) = \int_a^b \sum_{k=1}^n \varphi_k(t)\psi_k(s)x(s) \, ds = \sum_{k=1}^n \left\{ \int_a^b \psi_k(s)x(s) \, ds \right\} \varphi_k(t),$$

i.e., the operator A maps the whole space $L_2(a, b)$ into the finite-dimensional subspace spanned by the functions $\varphi_1(t), \ldots, \varphi_n(t)$.   Therefore, a *Fredholm integral operator with a degenerate kernel is completely continuous*.   On the other hand, every function $K(t, s)$ which is square integrable on the region $a \leqslant t, s \leqslant b$ can be expanded in a double Fourier series

$$K(t, s) = \sum_{k,m} a_{km} \sin k\pi \frac{s - a}{b - a} \sin m\pi \frac{t - a}{b - a},$$

which converges in the mean to $K(t, s)$.   (See Example 4 of Sec. 87.4, where $a = 0, b = \pi$.)   The degenerate kernels formed from the partial sums of this series define a sequence of completely continuous operators A_n.   Moreover, $\|A_n - A\| \to 0$ as $n \to \infty$, by the inequality (10).   Therefore, since the operators A_n are completely continuous, it follows from Lemma 2 that the operator A is also completely continuous, Q.E.D.

Remark.   In particular, all the conditions for this result are satisfied if the kernel $K(t, s)$ is continuous in the region $a \leqslant s, t \leqslant b$.   In this case, the Fredholm operator maps every function $x(t) \in L_2(a, b)$ into a *continuous function*.   In fact, if we set

$$y(t) = \int_a^b K(t, s)x(s) \, ds,$$

then, for any t' and t'', we have

$$|y(t') - y(t'')| \leqslant \int_a^b |K(t', s) - K(t'', s)| \, |x(s)| \, ds,$$

from which it follows at once that the function $y(t)$ is continuous.

89. Eigenvectors of a Symmetric Completely Continuous Operator

LEMMA 1.   *A symmetric completely continuous operator has a maximal vector* (Sec. 53).

Proof.   Since, by hypothesis, the set of all vectors $y = Ax$, where $\|x\| \leqslant 1$, is compact in the space R, the set is also bounded (Lemma 1 of Sec. 88.2), i.e., $\sup \|y\| = M < \infty$.   The number M equals the norm $\|A\|$ of the operator A (Sec. 53).   Thus, the completely continuous operator A is *bounded* and hence *continuous* (Sec. 88.1). Choose a sequence

$$y_n = Ax_n, \qquad (n = 1, 2, \ldots),$$

where $\|x_n\| = 1$, such that $\lim\limits_{n \to \infty} \|y_n\| = M$. By hypothesis, we can choose a convergent subsequence from the sequence y_n. By discarding superfluous vectors and renumbering the rest, we can assume that the sequence y_n itself converges as $n \to \infty$; let $y = \lim\limits_{n \to \infty} y_n$. Because of the continuity of the norm (Lemma 1 of Sec. 82),

$$\|y\| = \lim_{n \to \infty} \|y_n\| = M.$$

We now verify that the vector $z = y/M$ is the required maximal vector.

Because of the continuity of the operator A, we have

$$Az = \lim_{n \to \infty} A\left(\frac{y_n}{M}\right) = \lim_{n \to \infty} A\left(\frac{Ax_n}{M}\right).$$

The vectors Ax_n/M belong to the unit sphere, and therefore the lengths of the vectors $A(Ax_n/M)$ do not exceed M. Applying Lemma 3 of Sec. 66, we obtain

$$M \geqslant \left\|A\left(\frac{Ax_n}{M}\right)\right\| = \frac{1}{M}\|A^2x_n\| \geqslant \frac{1}{M}\|Ax_n\|^2 \to M,$$

which implies that

$$\|Az\| = \lim \left\|A\left(\frac{Ax_n}{M}\right)\right\| = M,$$

i.e., z is a maximal vector of the operator A, Q.E.D.

Applying Lemma 5 of Sec. 66, we obtain

LEMMA 2. *A symmetric completely continuous operator has an eigenvector with eigenvalue* $+M = +\|A\|$ *or* $-M = -\|A\|$.

By applying the process given in Theorem 33 (Sec. 66), we can construct other eigenvectors. Before doing this, we prove the following lemma:

LEMMA 3. *Let* A *be a completely continuous operator. Then, every orthonormal set of eigenvectors of* A *whose eigenvalues have absolute values greater than a positive number* δ *is finite.*

Proof. Suppose that we have found an infinite system F of such eigenvectors. Each of these eigenvectors is mapped by the operator A

into the same vector multiplied by a factor greater than δ. Let e_i and e_j be any two of these eigenvectors, i.e.,

$$\|e_i\| = \|e_j\| = 1, \quad (e_i, e_j) = 0, \quad Ae_i = \lambda_i e_i, \quad Ae_j = \lambda_j e_j.$$

Then we have

$$\|Ae_i - Ae_j\|^2 = \|\lambda_i e_i - \lambda_j e_j\|^2 = \lambda_i^2 + \lambda_j^2 > 2\delta^2,$$

which means that the distance between the vectors obtained by applying the operator A to the vectors of the system F must be greater than $\sqrt{2}\delta$. But then no convergent sequence can be selected from this set of vectors. This contradicts the complete continuity of the operator A and thereby proves the lemma. In particular, *there exists only a finite number of orthogonal vectors with a given eigenvalue $\lambda \neq 0$, i.e., every eigenspace corresponding to a nonzero eigenvalue of a symmetric completely continuous operator A is finite-dimensional.*

Lemma 3 allows us to draw certain definite conclusions about the set of all eigenvectors and eigenvalues of the operator A. Thus, consider the set of all eigenvalues of the operator A, represented by points on the real line. By Lemma 3, there exists only a finite number of eigenvalues whose absolute values exceed a given positive number δ. Therefore, if there is an infinite set of eigenvalues, they must form a sequence which converges to zero. Thus, we can give all the eigenvalues integral indices and then arrange them in the order of decreasing absolute value; in doing this, we agree to provide every eigenvalue with as many consecutive indices as there are dimensions in the corresponding eigenspace. Then, with the sequence of all nonzero eigenvalues

$$\lambda_1, \lambda_2, \lambda_3, \ldots, \lambda_n, \ldots$$

of the operator A, we can associate a sequence of eigenvectors

$$e_1, e_2, e_3, \ldots, e_n, \ldots,$$

where $Ae_n = \lambda_n e_n \ (n = 1, 2, \ldots)$. Moreover, we can assume that the vectors $e_1, e_2, \ldots$ are orthonormal. In fact, if $\lambda_n \neq \lambda_m$, then e_n and e_m are orthogonal by Lemma 1 of Sec. 66, while if $\lambda_n = \lambda_m$, then we can orthogonalize e_n and e_m within the finite-dimensional eigenspace corresponding to the eigenvalue $\lambda_n = \lambda_m$. Finally, we complete the construction by normalizing all the vectors obtained in this way.

We now show that *every vector z orthogonal to all the vectors $e_1, e_2, \ldots, e_n, \ldots$ just constructed is mapped into zero by the operator A.*

Proof. Consider the set P of all vectors z orthogonal to all the vectors $e_1, e_2, \ldots, e_n, \ldots$. The set P is a closed subspace, since it is the orthogonal complement of the linear manifold $L(e_1, e_2, \ldots)$ (Sec. 82.2). Since the linear manifold $L(e_1, e_2, \ldots)$ is obviously invariant with respect to the operator A, its orthogonal complement P is also invariant with respect to A, by Lemma 2 of Sec. 66. We denote by $M(P)$ the least upper bound of the values of $\|Ax\|$ on the unit sphere of the subspace P. By Lemma 2, there is an eigenvector e_0 in the subspace P with either the eigenvalue $\lambda_0 = M(P)$ or the eigenvalue $-\lambda_0 = -M(P)$. But, by the very construction of the subspace P, it cannot contain an eigenvector e_0 with a nonzero eigenvalue. Therefore, $\lambda_0 = M(P) = 0$, which means that $Az = 0$ for every vector $z \in P$, Q.E.D.

Let R' denote the closure of the linear manifold $L(e_1, e_2, \ldots)$ and let R'' denote the orthogonal complement of R'. Then, every vector x in the original space R can be written as a sum

$$x = x' + x'',$$

where $x' \in R'$, $x'' \in R''$. Moreover, the vector x' can be expanded in Fourier series with respect to the system $e_1, e_2, \ldots, e_n, \ldots$ which is complete in the space R', and by what has just been proved, the operator A maps the vector x'' into zero. Thus, we obtain the following theorem:

THEOREM 42. *Let R be a complete Euclidean space and let A be a symmetric completely continuous operator defined on R. Then every vector $x \in R$ can be represented in the form of a sum of orthogonal vectors*

$$x = x' + x'' = \xi_1 e_1 + \xi_2 e_2 + \cdots + x'',$$

where $e_1, e_2, \ldots$ are the eigenvectors of A corresponding to nonzero eigenvalues and $Ax'' = 0$.

90. Eigenvectors of a Fredholm Operator

In particular, the results of Sec. 89 are valid for a Fredholm integral operator A with a symmetric kernel $K(s, t) \equiv K(t, s)$, which is square integrable in the region $a \leqslant t, s \leqslant b$. The symmetry of the kernel guarantees the symmetry of the operator A (Sec. 67.2),[6] and the square integrability of the kernel guarantees the complete continuity of A (Sec. 88). Thus, every vector $f = f(t)$ in the space $L_2(a, b)$ can be represented in the form of a series

$$f(t) = \sum_{k=1}^{\infty} \xi_k e_k(t) + z(t)$$

---

[6] The proof given in Sec. 67 of the fact that the operator A is symmetric is based on the assumption that the kernel $K(t, s)$ is continuous, but the proof goes through in the general case, if we use Fubini's theorem (Sec. 85).

which converges in the mean, where $e_1(t), e_2(t), \ldots$ is the orthonormal system of eigenfunctions of A corresponding to nonzero eigenvalues, and $Az = 0$. The coefficients ξ_k, as always, are calculated by the formula $\xi_k = (x, e_k)$.

By using the concrete nature of the operator A, we can strengthen our results somewhat. First of all, we note that the equation defining the eigenfunctions of the Fredholm operator

$$\int_a^b K(t, s)e_n(s) \, ds = \lambda_n e_n(t) \tag{11}$$

shows that the quantity $\lambda_n e_n(t)$ is the Fourier coefficient of the function $K(t, s)$, for a fixed value of t. Therefore, applying Bessel's inequality (Sec. 87), we find that

$$\int_a^b K^2(t, s) \, ds \geqslant \sum_{n=0}^N \lambda_n^2 e_n^2(t) \tag{12}$$

for every value of N. Integrating this inequality with respect to t, we obtain

$$\int_a^b \int_a^b K^2(t, s) \, dt \, ds \geqslant \sum_{n=0}^N \lambda_n^2$$

for any integer N, which implies that *the sum of the squares of the eigenvalues of a symmetric Fredholm operator converges.* (We note that this result is not true in the general case of a symmetric completely continuous operator.)

A function of the form

$$g(t) = \int_a^b K(t, s)f(s) \, ds, \tag{13}$$

with an arbitrary $f(s) \in L_2(a, b)$, is said to be *source-derivable from the kernel* $K(t, s)$. The function $g(t)$ can be interpreted physically as follows: Consider, for example, the distribution of temperature in a rod lying along the segment $a \leqslant t \leqslant b$. Suppose that we know that the function $K(t, s)$ gives the distribution of temperature produced in the rod by a heat source of unit strength located at the point s. Then the integral (13) represents the distribution of temperature in the rod when every point s has a heat source of strength $f(s)$. Therefore, $g(t)$ can be derived from sources, which explains the term "source-derivable" (Russian: истокообразно представимый).

We now have the following important theorem, due to Hilbert:

THEOREM 43. *If the symmetric, square integrable kernel $K(t, s)$ satisfies the inequality*

$$\int_a^b K^2(t, s) \, ds \leqslant C, \tag{14}$$

where the constant C is independent of t; then every function $g(t)$ which *is source-derivable from $K(t, s)$ can be expanded in an absolutely and uniformly convergent Fourier series*

$$g = \sum_{k=1}^{\infty} \lambda_k(f, e_k)e_k,$$

in terms of the eigenfunctions of the Fredholm operator A with the kernel $K(t, s)$.

Proof. Applying the operator A to the expansion

$$f(t) = \sum_{k=1}^{\infty} \xi_k e_k(t) + z(t) \qquad [\xi_k = (f, e_k)],$$

and using the continuity of A and the fact that $Az = 0$, we obtain

$$g = Af = \sum_{k=1}^{\infty} \xi_k Ae_k = \sum_{k=1}^{\infty} \lambda_k(f, e_k)e_k(t). \tag{15}$$

It remains to verify that this series converges absolutely and uniformly. We have

$$(g, e_k) = (Af, e_k) = (f, Ae_k) = (f, \lambda_k e_k) = \lambda_k(f, e_k),$$

so that

$$\left[\sum_{k=n}^{n+m} |(g, e_k)e_k(t)| \right]^2 = \left[\sum_{k=n}^{n+m} |(f, e_k)\lambda_k e_k(t)| \right]^2$$
$$\leqslant \sum_{k=n}^{n+m} (f, e_k)^2 \sum_{k=n}^{n+m} \lambda_k^2 e_k^2(t), \tag{16}$$

where we have used the Schwarz inequality for a finite sum [see formula (9) of Sec. 50].   Since in the right-hand side of (16), the sum

$$\lambda_n^2 e_n^2(t) + \lambda_{n+1}^2 e_{n+1}^2(t) + \cdots$$

is uniformly bounded, according to the inequalities (12) and (14), while the sum

$$(f, e_n)^2 + (f, e_{n+1})^2 + \cdots$$

can be made arbitrarily small for sufficiently large n, by Bessel's inequality [see formula (6) of Sec. 87], it follows that the sum in the left-hand side of (16) can be made arbitrarily small for sufficiently large n.   Therefore, the Fourier series (15) is absolutely and uniformly convergent, Q.E.D.

Remark. In particular, everything we have said applies to the case of a continuous symmetric kernel $K(t, s)$.   We note that in this case, all the eigenfunctions with nonzero eigenvalues are also continuous; this follows from the remark at the end of Sec. 88.

91. Solution of Inhomogeneous Integral Equations

A natural and convenient basis in which to solve problems relating to a Fredholm operator is the system of eigenfunctions of the operator. An example of such a problem is finding the solution of the inhomogeneous integral equation

$$\varphi(t) = f(t) + \int_a^b K(t, s)\varphi(s)\,ds,$$

where $f(t)$ and $K(t, s) \equiv K(s, t)$ are given functions and the function $\varphi(t)$ is to be determined. In an abstract space, the corresponding equation is

$$\varphi = f + A\varphi,$$

where A is a symmetric completely continuous operator. Projecting this equation onto the axis defined by the eigenvector e_k (where $Ae_k = \lambda_k e_k$), we obtain

$$(\varphi, e_k) = (f, e_k) + (A\varphi, e_k) = (f, e_k) + (\varphi, Ae_k)$$
$$= (f, e_k) + (\varphi, \lambda_k e_k) = (f, e_k) + \lambda_k(\varphi, e_k), \qquad (17)$$

which gives

$$(\varphi, e_k) = \frac{(f, e_k)}{1 - \lambda_k} \qquad (18)$$

if $\lambda_k \neq 1$. This uniquely defines the Fourier coefficients of the vector φ. In the case where $\lambda_k = 1$ and $(f, e_k) \neq 0$, we obtain a contradictory formula, and therefore the required vector φ cannot exist. If $\lambda_k = 1$ and $(f, e_k) = 0$, then (17) does not impose any conditions on the quantity (φ, e_k), which can therefore be arbitrary. Thus, we arrive at the following result:

If none of the eigenvalues of the operator A equals 1, then all the Fourier coefficients of the required solution, with respect to the orthonormal system $e_1, e_2, \ldots$, are given uniquely by formula (18). If some of the eigenvalues of A equal 1, and if f is not orthogonal to the subspace spanned by the corresponding eigenvectors, then the solution φ does not exist. However, if f is orthogonal to this subspace, then the corresponding Fourier coefficients of the solution φ can be chosen arbitrarily.

If the solution φ exists, then the vector $\varphi - f = A\varphi$ is "source-derivable," and by Theorem 43 can be expanded in a Fourier series

$$\varphi - f = \sum_{k=1}^{\infty} c_k e_k, \qquad (19)$$

which converges not only in norm, but also absolutely and uniformly, if the condition (14) of Sec. 90 is met.   Calculating the Fourier coefficient c_k in (19), we obtain

$$c_k = (\varphi - f, e_k) = (\varphi, e_k) - (f, e_k) = \frac{(f, e_k)}{1 - \lambda_k} - (f, e_k) = (f, e_k)\frac{\lambda_k}{1 - \lambda_k}$$

if $\lambda_k \neq 1$, so that

$$\varphi = f + \sum_{k=1}^{\infty} (f, e_k)\frac{\lambda_k}{1 - \lambda_k} e_k. \tag{20}$$

On the other hand, the series

$$\sum_{k=1}^{\infty} (f, e_k)\frac{\lambda_k}{1 - \lambda_k} e_k \qquad (\lambda_k \neq 1),$$

constructed without reference to these considerations, converges in norm in a complete space R, since the sum of squares of its coefficients obviously converges, by Bessel's inequality and the fact that $|\lambda_k/(1 - \lambda_k)|$ is bounded.[7] For a Fredholm operator satisfying the condition (14), this series also converges absolutely and uniformly, since, denoting the largest of the values $1/|1 - \lambda_k|$ by K, and making a calculation like that in Sec. 90, we have

$$\left[\sum_{k=n}^{n+m} |(f, e_k)\frac{\lambda_k}{1 - \lambda_k} e_k(t)|\right]^2 \leqslant K^2\left[\sum_{k=n}^{n+m} |(f, e_k)\lambda_k e_k(t)|\right]^2$$

$$\leqslant K^2 \sum_{k=n}^{n+m} (f, e_k)^2 \sum_{k=n}^{n+m} \lambda_k^2 e_k^2(t).$$

Thus, the vector φ, *defined* by formula (20), exists.   We now verify that φ is actually a solution of the equation $\varphi = f + A\varphi$.   In fact, we have

$$A\varphi = Af + \sum_{k=1}^{\infty} (f, e_k)\frac{\lambda_k}{1 - \lambda_k} \lambda_k e_k$$

$$= \sum_{k=1}^{\infty} \lambda_k(f, e_k)e_k + \sum_{k=1}^{\infty} (f, e_k)\frac{\lambda_k}{1 - \lambda_k} \lambda_k e_k$$

$$= \sum_{k=1}^{\infty} \left(1 + \frac{\lambda_k}{1 - \lambda_k}\right) \lambda_k(f, e_k)e_k = \sum_{k=1}^{\infty} \frac{\lambda_k}{1 - \lambda_k} (f, e_k)e_k = \varphi - f.$$

Finally, if some of the eigenvalues of the operator A equal 1, and the corresponding coefficients (f, e_k) are zero (which, as we have seen, is a

---

[7] The quantities

$$\frac{1}{|1 - \lambda_k|}, \qquad \left|\frac{\lambda_k}{1 - \lambda_k}\right|$$

$(\lambda_k \neq 0)$ are uniformly bounded, since $\lambda_k \to 0$ (Sec. 89).

necessary condition for the equation $\varphi = f + A\varphi$ to have a solution), then the series (20) also defines a solution, where we can now replace a finite number of coefficients

$$(f, e_k) \frac{\lambda_k}{1 - \lambda_k} \qquad (\lambda_k = 1)$$

(which have no meaning for $\lambda_k = 1$), by *arbitrary numbers*. This can be proved by using a calculation like that just given.

*92. Unbounded Operators with Inverses that are Symmetric and Completely Continuous

Let L be an unbounded linear operator in a Euclidean space R, with a domain of definition $R' \subset R$ which is a linear manifold. A bounded operator A defined on the whole space R is said to be the *inverse* of the operator L if $Ax \in R'$ and $LAx = x$ for any $x \in R$, and if $ALy = y$ for any $y \in R'$.

Example. Consider the space $C_2(a, b)$ and let the operator L be the differentiation operator

$$Lx(t) = x'(t),$$

defined for the functions $x(t) \in C_2(a, b)$ which equal zero at $x = a$ and have a continuous derivative. Let the operator A be the integral

$$Ax(t) = \int_a^t x(s) \, ds,$$

with a variable upper limit. It is easily seen (cf. the proof of the boundedness of the Fredholm operator given in Sec. 88) that the operator A is bounded on $C_2(a, b)$. Obviously, $LAx(t) = x(t)$ for any $x(t) \in C_2(a, b)$ and $ALx(t) = x(t)$ for any $x(t)$ in the domain of definition of the operator L. Therefore, A is the inverse operator of L.

Suppose that a given unbounded operator L has an inverse operator A. Then, it can be asserted that *all the eigenvalues of A are different from zero*. In fact, if $Ax = 0$ for some x, then $x = LAx$ would be the zero vector, which is impossible. Moreover, *every eigenvector of A with the eigenvalue λ is simultaneously an eigenvector of L with the eigenvalue $1/\lambda$*. In fact, if $Ax = \lambda x$, then $x = LAx = L\lambda x = \lambda Lx$, so that $Lx = (1/\lambda)x$.

We now assume that the operator A which is the inverse of L is symmetric and completely continuous. Applying the above result and Theorem 42 (Sec. 89), we find that *there exists a complete orthogonal system of eigenvectors of the operator A with nonzero eigenvalues, from which it follows that the operator L also has a complete orthogonal system of eigenvectors with nonzero eigenvalues*.

As an example, we take the operator L to be the Sturm-Liouville differential operator (Sec. 67). In the theory of differential equations, there is the following theorem, which we cite without proof:[8] Consider the Sturm-Liouville equation $Lx(t) = 0$, on the interval $a \leqslant t \leqslant b$. Suppose this equation has no nonzero solutions satisfying the boundary conditions guaranteeing the symmetry of the Sturm-Liouville operator (Sec. 67). (In this case, the equation and the corresponding Sturm-Liouville operator are called *nonsingular*.) Then, there exists a symmetric continuous function of two variables $K(t, s)$, the *Green's function*, with the property that if $\varphi(t) \in C_2(a, b)$, then the function $x(t)$ defined by

$$x(t) = \int_a^b K(t, s)\varphi(s)\, ds \tag{21}$$

satisfies the boundary conditions and the differential equation

$$Lx(t) = \varphi(t). \tag{22}$$

Conversely, every function $x(t)$ satisfying the boundary conditions and (22) can be represented by formula (21). In other words, a nonsingular Sturm-Liouville operator L has a bounded inverse operator, which is symmetric and completely continuous. Thus, the following important result follows from what was proved earlier:

A nonsingular Sturm-Liouville operator has a complete orthogonal system of eigenfunctions $e_1(t), e_2(t), \ldots, e_n(t), \ldots$.

Moreover, suppose that the twice-differentiable function $x(t)$ satisfies the boundary conditions, and set $\varphi(t) = Lx(t)$. Then, by the theorem just cited, $x(t)$ is source-derivable from the kernel $K(t, s)$, and hence, by Theorem 43 (Sec. 90), $x(t)$ can be expanded as an absolutely and uniformly convergent Fourier series with respect to the eigenfunctions $e_1(t), e_2(t), \ldots$. Thus, we obtain the following theorem:

THEOREM 44. (*Steklov's theorem*). *Let L be a nonsingular Sturm-Liouville operator with eigenfunctions $e_1(t), e_2(t), \ldots$. Then, the Fourier series with respect to $e_1(t), e_2(t), \ldots$ of any twice-differentiable function $x(t)$ satisfying the boundary conditions converges absolutely and uniformly.*

Example 1. Consider the Sturm-Liouville equation $y'' = 0$ on the interval $0 \leqslant t \leqslant \pi$, with boundary conditions $y(0) = 0$, $y(\pi) = 0$. The unique solution of this equation satisfying the given boundary conditions is $y = 0$, so that the hypothesis of the theorem cited above is met. In this case, the theorem states that the operator $Ly = y''$ has a bounded inverse, which is

---

[8] A proof can be found, e.g., in I. G. Petrovsky, *Lectures on Partial Differential Equations* (translated by A. Shenitzer), Interscience Publishers, Inc., New York, Sec. 24 (1957).

symmetric and completely continuous. Moreover, the operator has a complete orthonormal system of eigenfunctions $e_1(t)$, $e_2(t)$, ... with nonzero eigenvalues, and any twice-differentiable function $f(t) \in C_2(0, \pi)$ which satisfies the boundary conditions can be expanded in an absolutely and uniformly convergent Fourier series with respect to these eigenfunctions. Obviously, in this case, the eigenfunctions are

$$e_n(t) = \sin nt \qquad (n = 1, 2, \ldots).$$

Since they form a complete orthogonal system (Sec. 87), there are no other eigenfunctions. In this case, the expansion theorem asserts that every twice-differentiable function $f(t) \in C_2(0, \pi)$, which satisfies the conditions $f(0) = f(\pi) = 0$, can be expanded in an absolutely and uniformly convergent series

$$f(t) = \sum_{n=1}^{\infty} b_n \sin nt.$$

Of course, this result can also be obtained from a general theorem on the expansion of functions in trigonometric Fourier series.

Example 2. If we wish to apply Theorem 44 to the Sturm-Liouville operator $Ly = y''$ in the space $C_2(-\pi, \pi)$, with the boundary conditions $y(-\pi) = y(\pi)$, $y'(-\pi) = y'(\pi)$, or to the operator $Ly = [(t^2 - 1)y']'$ in the space $C_2(-1, 1)$, which we considered previously (Secs. 67, 87), then we encounter a difficulty associated with the fact that the corresponding equation $Ly = 0$ has a nonzero solution $y(t) \equiv 1$, so that the hypothesis of Theorem 44 is not met. The following lemma allows us to obviate this difficulty.

LEMMA. *Suppose that the linear operator A has eigenvectors e_1, e_2, ..., e_n, ... with corresponding eigenvalues λ_1, λ_2, ..., λ_n, ... Consider the linear operator $A - \lambda_0 E$, where λ_0 is a fixed number. Then the vectors e_1, e_2, ..., e_n, ... are also eigenvectors of the operator $A - \lambda_0 E$, with corresponding eigenvalues $\lambda_1 - \lambda_0$, $\lambda_2 - \lambda_0$, ..., $\lambda_n - \lambda_0$, ...*

Proof. Applying the operator $A - \lambda_0 E$ to the vector e_n, we obtain

$$(A - \lambda_0 E)e_n = Ae_n - \lambda_0 e_n = (\lambda_n - \lambda_0)e_n,$$

i.e., the vector e_n is also an eigenvector of the operator $A - \lambda_0 E$, and the corresponding eigenvalue is $\lambda_n - \lambda_0$, Q.E.D.

We now consider the Sturm-Liouville operators just mentioned. As we already know, the operator $Ly = y''$ in the space $C_2(-\pi, \pi)$ has the eigenfunctions 1, $\cos t$, $\sin t$, ..., $\cos nt$, $\sin nt$, ..., with corresponding eigenvalues 0, -1, -1, ..., $-n^2$, $-n^2$, ..., and moreover this operator has no other eigenfunctions with different eigenvalues (Sec. 87). Choose a number λ_0

which is different from all the numbers $-n^2$ $(n = 0, 1, 2, \ldots)$. Then, by our lemma, the operator $Ly - \lambda_0 y$ has the same system of eigenfunctions with eigenvalues $-n^2 - \lambda_0$, which are all different from zero. The operator $Ly - \lambda_0 y$ has no other eigenvalues, since the system of eigenfunctions 1, $\cos t$, $\sin t$, ... is complete in the space $C_2(-1, 1)$. Applying Theorem 44, we find that any twice-differentiable function $f(t)$ satisfying the conditions $f(-\pi) = f(\pi)$, $f'(-\pi) = f'(\pi)$, can be expanded in an absolutely and uniformly convergent Fourier series

$$f(t) = \frac{a_0}{2} + \sum_{n=1}^{\infty} (a_n \cos nt + b_n \sin nt),$$

for $-\pi \leqslant t \leqslant \pi$. Similarly, applying our lemma in the case of the operator $Ly = [(t^2 - 1)y']'$ in the space $C_2(-1, 1)$, we find that any twice-differentiable function $f(t) \in C_2(-1, 1)$ has a series expansion in terms of the Legendre polynomials.

More generally, a theorem on expansion in trigonometric Fourier series can be proved for a function $f(t) \in C_2(-\pi, \pi)$ which satisfies less restrictive conditions (e.g., piecewise continuity of $f'(t)$). As we see, the theorem obtained here imposes more restrictive conditions on the function $f(t)$, but as a result the theorem is of a much more general nature. However, even in the general case, we can avoid imposing too restrictive conditions on the function $f(t)$. In fact, it turns out that any condition guaranteeing the uniform convergence of the trigonometric Fourier series of a function $f(t)$ at the same time also guarantees the uniform convergence of the series expansion of $f(t)$ in terms of the eigenfunctions of any Sturm-Liouville operator.[9]

93. Calculation of Eigenfunctions and Eigenvalues

To make practical application of our results, we have to know the eigenfunctions $e_k(t)$ $(k = 1, 2, \ldots)$ of the Fredholm operator under consideration. We now discuss the problem of finding these eigenfunctions and the corresponding eigenvalues.

93.1. If the kernel $K(t, s)$ of the Fredholm operator A is degenerate, i.e.,

$$K(t, s) = \sum_{k=1}^{m} p_k(t)q_k(s),$$

then, as we have seen (Sec. 88.3), the operator A maps the whole space L_2

---

[9] See B. M. Levitan, Разложение по Собственным Функциям Дифференциальных Уравнений Второго Порядка (*Expansion in Eigenfunctions of Second-Order Differential Equations*), Gostekhizdat, Moscow-Leningrad, Chap. 1 (1950).

into a finite-dimensional subspace spanned by the vectors $p_k(t)$ ($k = 1, 2, \ldots, m$).   Thus, to find eigenfunctions with nonzero eigenvalues, we need only look in this subspace, and the eigenfunctions have the form

$$e(t) = \sum_{k=1}^{m} c_k p_k(t). \tag{23}$$

To determine the coefficients c_k, we substitute the function (23) into equation (11) of Sec. 90, obtaining

$$\lambda e(t) = \sum_{k=1}^{m} \lambda c_k p_k(t) = \int_a^b K(t, s) \sum_{i=1}^{m} c_i p_i(s)\, ds$$
$$= \sum_{k=1}^{m} \sum_{i=1}^{m} c_i \left(\int_a^b p_i(s) q_k(s)\, ds \right) p_k(t)$$
$$= \sum_{k=1}^{m} \sum_{i=1}^{m} c_i p_{ik} p_k(t),$$

where

$$p_{ik} = \int_a^b p_i(s) q_k(s)\, ds. \tag{24}$$

Therefore, we have

$$\lambda c_k = \sum_{i=1}^{m} c_i p_{ik} \qquad (k = 1, 2, \ldots, m). \tag{25}$$

This system can be solved in the usual way to find λ and the constants c_k.

The system (25) takes a particularly simple form when $p_i(s) \equiv q_i(s)$ and $(p_i, q_k) = 0$ for $i \neq k$. Then, the quantities p_{ik} vanish for $i \neq k$, and the system (25) has the obvious solution $c_k = 1$ for some k, $c_i = 0$ for $i \neq k$, $\lambda = p_{kk}$. By (23), the eigenfunction corresponding to this solution is the function $p_k(t) \equiv q_k(t)$. Thus, in this case, the eigenfunctions are just the functions $p_k(t)$ ($k = 1, 2, \ldots, m$), and the eigenvalues are the numbers p_{kk}, i.e., the squared norms of the functions $p_k(t)$.

The following problems are to be solved by the method just given.

Solve the following four integral equations:

Problem 1. $\varphi(t) = 3 \int_0^2 st\varphi(s)\, ds + 3t - 2.$ *Ans.* $\varphi(t) = \frac{9}{7}t - 2.$

Problem 2. $\varphi(t) = 3 \int_0^1 st\varphi(s)\, ds + 3t - 2.$
Ans. $\varphi(t) = Ct - 2$, where C is arbitrary.

Problem 3. $\varphi(t) = \int_0^1 (s + t)\varphi(s)\, ds + 18t^2 - 9t - 4.$
Ans. $\varphi(t) = 18t^2 + 12t + 9.$

Problem 4. $\varphi(t) = \int_0^\pi \cos(s + t)\varphi(s)\,ds + 1.$

$$Ans. \quad \varphi(t) = 1 - \frac{2\sin t}{1 - (\pi/2)}.$$

93.2. If the kernel $K(t, s)$ of the Fredholm operator A is nondegenerate, then one often uses the following approximate method for calculating eigenvectors and eigenvalues: The given kernel $K(t, s)$ is replaced by degenerate kernels $K_n(t, s)$ which approximate $K(t, s)$ (e.g., partial sums of the Fourier series), and then the eigenfunctions and eigenvalues of the operator A_n corresponding to $K_n(t, s)$ are found by the method described above. It turns out that under certain assumptions concerning the smoothness of the kernel $K(t, s)$, the eigenfunctions and eigenvalues of the operator A_n converge to the corresponding eigenfunctions and eigenvalues of the operator A. Since we do not have the space to go into these matters here, we refer the reader to the special literature of the subject.[10]

94. Integral Equations with Nonsymmetric Kernels. The Fredholm Alternative

94.1. Consider the integral equation

$$\varphi(t) - \int_a^b K(t, s)\varphi(s)\,ds = f(t), \tag{26a}$$

where the kernel $K(t, s)$ is no longer assumed to be symmetric, but it is still assumed that $K(t, s)$ is square integrable in the region $a \leqslant t, s \leqslant b$. The function $f(t)$ is assumed to belong to the space $L_2\,(a, b)$, and we look for a solution $\varphi(t)$ in the same space. For $f(t) \equiv 0$, we obtain a homogeneous equation, in which we denote the unknown function by $\varphi_0(t)$:

$$\varphi_0(t) - \int_a^b K(t, s)\varphi_0(s)\,ds = 0. \tag{26b}$$

Together with equations (26a) and (26b), it turns out to be natural to study the "adjoint" equations, with the kernel $K(s, t)$, obtained from the original kernel by transposing the arguments:

$$\psi(t) - \int_a^b K(s, t)\psi(s)\,ds = g(t), \tag{26c}$$

$$\psi_0(t) - \int_a^b K(s, t)\psi_0(s)\,ds = 0. \tag{26d}$$

---

[10] See R. Courant and D. Hilbert, *Methods of Mathematical Physics*, Interscience Publishers, Inc., New York, vol. 1, Chap. 3, Sec. 8.4 (1953); and L. V. Kantorovich and V.. I. Krylov, *Approximate Methods of Higher Analysis* (translated by C. D. Benster), Interscience Publishers, Inc., New York, Chap. 2, Sec. 4 (1958).

The following basic theorem, called the *Fredholm alternative*, establishes the relation between the solutions of equations (26a), (26b), (26c), and (26d):

THEOREM 45. *Two cases are possible:* I) *Either* (26b) *has only the solution* $\varphi_0(t) \equiv 0$, *or* II) (26b) *has a solution* $\varphi_0(t) \not\equiv 0$.

In case I, (26a) *has a unique solution for every* $f(t) \in L_2$, (26d) *has the unique solution* $\psi_0(t) \equiv 0$, *and* (26c) *has a unique solution for every* $g(t) \in L_2$.

In case II, *the number of linearly independent solutions of* (26b) *is finite; let this number be* ν. *Then* (26d) *also has* ν *linearly independent solutions. Equation* (26a) *has a solution if and only if the function* $f(t)$ *is orthogonal to all* ν *solutions of* (26d); *this solution is not unique, but rather is defined only to within a term which is a solution of* (26b). *Moreover, only one of the solutions of* (26a) *is orthogonal to all the solutions of* (26b). *Analogous assertions hold for the solutions of* (26c).

94.2. First, we consider the analog of Theorem 45 for the case of a system of algebraic equations:

$$\sum_{j=1}^{m} a_{ij}\xi_j = b_i \qquad (i = 1, 2, \ldots, m). \tag{27a}$$

We write the corresponding homogeneous system

$$\sum_{=1}^{m} a_{ij}\xi_j^0 = 0, \tag{27b}$$

and the adjoint systems

$$\sum_{j=1}^{m} a_{ji}\eta_j = c_i, \tag{27c}$$

$$\sum_{j=1}^{m} a_{ji}\eta_j^0 = 0, \tag{27d}$$

whose matrix is obtained by transposing the matrix of the systems (27a) and (27b). We now examine the algebraic analogs of the statements made in Theorem 45.

I) Assume that the system (27b) has only the trivial solution. This means that the rank of the matrix $A = \|a_{ij}\|$ equals m (Sec. 20), i.e., $\det A \neq 0$. Therefore the system (21a) has a solution for arbitrary constant terms b_i (Sec. 7). The determinant of the system (27d) is

$$\det \|a_{ji}\| = \det \|a_{ij}\|$$

(Sec. 3), which is also nonvanishing. Therefore, the system (27c) has a unique solution for any constant terms c_i. In particular, the unique solution of

(27d) [corresponding to setting $c_i = 0$ in (27c)] is $\eta_j^0 = 0$. This verifies the analogs of all the statements made in case I of Theorem 45.

II) Suppose that the system (27b) has a nontrivial solution ξ_j^0. This means that the rank r of the matrix A is less than m (Sec. 20). The number ν of linearly independent solutions of (27a) equals $m - r$ (Sec. 23). Since the rank of a matrix remains unchanged when the matrix is transposed (Sec. 9), the number of linearly independent solutions of the system (27d) also equals $m - r = \nu$. The system (27a) no longer has solutions for all values of the constant terms b_i. To discover the conditions which have to be imposed on the b_i if the system (27a) is to have a solution, we interpret (27a) geometrically, regarding the set of numbers $(\xi_1, \xi_2, \ldots, \xi_m)$ as a vector in an m-dimensional Euclidean space T_m (Sec. 49). The existence of a solution of (27a) is equivalent to the assertion that the vector $b = (b_1, b_2, \ldots, b_m)$ is an element of the linear manifold spanned by the vectors

$$
\begin{aligned}
a_1 &= (a_{11}, a_{21}, \ldots, a_{m1}), \\
a_2 &= (a_{12}, a_{22}, \ldots, a_{m2}), \\
&\quad \ldots \\
a_m &= (a_{1m}, a_{2m}, \ldots, a_{mm}).
\end{aligned}
$$

If Z is the orthogonal complement of this linear manifold, then what we have just said can also be expressed as follows: The system (27a) has a solution if and only if the vector b is orthogonal to the subspace Z. The condition for a vector η^0 to belong to the subspace Z can be written in the form of the system (27d). It follows that the system (27a) has a solution if and only if the vector b is orthogonal to any solution of the system (27d). Moreover, in this case, the system (27a) has a whole set of solutions, the geometric model of which is the hyperplane parallel to the subspace of solutions of the system (27b) (Sec. 23). The perpendicular dropped from the origin of coordinates onto this hyperplane (Sec. 56) is uniquely distinguished from all the other solutions by the fact that it is orthogonal to all the solutions of the system (27b). Thus, we have verified the analogs of the statements made in case II of Theorem 45.

94.3. Turning next to integral equations, we first consider the equations with the degenerate kernel

$$
K(t, s) = \sum_{k=1}^{m} p_k(t) q_k(s),
$$

$$
K(s, t) = \sum_{k=1}^{m} p_k(s) q_k(t).
$$

We can regard both the functions $p_k(t)$ and the functions $q_k(s)$ ($k = 1, 2$,

$\ldots, m$) as being linearly independent. Equations (26a) through (26b) become

$$\varphi(t) - \sum_{k=1}^{m} p_k(t) \int_a^b q_k(s)\varphi(s) \, ds = f(t), \tag{28a}$$

$$\varphi_0(t) - \sum_{k=1}^{m} p_k(t) \int_a^b q_k(s)\varphi_0(s) \, ds = 0, \tag{28b}$$

$$\psi(t) - \sum_{k=1}^{m} q_k(t) \int_a^b p_k(s)\psi(s) \, ds = g(t), \tag{28c}$$

$$\psi_0(t) - \sum_{k=1}^{m} q_k(t) \int_a^b p_k(s)\psi_0(s) \, ds = 0. \tag{28d}$$

These equations can be written in abstract form as

$$\varphi - \Sigma p_k(q_k, \varphi) = f, \tag{29a}$$

$$\varphi_0 - \Sigma p_k(q_k, \varphi_0) = 0, \tag{29b}$$

$$\psi - \Sigma q_k(p_k, \psi) = g, \tag{29c}$$

$$\psi_0 - \Sigma q_k(p_k, \psi_0) = 0, \tag{29d}$$

where the vectors φ, f, p_k, q_k, etc., belong to some Euclidean space E.

The operators appearing in these equations belong to the category of *degenerate operators*; this is the name given to operators defined by formulas of the type

$$B\varphi = \sum_{k=1}^{m} p_k(q_k, \varphi).$$

Obviously, a degenerate operator B maps the whole space into the finite-dimensional subspace spanned by the vectors $p_1, p_2, \ldots, p_m$. It is clear from (29a) that if a solution exists, it has the form

$$\varphi = f + \Sigma \xi_k p_k, \tag{30a}$$

where the ξ_k are certain unknown coefficients. Similarly, the solutions of the other equations have the form

$$\varphi_0 = \Sigma \xi_k^0 p_k, \tag{30b}$$

$$\psi = g + \Sigma \eta_k q_k, \tag{30c}$$

$$\psi_0 = \Sigma \eta_k^0 q_k. \tag{30d}$$

Substituting (30a) in (29a), we find that the numbers ξ_k have to satisfy the equation

$$\sum_{k=1}^{m} \xi_k p_k - \sum_{k=1}^{m} p_k(q_k, f) - \sum_{k=1}^{m} p_k \left(q_k, \sum_{i=1}^{m} \xi_i p_i \right) = 0;$$

since the vectors p_k are linearly independent, this equation is equivalent to the system

$$\xi_k - \sum_{i=1}^{m} \xi_i(p_i, q_k) = (f, q_k) \qquad (k = 1, 2, \ldots, m).$$

Writing $a_{ik} = -(p_k, q_i)$ $(i \neq k)$, $a_{ii} = 1 - (p_i, q_i)$, $b_i = (f, q_i)$, we reduce this system to the form

$$\sum_{j=1}^{m} a_{ij}\xi_j = b_i \qquad (i = 1, 2, \ldots, m). \tag{31a}$$

Similarly, equations (29b), (29c), and (29d) become

$$\sum_{j=1}^{m} a_{ij}\xi_j^0 = 0, \tag{31b}$$

$$\sum_{j=1}^{m} a_{ji}\eta_j = c_i, \tag{31c}$$

$$\sum_{j=1}^{m} a_{ji}\eta_j^0 = 0, \tag{31d}$$

where $c_i = (g, p_i)$. If any of the systems (31a) to (31d) has a solution, then by using the corresponding formula from the set (30a) to (30d), we can construct a solution of the corresponding equation from the set (29a) to (29d), i.e., a solution of the corresponding equation from the set (28a) to (28d). It remains only to verify (for example) that the scalar product of the vector f and a solution ψ_0 of (29d), in the sense of the metric of the abstract Euclidean space E, is the same as the scalar product of the vector b with components $b_k = (f, q_k)$ and the vector $\eta^0 = (\eta_1^0, \eta_2^0, \ldots, \eta_m^0)$, in the finite-dimensional Euclidean space T_m. This can be verified by the following simple calculation:

$$(f, \psi_0) = \left(f, \sum_{k=1}^{m} \eta_k^0 q_k \right) = \sum_{k=1}^{m} \eta_k^0(f, q_k) = \sum_{k=1}^{m} b_k \eta_k^0 = (b, \eta^0).$$

Thus, we have proved that the Fredholm alternative holds in the case of a degenerate kernel $K(t, s)$.

94.4. We now turn to the general case, where $K(t, s)$ is an arbitrary function which is square integrable in the region $a \leqslant t, s \leqslant b$. As shown in Sec. 88.3, the integral operator

$$A\varphi = \int_a^b K(t, s)\varphi(s)\, ds$$

can be represented as the limit (in norm) of certain integral operators

$$A_n\varphi = \int_a^b K_n(t, s)\varphi(s)\, ds$$

with degenerate kernels $K_n(t, s)$. It is obvious that at the same time, the adjoint integral operator

$$A^*\psi = \int_a^b K(s, t)\psi(s)\, ds$$

can be represented as the limit of the integral operators

$$A_n^*\psi = \int_a^b K_n(s, t)\psi(s)\, ds,$$

whose kernels are also degenerate.

We now write equations (26a) to (26d) in abstract form, regarding the vectors φ, f, φ_0, etc., as elements of a Euclidean space E:

$$\varphi - A\varphi = f, \tag{32a}$$

$$\varphi_0 - A\varphi_0 = 0, \tag{32b}$$

$$\psi - A^*\psi = g, \tag{32c}$$

$$\psi_0 - A^*\psi_0 = 0. \tag{32d}$$

Henceforth, the term *eigenvector* will refer to a solution of an equation of the type (32b) (with eigenvalue 1). We note that the adjoint operator A^* is connected with the operator A by the relation

$$(A^*p, q) = (p, Aq) \tag{33}$$

where p and q are any vectors. To see this, we write

$$(A^*p, q) = \int_a^b \left\{ \int_a^b K(s, t)p(s)\, ds \right\} q(t)\, dt,$$

$$(p, Aq) = \int_a^b p(t) \left\{ \int_a^b K(t, s)q(s)\, ds \right\} dt,$$

and observe that the first integral can be converted into the second integral by interchanging the variables s and t, and then changing the order of integration, which in the general case is justified by Fubini's theorem (Sec. 85.4). The operators corresponding to the degenerate kernels $K_n(t, s)$ and $K_n(s, t)$ are denoted by A_n and A_n^*, respectively.

LEMMA 1. *If the homogeneous equation*

$$\varphi_n^0 - A_n\varphi_n^0 = 0 \tag{34}$$

has a solution $\varphi_n^0 \neq 0$ for every n, then equation (32b) *also has a nonzero solution.*

Proof. The solutions φ_n^0 of (34) can always be regarded as normalized, i.e., $\|\varphi_n^0\| = 1$, $n = 1, 2, \ldots$. Since the operator A is completely continuous, the sequence $A\varphi_n^0$ contains a convergent subsequence. Discarding superfluous vectors and renumbering the rest, we can assume that the sequence $A\varphi_n^0$ itself converges. Then $A_n\varphi_n^0$ also converges, since

$$A_n\varphi_n^0 = (A_n - A)\varphi_n^0 + A\varphi_n^0$$

and

$$\|(A_n - A)\varphi_n^0\| \leqslant \|A_n - A\| \, \|\varphi_n^0\| \to 0.$$

Moreover, if $A_n\varphi_n^0$ converges, so does the sequence $\varphi_n^0 = A_n\varphi_n^0$; we set

$$\varphi_0 = \lim_{n \to \infty} \varphi_n^0.$$

Since the vectors φ_n^0 have norm 1, so does the vector φ_0, and

$$A\varphi_0 = \lim_{n \to \infty} A\varphi_n^0 = \lim_{n \to \infty} A_n\varphi_n^0 = \lim_{n \to \infty} \varphi_n^0 = \varphi_0.$$

Thus, equation (32b) does in fact have a nonzero solution φ_0, Q.E.D.

LEMMA 2. *If, for every n, (34) has k linearly independent solutions $\varphi_{1n}^0, \varphi_{2n}^0, \ldots, \varphi_{kn}^0$, then (32b) also has k linearly independent solutions $\varphi_1^0, \varphi_2^0, \ldots, \varphi_n^0$.*

Proof. We can assume that the solutions $\varphi_{1n}^0, \varphi_{2n}^0, \ldots, \varphi_{kn}^0$ of (34) are orthonormal. We form the sequence

$$\begin{array}{cccc}
\varphi_{11}^0, & \varphi_{12}^0, & \ldots, & \varphi_{1n}^0, \ldots \\
\varphi_{21}^0, & \varphi_{22}^0, & \ldots, & \varphi_{2n}^0, \ldots \\
\cdot & \cdot & \ldots & \cdot \quad \ldots \\
\varphi_{k1}^0, & \varphi_{k2}^0, & \ldots, & \varphi_{kn}^0, \ldots
\end{array}$$

As proved in Lemma 1, each row contains a convergent subsequence. Discarding superfluous vectors and renumbering the rest, we can assume that the sequences themselves converge, and that their limits $\varphi_1^0, \varphi_2^0, \ldots, \varphi_k^0$ are nonzero solutions of equation (32b). Moreover, since the functions $\varphi_{1n}^0, \varphi_{2n}^0, \ldots, \varphi_{kn}^0$ are orthogonal for every n, their limits $\varphi_1^0, \varphi_2^0, \ldots, \varphi_k^0$ are also orthogonal, and hence also linearly independent, Q.E.D.

The set of all solutions of (32b) is a subspace which we denote by Φ_0. We now show that Φ_0 is finite-dimensional. If Φ_0 were infinite-dimensional, then we could construct an infinite-dimensional orthonormal system $e_1, e_2, \ldots, e_n, \ldots$ in Φ_0. Since the operator A is completely continuous, the vectors Ae_n, $n = 1, 2, \ldots$, would then form a compact set. But the e_n are solutions of (32b), so that $Ae_n = e_n$, and clearly, we cannot select a convergent subsequence from the sequence $e_1, e_2, \ldots, e_n, \ldots$, since $\|e_n - e_m\| =$

$\sqrt{2}$, for any n and m. This contradiction shows that the subspace Φ_0 is finite-dimensional; we denote its dimension by ν.

LEMMA 3. *Suppose that the completely continuous operator A can be represented as the limit of a sequence of degenerate operators A_n. Then A can also be represented as the limit of a sequence of degenerate operators $\tilde{A}_n$, where the space spanned by the eigenvectors of each of the operators $\tilde{A}_n$ coincides with the space Φ_0.*

Proof. Let $\varphi_1^0, \varphi_2^0, \ldots, \varphi_\nu^0$ be an orthonormal system in the subspace Φ_0, and write

$$h_i^n = \varphi_i^0 - A_n\varphi_i^0 \qquad (i = 1, 2, \ldots, \nu).$$

We define the operator $\tilde{A}_n$ by the formula

$$\tilde{A}_n\varphi = A_n\varphi + \sum_{i=1}^{\nu} h_i^n(\varphi, \varphi_i^0).$$

Obviously, $\tilde{A}_n$ is a degenerate operator. It is clear that $\|\tilde{A}_n - A_n\| \to 0$, since $h_i^n \to 0$. It follows that $\|\tilde{A}_n - A\| \to 0$ also. The vectors φ_j^0 are eigenvectors of the operator $\tilde{A}_n$, since

$$\tilde{A}_n\varphi_j^0 = A_n\varphi_j^0 + \sum_{i=1}^{\nu} (\varphi_i^0 - A_n\varphi_i^0)(\varphi_j^0, \varphi_i^0) = A_n\varphi_j^0 + \varphi_j^0 - A_n\varphi_j^0 = \varphi_j^0.$$

In general, the operators $\tilde{A}_n$ can have more than ν linearly independent eigenvectors, but we maintain that this cannot be the case for an infinite number of indices, for otherwise, considering only those values of n for which the operators $\tilde{A}_n$ have at least $\nu + 1$ linearly independent eigenvectors and applying Lemma 2, we would find that the operator A also has at least $\nu + 1$ linearly independent eigenvectors, contrary to assumption. Thus, we find that only a finite number of the operators $\tilde{A}_n$ can have more than ν linearly independent eigenvectors. Discarding superfluous vectors and renumbering the rest, we obtain a sequence of operators $\tilde{A}_n$ satisfying the conditions of Lemma 3.

We can now prove that *the subspace Ψ_0 of solutions of equation* (32d) *has the same dimension as the subspace Φ_0*. Consider the sequence of degenerate operators $\tilde{A}_n$ figuring in Lemma 3, and also the sequence of adjoint operators $\tilde{A}_n^*$. Since the Fredholm alternative holds for degenerate operators, the dimension of the space spanned by the eigenvectors of each of the operators $\tilde{A}_n^*$ equals ν. Since $\tilde{A}_n \to A$, we also have $\tilde{A}_n^* \to A^*$. By Lemma 2, there exists a system of ν orthonormal solutions of the equation $A^*\psi_0 = \psi_0$. There can be no more than ν such solutions, since by using $\nu + 1$ solutions and arguing in the reverse direction, from the operator A^* to the operator

A, we would obtain $\nu + 1$ linearly independent solutions for the equation $A\varphi_0 = \varphi_0$, which is contrary to hypothesis. Thus, the equations $A\varphi_0 = \varphi_0$ and $A^*\psi_0 = \psi_0$ always have the same number of linearly independent solutions.

Next, we consider the question of the existence of solutions for equation (32a). Let the vector φ be a solution of (32a), and let ψ_0 be a solution of (32d). Multiplying (32a) by ψ_0 and using (33), we obtain

$$(\varphi, \psi_0) - (A\varphi, \psi_0) = (\varphi, \psi_0) - (\varphi, A^*\psi_0) = (f, \psi_0).$$

But, by (32d), we have

$$(\varphi, \psi_0) - (\varphi, A^*\psi_0) = (\varphi, \psi_0) - (\varphi, \psi_0) = 0,$$

so that any solution ψ_0 of (32d) satisfies

$$(f, \psi_0) = 0.$$

Thus, (32a) can have a solution only under the condition that the vector f be orthogonal to all the solutions of (32d). We now show that if this condition is met, there always exists a solution of (32a).

Consider a sequence of degenerate operators $\tilde{A}_n \to A$ such that the space spanned by the eigenvectors of each operator $\tilde{A}_n$ is the same as the subspace Φ_0 spanned by the eigenvectors of the operator A; let the dimension of Φ_0 be ν. (The existence of such a sequence of operators $\tilde{A}_n$ was proved in Lemma 3.) Each of the operators $\tilde{A}_n^*$ has the same number ν of orthonormal eigenvectors ψ_i^n, since the Fredholm alternative is valid for degenerate operators. By Lemma 2, we can assume that as $n \to \infty$, these vectors converge to an orthonormal system of eigenvectors ψ_i^0 $(i = 1, 2, \ldots, \nu)$ of the operator A^*. Now let the right-hand side of equation (32a) be

$$f_n = f - \sum_{i=1}^{\nu} (f, \psi_i^n)\psi_i^n.$$

The vector f_n is orthogonal to all the vectors ψ_i^n, since f_n is the perpendicular dropped from the end of the vector f onto the subspace $L(\psi_1^n, \psi_2^n, \ldots, \psi_\nu^n)$. Therefore, applying the Fredholm alternative for the degenerate operator $\tilde{A}_n$, we establish the existence of a vector φ_n satisfying the equation

$$\varphi_n - \tilde{A}_n\varphi_n = f_n.$$

As $n \to \infty$, the vectors f_n converge to the vector f, since $\psi_i^n \to \psi_i^0$ and $(f, \psi_i^0) = 0$ $(i = 1, 2, \ldots, \nu)$. Moreover, for any n, the vector φ_n can be chosen to be orthogonal to the subspace spanned by the eigenvectors of the operator $\tilde{A}_n$, i.e., to the subspace Φ_0.

Next, we show that the vectors φ_n obtained in this way have *bounded norms.* Assume the contrary, i.e., let the norms of the vectors φ_n be unbounded. Then, discarding certain superfluous vectors, we can assume that $\|\varphi_n\| \to \infty$. Setting

$$\tilde{\varphi}_n = \frac{\varphi_n}{\|\varphi_n\|},$$

we obtain a sequence of normalized vectors satisfying the equation

$$\tilde{\varphi}_n - \tilde{A}_n \tilde{\varphi}_n = \frac{f}{\|\varphi_n\|}. \tag{35}$$

The right-hand side of (35) converges to zero as $n \to \infty$. By a previous argument, the sequence of vectors $\tilde{A}_n \tilde{\varphi}_n$ can be assumed to be convergent, and then the sequence of vectors $\tilde{\varphi}_n$ will also be convergent. Suppose that

$$\varphi_0 = \lim_{n \to \infty} \tilde{\varphi}_n;$$

$\|\varphi_0\| = 1$, since $\|\tilde{\varphi}_n\| = 1$. Passing to the limit in (35), we find that the vector φ_0 satisfies the equation

$$\varphi_0 - A\varphi_0 = 0.$$

which implies that φ_0 is an element of the subspace Φ_0. On the other hand, since all the vectors $\tilde{\varphi}_n$ are orthogonal to Φ_0, the limit vector φ_0 is also orthogonal to Φ_0. This contradiction shows that the vectors φ_n actually have bounded norms.

Since the φ_n are bounded, then, just as before, the sequence $A\varphi_n$ can be regarded as convergent, and then the sequences

$$\tilde{A}_n \varphi_n = (\tilde{A}_n - A)\varphi_n + A\varphi_n$$

and

$$\varphi_n = f_n + \tilde{A}_n \varphi_n$$

will also be convergent. Denoting the limit of the sequence φ_n by φ, and passing to the limit in the relation

$$\varphi_n = f_n + (\tilde{A}_n - A)\varphi_n + A\varphi_n,$$

we obtain

$$\varphi = f + A\varphi,$$

i.e., the vector φ is a solution of (32a). Thus, *if the right-hand side f of* (32a) *is orthogonal to any solution of* (32d), *equation* (32a) *has a solution*; the solution φ is determined only to within any solution φ_0 of the homogeneous equation (32b), and since the set of all solutions of (32b) is finite-dimensional, φ can be chosen to be orthogonal to this whole set. This choice singles out a unique solution of (32a).

Thus, we have verified all the statements in case II of the Fredholm alternative. It remains only to verify the statements in case I.

In case I, equation (32a) has no nonzero solutions, and the subspace Φ_0 contains only the zero vector. Then, by what has been proved, the subspace Ψ_0 of solutions of (32d) contains only the zero vector. As we have shown, (32a) has a solution if f is orthogonal to the subspace Ψ_0; in the present case, any vector f satisfies this condition, and hence (32a) has a solution for any $f \in E$. This solution is unique, since the difference between any two solutions of (32a) is a solution of (32b), and hence is zero, by assumption. This completes the proof of Theorem 45.

We note that if the kernel $K(t, s)$ and the function $f(t)$ are continuous, then it follows from the last remark of Sec. 88.3 that the solution of the integral equation (26a) is also continuous.

*95. Applications to Potential Theory

95.1. We shall assume that the reader is familiar with the following facts from potential theory:

a) A function $u(x, y)$ satisfying the equation

$$\Delta u \equiv \frac{\partial^2 u}{\partial x^2} + \frac{\partial^2 u}{\partial y^2} = 0$$

in a region G of the (x, y) plane is said to be a *harmonic function*. An example is the function

$$\log \frac{1}{\sqrt{(x - \xi)^2 + (y - \eta)^2}}, \tag{36}$$

which depends on the parameters ξ, η and is harmonic wherever the expression under the radical is different from zero, i.e., everywhere except at the point $x = \xi, y = \eta$. Setting $(x, y) = P$, $(\xi, \eta) = Q$, we abbreviate the expression (36) as

$$\log \frac{1}{r(P, Q)}.$$

The partial derivatives of the function (36) are also harmonic functions for $P \neq Q$.

b) Let C be a simple smooth closed contour, which divides the plane into two regions, an interior region G_i and an exterior region G_e. Consider a function v which is continuous and differentiable in the regions G_i and G_e, but which possibly suffers a discontinuity in traversing the points of the contour C. We denote by v_i the limiting values of the function v when the contour C is approached from the inside, and by v_e the limiting values of the function v when the contour C is approached from the outside. The normal derivatives $\partial v_i / \partial n$ and $\partial v_e / \partial n$ are defined similarly; it is assumed that the positive direction of the normal points in the outward direction.

We have the *Green's formulas*

$$\iint_{G_i} \left[\left(\frac{\partial v}{\partial x}\right)^2 + \left(\frac{\partial v}{\partial y}\right)^2\right] dx\,dy = \int_C v_i \frac{\partial v_i}{\partial n}\,dl, \tag{37}$$

if v is a function which is harmonic in the region G_i and

$$\iint_{G_e} \left[\left(\frac{\partial v}{\partial x}\right)^2 + \left(\frac{\partial v}{\partial y}\right)^2\right] dx\,dy = -\int_C v_e \frac{\partial v_e}{\partial n}\,dl, \tag{38}$$

if v is a function which is harmonic in the region G_e, provided that the integrals in the right-hand side exist; for example, this is guaranteed by the requirement that $\partial v/\partial x$, $\partial v/\partial y$ fall off at infinity no slower than $r^{-(1+\epsilon)}$.

c) Let C be the contour used in the constructions of paragraph b), and let l denote a parameter of the contour, e.g., the arc length measured from a fixed initial point Q_0 to the variable point Q. Suppose that we are given a continuous function $\rho(Q)$. Then the function

$$v(P) = \int_C \rho(Q) \frac{\partial}{\partial n_Q} \log \frac{1}{r(P, Q)}\,dl \tag{39}$$

[the potential of a double layer with density $\rho(Q)$] is a function which is harmonic in both regions G_i and G_e. The integral (39) exists even for points P on the contour C itself.

In particular, in the case $\rho(Q) \equiv 1$, a simple construction shows that the function $v(P)$ is the total change in the angle swept out by the ray PQ as Q describes the contour C in the negative direction, and therefore

$$\int_C \frac{\partial}{\partial n_Q} \log \frac{1}{r(P, Q)}\,dl = -2\pi \quad \text{if } P \in G_i, \tag{40}$$

$$\int_C \frac{\partial}{\partial n_Q} \log \frac{1}{r(P, Q)}\,dl = -\pi \quad \text{if } P \in C, \tag{41}$$

$$\int_C \frac{\partial}{\partial n_Q} \log \frac{1}{r(P, Q)}\,dl = 0 \quad \text{if } P \in G_e. \tag{42}$$

Thus, in this case, the values of the function $v(P)$ in the region G_i are π less than its values on the boundary, and its values in the region G_e are π more than its values on the boundary. In the general case, where we have an arbitrary continuous density $\rho(Q)$, the following formulas hold:

$$v_i(P) = v(P) - \pi\rho(P), \tag{43}$$

$$v_e(P) = v(P) + \pi\rho(P), \tag{44}$$

$$\frac{\partial v_i(P)}{\partial n} = \frac{\partial v_e(P)}{\partial n}, \tag{45}$$

where $P \in C$.

The function

$$u(P) = \int_C \rho(Q) \log \frac{1}{r(P, Q)} \, dl \tag{46}$$

[the potential of a single layer with density $\rho(Q)$] is also harmonic in both regions G_i and G_e, and the following formulas are valid:

$$u_i(P) = u_e(P) = u(P), \tag{47}$$

$$\frac{\partial u_i(P)}{\partial n} = \int_C \rho(Q) \frac{\partial}{\partial n_P} \log \frac{1}{r(P, Q)} \, dl + \pi\rho(P), \tag{48}$$

$$\frac{\partial u_e(P)}{\partial n} = \int_C \rho(Q) \frac{\partial}{\partial n_P} \log \frac{1}{r(P, Q)} \, dl - \pi\rho(P), \tag{49}$$

where $P \in C$.

95.2. We now pose the following problems:

1. The first boundary value problem (the *Dirichlet problem*): Find a function $v(P)$ harmonic in the region G_i (the *interior* problem) or in G_e (the *exterior* problem), which takes preassigned boundary values $f(Q)$ on the contour C.

2. The second boundary value problem (the *Neumann problem*): Find a function $u(P)$ harmonic in the region G_i (the *interior* problem) or in G_e (the *exterior* problem), whose normal derivative takes preassigned boundary values $g(Q)$ on the contour C.

Green's formulas (37) and (38) allow us to determine the degree of arbitrariness of the solutions of these problems, provided the solutions exist. For example, suppose that we have two solutions of the interior Dirichlet problem, which correspond to the same boundary function $f(Q)$. Then $v(P)$, the difference between these two solutions, is a harmonic function in the region G_i, for which $v_i = 0$. By Green's formula (37), $\partial v/\partial x = 0$, $\partial v/\partial y = 0$, and hence the function is constant in G_i; but then $v = 0$, since $v_i = 0$. Thus, the interior Dirichlet problem can have only one solution, Similarly, the interior Neumann problem has a unique solution to within an additive constant. Moreover, both exterior problems have only one solution for which the integral in the right-hand side of (38) converges.

We now give the solution due to Fredholm of both of these problems for the case of a contour C with continuous curvature.

95.3. We look for a solution of the Dirichlet problem in the form of the potential of a double layer

$$v(P) = \int_C \rho(Q) \frac{\partial}{\partial n_Q} \log \frac{1}{r(P, Q)} \, dl, \tag{50}$$

with an unknown continuous function $\rho(Q)$. By (43) and the statement of the problem, we have

$$v_i(P) = \int_C \rho(Q) \frac{\partial}{\partial n_Q} \log \frac{1}{r(P, Q)} \, dl - \pi\rho(P) = f(P),$$

where $P \in C$. Thus, the function $\rho(P)$ is the solution of a Fredholm integral equation of the second kind with the kernel

$$K(P, Q) = \frac{\partial}{\partial n_Q} \log \frac{1}{r(P, Q)}.$$

This kernel is continuous for all P and Q on the contour C, and it can be shown that as $P \to Q$, it approaches a limit equal to the curvature of the contour C at the point Q.

By the Fredholm alternative, proved in Sec. 94, to prove that the Dirichlet problem has a solution for any $f(P)$, it suffices to show that the corresponding homogeneous equation

$$\int_C \rho(Q)K(P, Q) \, dl - \pi\rho(P) = 0 \qquad (P \in C)$$

has only the zero solution. We now prove that this is the case. Assume the contrary, and let $\rho_0(P)$ be a nonzero solution of this homogeneous equation. Then, for the harmonic function

$$v_0(P) = \int_C K(P, Q)\rho_0(Q) \, dl,$$

we have

$$v_{i0}(P) = v_0(P) - \pi\rho_0(P) \equiv 0,$$

so that $v_0(P) \equiv 0$ in the region G_i, by Green's formula (37). But then $\partial v_{i0}(P)/\partial n \equiv 0$, and by (45), we also have $\partial v_{e0}(P)/\partial n \equiv 0$. Since the integral in the left-hand side of (38) exists for the function $v_0(P)$, applying this formula, we find that $v_0(P) = $ const in the region G_e. Moreover, since $v_0(P) = 0$ at infinity, we have $v_0(P) \equiv 0$ in G_e, and hence $v_{e0}(P) = 0$. We now find from (43) and (44) that $\rho_0(P) \equiv 0$ where $P \in C$, contrary to our assumption, Q.E.D.

Applying the results of Sec. 94, we find that the integral equation (50) has a solution for every function $f(P)$. If $f(P)$ is continuous, then by the continuity of the kernel $K(P, Q)$ and the last remark of Sec. 94.4, the solution $\rho(P)$ is also a continuous function. Therefore, formulas (43) to (45) are valid for the function $\rho(P)$, so that it is legitimate to reduce the Dirichlet problem to the potential (50). Thus, finally *the interior Dirichlet problem has a solution for any continuous boundary function $f(P)$.*

We now look for a solution to the exterior Dirichlet problem, of the same form (50). In this case, we obtain the equation

$$v_e(P) = \int_C \rho(Q) \frac{\partial}{\partial n_Q} \log \frac{1}{r(P, Q)} \, dl + \pi\rho(P) = f(P). \tag{51}$$

However, this time the corresponding homogeneous equation

$$\int_C \rho(Q) K(P, Q) \, dl + \pi \rho(P) = 0 \qquad (P \in C)$$

has a nonzero solution $\rho(P) \equiv 1$ [see (41)]. This solution is unique to within a numerical factor. In fact, arguing in just the same way as above, but interchanging the subscripts i and e everywhere, we arrive at the relation $\partial v_i(P)/\partial n = 0$, from which it follows by Green's formula that $v(P) = \text{const}$ in G_i. Then, we find from (43) and (44) that $\rho(P) = \text{const}$. Therefore, by the Fredholm alternative, equation (51) does not have a solution for all f, but only for those which are orthogonal to some fixed function $\rho_0(P)$, i.e., a solution (unique to within a numerical factor) of the adjoint homogeneous equation.

However, we are able to solve the problem for *any* boundary function f, if beside the solutions given by equation (50), which obviously converge to zero at infinity, we also consider solutions which are obtained from solutions of (50) by adding a constant. In fact, if f is any function specified on the boundary, we can always find a constant c such that the function $f - c$ is orthogonal to the function ρ_0. Then, by what has been proved, there exists a solution $v(P)$ of the exterior problem with boundary values $f - c$. On the other hand, $v_0(P) \equiv c$ is a solution of the exterior problem with boundary values equal to c. Therefore, $v(P) + v_0(P)$ is a solution of the exterior Dirichlet problem with boundary values f. Thus, finally *the exterior Dirichlet problem has a solution for any continuous boundary function f.*

95.4. Next, we look for a solution of the interior Neumann problem, in the form of the potential of a simple layer

$$u(P) = \int_C \rho(Q) \log \frac{1}{r(P, Q)} \, dl, \tag{52}$$

with an unknown function $\rho(Q)$. By (48) and the statement of the problem, we have

$$\frac{\partial u_i(P)}{\partial n} = \int_C \rho(Q) \frac{\partial}{\partial n_P} \log \frac{1}{r(P, Q)} \, dl + \pi \rho(P) = g(P), \tag{53}$$

so that the function $\rho(P)$ is again the solution of a Fredholm integral equation, this time with a kernel

$$K_1(P, Q) = \frac{\partial}{\partial n_P} \log \frac{1}{r(P, Q)}$$

which is the transpose of the kernel $K(P, Q)$ appearing in the Dirichlet problem. As we have already proved, the homogeneous adjoint equation

$$\int_C \rho(P) \frac{\partial}{\partial n_Q} \log \frac{1}{r(P, Q)} \, dl + \pi \rho(P) = 0$$

has only a constant solution. Therefore, by the Fredholm alternative, (53) has a solution if and only if the function $g(P)$ is orthogonal to 1, i.e.,

$$\int_C g(Q) \, dl = 0. \tag{54}$$

But it is well known that any function $u(P)$ which is harmonic in the region G_i, whether it has the form (46) or not, satisfies the equation

$$\int_C \frac{\partial u(Q)}{\partial n} \cdot dl = 0.$$

Thus, we see that *the condition* (54) *is necessary and sufficient for the interior Neumann problem to have a solution.*

Finally, we look for a solution of the exterior Neumann problem in the same form (52), and reduce it to the integral equation

$$\frac{\partial u_e(P)}{\partial n} = \int_C \rho(Q) \frac{\partial}{\partial n_P} \log \frac{1}{r(P, Q)} \, dl - \pi \rho(P) = g(P).$$

As we have proved, the homogeneous adjoint equation

$$\int_C \rho(Q) K(P, Q) \, dl - \pi \rho(P) = 0 \qquad (P \in C)$$

has no nonzero solutions. Therefore, *the exterior Neumann problem has a solution for any continuous boundary function* $g(P)$.

Remark. The condition used by Fredholm to guarantee the continuity of the kernel $K(P, Q)$, namely, that the contour C has continuous curvature, has subsequently been weakened considerably. Since we do not have space to go into these matters here, we refer the reader to the special literature on the subject.[11]

---

[11] Smirnov, V. I., "Курс Высшей Математики", translated into German as *Lehrgang der Höheren Mathematik*, Deutscher Verlag der Wissenschaften, Berlin, vol. 4 (1958).

BIBLIOGRAPHY[1]

Aitken, A. C., *Determinants and Matrices*, Oliver and Boyd, London (1959).

Bellman, R., *Introduction to Matrix Algebra*, McGraw-Hill Book Co., Inc., New York (1960).

Gantmakher, F. R., *The Theory of Matrices*, 2 vols., translated by K. A. Hirsch, Chelsea Publishing Co., New York (1959).

Gelfand, I. M., *Lectures on Linear Algebra*, translated by A. Shenitzer, Interscience Publishers, Inc., New York (1961).

Goldstein, H., *Classical Mechanics*, Addison-Wesley Publishing Co., Reading, Mass. (1951). (*Re* Sec. 76.)

Halmos, P. R., *Finite-Dimensional Vector Spaces*, 2d ed., D. Van Nostrand Co., Inc., Princeton, N. J. (1958).

Halmos, P. R., *Introduction to Hilbert Space*, Chelsea Publishing Co., New York (1951).

Hamburger, H. L. and M. E. Grimshaw, *Linear Transformations*, Cambridge University Press, New York (1951).

Hoffman, K. and R. Kunze, *Linear Algebra*, Prentice-Hall, Inc., Englewood Cliffs, N.J. (1961).

Jacobson, N., *Lectures in Abstract Algebra, Vol. 2, Linear Algebra*, D. Van Nostrand Co., Inc., Princeton, N. J. (1953).

Kellogg, O. D., *Foundations of Potential Theory*, Dover Publications, Inc., New York (1953). (*Re* Sec. 95.)

Mirsky, L., *An Introduction to Linear Algebra*, Oxford University Press, New York (1955).

Perlis, S., *The Theory of Matrices*, Addison-Wesley Publishing Co., Reading, Mass. (1952).

Schreier, O. and E. Sperner, *Introduction to Modern Algebra and Matrix Theory*, translated by M. Davis and M. Hauser, Chelsea Publishing Co., New York (1959).

Smirnov, V. I., *Linear Algebra and Group Theory*, translated by R. A. Silverman, McGraw-Hill Book Co., Inc., New York (1961).

Spain, B., *Analytical Quadrics*, Pergamon Press, London (1960). (*Re* Chap. 11.)

Stoll, R. R., *Linear Algebra and Matrix Theory*, McGraw-Hill Book Co., Inc., New York (1952).

Struik, D. J., *Lectures on Classical Differential Geometry*, Addison-Wesley Publishing Co., Reading, Mass. (1950). (*Re* Sec. 75.)

Synge, J. L. and A. Schild, *Tensor Calculus*, University of Toronto Press, Toronto (1956). (*Re* Sec. 39.)

Thrall, R. M. and L. Tornheim, *Vector Spaces and Matrices*, John Wiley and Sons, Inc., New York (1957).

Tricomi, F. G., *Integral Equations*, Interscience Publishers, Inc., New York (1957). (*Re* Chap. 12.)

---

[1] See also books cited in Secs. 53, 70, 82, 85, 92, 93, and 95.

INDEX

A CATALOGUE OF SELECTED DOVER BOOKS
IN ALL FIELDS OF INTEREST

A CATALOGUE OF SELECTED DOVER BOOKS
IN ALL FIELDS OF INTEREST

AMERICA'S OLD MASTERS, James T. Flexner. Four men emerged unexpectedly from provincial 18th century America to leadership in European art: Benjamin West, J. S. Copley, C. R. Peale, Gilbert Stuart. Brilliant coverage of lives and contributions. Revised, 1967 edition. 69 plates. 365pp. of text.

21806-6 Paperbound $3.00

FIRST FLOWERS OF OUR WILDERNESS: AMERICAN PAINTING, THE COLONIAL PERIOD, James T. Flexner. Painters, and regional painting traditions from earliest Colonial times up to the emergence of Copley, West and Peale Sr., Foster, Gustavus Hesselius, Feke, John Smibert and many anonymous painters in the primitive manner. Engaging presentation, with 162 illustrations. xxii + 368pp.

22180-6 Paperbound $3.50

THE LIGHT OF DISTANT SKIES: AMERICAN PAINTING, 1760-1835, James T. Flexner. The great generation of early American painters goes to Europe to learn and to teach: West, Copley, Gilbert Stuart and others. Allston, Trumbull, Morse; also contemporary American painters—primitives, derivatives, academics—who remained in America. 102 illustrations. xiii + 306pp.   22179-2 Paperbound $3.50

A HISTORY OF THE RISE AND PROGRESS OF THE ARTS OF DESIGN IN THE UNITED STATES, William Dunlap. Much the richest mine of information on early American painters, sculptors, architects, engravers, miniaturists, etc. The only source of information for scores of artists, the major primary source for many others. Unabridged reprint of rare original 1834 edition, with new introduction by James T. Flexner, and 394 new illustrations. Edited by Rita Weiss. 6⅝ x 9⅝.

21695-0, 21696-9, 21697-7 Three volumes, Paperbound $13.50

EPOCHS OF CHINESE AND JAPANESE ART, Ernest F. Fenollosa. From primitive Chinese art to the 20th century, thorough history, explanation of every important art period and form, including Japanese woodcuts; main stress on China and Japan, but Tibet, Korea also included. Still unexcelled for its detailed, rich coverage of cultural background, aesthetic elements, diffusion studies, particularly of the historical period. 2nd, 1913 edition. 242 illustrations. lii + 439pp. of text.

20364-6, 20365-4 Two volumes, Paperbound $6.00

THE GENTLE ART OF MAKING ENEMIES, James A. M. Whistler. Greatest wit of his day deflates Oscar Wilde, Ruskin, Swinburne; strikes back at inane critics, exhibitions, art journalism; aesthetics of impressionist revolution in most striking form. Highly readable classic by great painter. Reproduction of edition designed by Whistler. Introduction by Alfred Werner. xxxvi + 334pp.

21875-9 Paperbound $3.00

VISUAL ILLUSIONS: THEIR CAUSES, CHARACTERISTICS, AND APPLICATIONS, Matthew Luckiesh. Thorough description and discussion of optical illusion, geometric and perspective, particularly; size and shape distortions, illusions of color, of motion; natural illusions; use of illusion in art and magic, industry, etc. Most useful today with op art, also for classical art. Scores of effects illustrated. Introduction by William H. Ittleson. 100 illustrations. xxi + 252pp.

21530-X Paperbound $2.00

A HANDBOOK OF ANATOMY FOR ART STUDENTS, Arthur Thomson. Thorough, virtually exhaustive coverage of skeletal structure, musculature, etc. Full text, supplemented by anatomical diagrams and drawings and by photographs of undraped figures. Unique in its comparison of male and female forms, pointing out differences of contour, texture, form. 211 figures, 40 drawings, 86 photographs. xx + 459pp. 5⅜ x 8⅜.

21163-0 Paperbound $3.50

150 MASTERPIECES OF DRAWING, Selected by Anthony Toney. Full page reproductions of drawings from the early 16th to the end of the 18th century, all beautifully reproduced: Rembrandt, Michelangelo, Dürer, Fragonard, Urs, Graf, Wouwerman, many others. First-rate browsing book, model book for artists. xviii + 150pp. 8⅜ x 11¼.

21032-4 Paperbound $2.50

THE LATER WORK OF AUBREY BEARDSLEY, Aubrey Beardsley. Exotic, erotic, ironic masterpieces in full maturity: Comedy Ballet, Venus and Tannhauser, Pierrot, Lysistrata, Rape of the Lock, Savoy material, Ali Baba, Volpone, etc. This material revolutionized the art world, and is still powerful, fresh, brilliant. With *The Early Work,* all Beardsley's finest work. 174 plates, 2 in color. xiv + 176pp. 8⅛ x 11.

21817-1 Paperbound $3.00

DRAWINGS OF REMBRANDT, Rembrandt van Rijn. Complete reproduction of fabulously rare edition by Lippmann and Hofstede de Groot, completely reedited, updated, improved by Prof. Seymour Slive, Fogg Museum. Portraits, Biblical sketches, landscapes, Oriental types, nudes, episodes from classical mythology—All Rembrandt's fertile genius. Also selection of drawings by his pupils and followers. "Stunning volumes," *Saturday Review.* 550 illustrations. lxxviii + 552pp. 9⅛ x 12¼.

21485-0, 21486-9 Two volumes, Paperbound $10.00

THE DISASTERS OF WAR, Francisco Goya. One of the masterpieces of Western civilization—83 etchings that record Goya's shattering, bitter reaction to the Napoleonic war that swept through Spain after the insurrection of 1808 and to war in general. Reprint of the first edition, with three additional plates from Boston's Museum of Fine Arts. All plates facsimile size. Introduction by Philip Hofer, Fogg Museum. v + 97pp. 9⅜ x 8¼.

21872-4 Paperbound $2.00

GRAPHIC WORKS OF ODILON REDON. Largest collection of Redon's graphic works ever assembled: 172 lithographs, 28 etchings and engravings, 9 drawings. These include some of his most famous works. All the plates from *Odilon Redon: oeuvre graphique complet,* plus additional plates. New introduction and caption translations by Alfred Werner. 209 illustrations. xxvii + 209pp. 9⅛ x 12¼.

21966-8 Paperbound $4.50

DESIGN BY ACCIDENT; A BOOK OF "ACCIDENTAL EFFECTS" FOR ARTISTS AND DESIGNERS, James F. O'Brien. Create your own unique, striking, imaginative effects by "controlled accident" interaction of materials: paints and lacquers, oil and water based paints, splatter, crackling materials, shatter, similar items. Everything you do will be different; first book on this limitless art, so useful to both fine artist and commercial artist. Full instructions. 192 plates showing "accidents," 8 in color. viii + 215pp. 8⅜ x 11¼.   21942-9 Paperbound $3.50

THE BOOK OF SIGNS, Rudolf Koch. Famed German type designer draws 493 beautiful symbols: religious, mystical, alchemical, imperial, property marks, runes, etc. Remarkable fusion of traditional and modern. Good for suggestions of timelessness, smartness, modernity. Text. vi + 104pp. 6⅛ x 9¼.

20162-7 Paperbound $1.25

HISTORY OF INDIAN AND INDONESIAN ART, Ananda K. Coomaraswamy. An unabridged republication of one of the finest books by a great scholar in Eastern art. Rich in descriptive material, history, social backgrounds; Sunga reliefs, Rajput paintings, Gupta temples, Burmese frescoes, textiles, jewelry, sculpture, etc. 400 photos. viii + 423pp. 6⅜ x 9¾.   21436-2 Paperbound $5.00

PRIMITIVE ART, Franz Boas. America's foremost anthropologist surveys textiles, ceramics, woodcarving, basketry, metalwork, etc.; patterns, technology, creation of symbols, style origins. All areas of world, but very full on Northwest Coast Indians. More than 350 illustrations of baskets, boxes, totem poles, weapons, etc. 378 pp.

20025-6 Paperbound $3.00

THE GENTLEMAN AND CABINET MAKER'S DIRECTOR, Thomas Chippendale. Full reprint (third edition, 1762) of most influential furniture book of all time, by master cabinetmaker. 200 plates, illustrating chairs, sofas, mirrors, tables, cabinets, plus 24 photographs of surviving pieces. Biographical introduction by N. Bienenstock. vi + 249pp. 9⅞ x 12¾.   21601-2 Paperbound $4.00

AMERICAN ANTIQUE FURNITURE, Edgar G. Miller, Jr. The basic coverage of all American furniture before 1840. Individual chapters cover type of furniture—clocks, tables, sideboards, etc.—chronologically, with inexhaustible wealth of data. More than 2100 photographs, all identified, commented on. Essential to all early American collectors. Introduction by H. E. Keyes. vi + 1106pp. 7⅞ x 10¾.

21599-7, 21600-4 Two volumes, Paperbound $11.00

PENNSYLVANIA DUTCH AMERICAN FOLK ART, Henry J. Kauffman. 279 photos, 28 drawings of tulipware, Fraktur script, painted tinware, toys, flowered furniture, quilts, samplers, hex signs, house interiors, etc. Full descriptive text. Excellent for tourist, rewarding for designer, collector. Map. 146pp. 7⅞ x 10¾.

21205-X Paperbound $2.50

EARLY NEW ENGLAND GRAVESTONE RUBBINGS, Edmund V. Gillon, Jr. 43 photographs, 226 carefully reproduced rubbings show heavily symbolic, sometimes macabre early gravestones, up to early 19th century. Remarkable early American primitive art, occasionally strikingly beautiful; always powerful. Text. xxvi + 207pp. 8⅜ x 11¼.   21380-3 Paperbound $3.50

ALPHABETS AND ORNAMENTS, Ernst Lehner. Well-known pictorial source for decorative alphabets, script examples, cartouches, frames, decorative title pages, calligraphic initials, borders, similar material. 14th to 19th century, mostly European. Useful in almost any graphic arts designing, varied styles. 750 illustrations. 256pp. 7 x 10. 21905-4 Paperbound $4.00

PAINTING: A CREATIVE APPROACH, Norman Colquhoun. For the beginner simple guide provides an instructive approach to painting: major stumbling blocks for beginner; overcoming them, technical points; paints and pigments; oil painting; watercolor and other media and color. New section on "plastic" paints. Glossary. Formerly *Paint Your Own Pictures*. 221pp. 22000-1 Paperbound $1.75

THE ENJOYMENT AND USE OF COLOR, Walter Sargent. Explanation of the relations between colors themselves and between colors in nature and art, including hundreds of little-known facts about color values, intensities, effects of high and low illumination, complementary colors. Many practical hints for painters, references to great masters. 7 color plates, 29 illustrations. x + 274pp.
20944-X Paperbound $2.75

THE NOTEBOOKS OF LEONARDO DA VINCI, compiled and edited by Jean Paul Richter. 1566 extracts from original manuscripts reveal the full range of Leonardo's versatile genius: all his writings on painting, sculpture, architecture, anatomy, astronomy, geography, topography, physiology, mining, music, etc., in both Italian and English, with 186 plates of manuscript pages and more than 500 additional drawings. Includes studies for the Last Supper, the lost Sforza monument, and other works. Total of xlvii + 866pp. 7⅞ x 10¾.
22572-0, 22573-9 Two volumes, Paperbound $10.00

MONTGOMERY WARD CATALOGUE OF 1895. Tea gowns, yards of flannel and pillow-case lace, stereoscopes, books of gospel hymns, the New Improved Singer Sewing Machine, side saddles, milk skimmers, straight-edged razors, high-button shoes, spittoons, and on and on . . . listing some 25,000 items, practically all illustrated. Essential to the shoppers of the 1890's, it is our truest record of the spirit of the period. Unaltered reprint of Issue No. 57, Spring and Summer 1895. Introduction by Boris Emmet. Innumerable illustrations. xiii + 624pp. 8½ x 11⅝.
22377-9 Paperbound $6.95

THE CRYSTAL PALACE EXHIBITION ILLUSTRATED CATALOGUE (LONDON, 1851). One of the wonders of the modern world—the Crystal Palace Exhibition in which all the nations of the civilized world exhibited their achievements in the arts and sciences—presented in an equally important illustrated catalogue. More than 1700 items pictured with accompanying text—ceramics, textiles, cast-iron work, carpets, pianos, sleds, razors, wall-papers, billiard tables, beehives, silverware and hundreds of other artifacts—represent the focal point of Victorian culture in the Western World. Probably the largest collection of Victorian decorative art ever assembled—indispensable for antiquarians and designers. Unabridged republication of the Art-Journal Catalogue of the Great Exhibition of 1851, with all terminal essays. New introduction by John Gloag, F.S.A. xxxiv + 426pp. 9 x 12.
22503-8 Paperbound $5.00

A HISTORY OF COSTUME, Carl Köhler. Definitive history, based on surviving pieces of clothing primarily, and paintings, statues, etc. secondarily. Highly readable text, supplemented by 594 illustrations of costumes of the ancient Mediterranean peoples, Greece and Rome, the Teutonic prehistoric period; costumes of the Middle Ages, Renaissance, Baroque, 18th and 19th centuries. Clear, measured patterns are provided for many clothing articles. Approach is practical throughout. Enlarged by Emma von Sichart. 464pp.   21030-8 Paperbound $3.50.

ORIENTAL RUGS, ANTIQUE AND MODERN, Walter A. Hawley. A complete and authoritative treatise on the Oriental rug—where they are made, by whom and how, designs and symbols, characteristics in detail of the six major groups, how to distinguish them and how to buy them. Detailed technical data is provided on periods, weaves, warps, wefts, textures, sides, ends and knots, although no technical background is required for an understanding. 11 color plates, 80 halftones, 4 maps. vi + 320pp. 6⅛ x 9⅛.   22366-3 Paperbound $5.00

TEN BOOKS ON ARCHITECTURE, Vitruvius. By any standards the most important book on architecture ever written. Early Roman discussion of aesthetics of building, construction methods, orders, sites, and every other aspect of architecture has inspired, instructed architecture for about 2,000 years. Stands behind Palladio, Michelangelo, Bramante, Wren, countless others. Definitive Morris H. Morgan translation. 68 illustrations. xii + 331pp.   20645-9 Paperbound $3.00

THE FOUR BOOKS OF ARCHITECTURE, Andrea Palladio. Translated into every major Western European language in the two centuries following its publication in 1570, this has been one of the most influential books in the history of architecture. Complete reprint of the 1738 Isaac Ware edition. New introduction by Adolf Placzek, Columbia Univ. 216 plates. xxii + 110pp. of text. 9½ x 12¾.
21308-0 Clothbound $12.50

STICKS AND STONES: A STUDY OF AMERICAN ARCHITECTURE AND CIVILIZATION, Lewis Mumford.One of the great classics of American cultural history. American architecture from the medieval-inspired earliest forms to the early 20th century; evolution of structure and style, and reciprocal influences on environment. 21 photographic illustrations. 238pp.   20202-X Paperbound $2.00

THE AMERICAN BUILDER'S COMPANION, Asher Benjamin. The most widely used early 19th century architectural style and source book, for colonial up into Greek Revival periods. Extensive development of geometry of carpentering, construction of sashes, frames, doors, stairs; plans and elevations of domestic and other buildings. Hundreds of thousands of houses were built according to this book, now invaluable to historians, architects, restorers, etc. 1827 edition. 59 plates. 114pp. 7⅞ x 10¾.
22236-5 Paperbound $3.50

DUTCH HOUSES IN THE HUDSON VALLEY BEFORE 1776, Helen Wilkinson Reynolds. The standard survey of the Dutch colonial house and outbuildings, with constructional features, decoration, and local history associated with individual homesteads. Introduction by Franklin D. Roosevelt. Map. 150 illustrations. 469pp. 6⅝ x 9¼.   21469-9 Paperbound $5.00

THE ARCHITECTURE OF COUNTRY HOUSES, Andrew J. Downing. Together with Vaux's *Villas and Cottages* this is the basic book for Hudson River Gothic architecture of the middle Victorian period. Full, sound discussions of general aspects of housing, architecture, style, decoration, furnishing, together with scores of detailed house plans, illustrations of specific buildings, accompanied by full text. Perhaps the most influential single American architectural book. 1850 edition. Introduction by J. Stewart Johnson. 321 figures, 34 architectural designs. xvi + 560pp.

22003-6 Paperbound $4.00

LOST EXAMPLES OF COLONIAL ARCHITECTURE, John Mead Howells. Full-page photographs of buildings that have disappeared or been so altered as to be denatured, including many designed by major early American architects. 245 plates. xvii + 248pp. 7⅞ x 10¾.

21143-6 Paperbound $3.50

DOMESTIC ARCHITECTURE OF THE AMERICAN COLONIES AND OF THE EARLY REPUBLIC, Fiske Kimball. Foremost architect and restorer of Williamsburg and Monticello covers nearly 200 homes between 1620-1825. Architectural details, construction, style features, special fixtures, floor plans, etc. Generally considered finest work in its area. 219 illustrations of houses, doorways, windows, capital mantels. xx + 314pp. 7⅞ x 10¾.

21743-4 Paperbound $4.00

EARLY AMERICAN ROOMS: 1650-1858, edited by Russell Hawes Kettell. Tour of 12 rooms, each representative of a different era in American history and each furnished, decorated, designed and occupied in the style of the era. 72 plans and elevations, 8-page color section, etc., show fabrics, wall papers, arrangements, etc. Full descriptive text. xvii + 200pp. of text. 8⅜ x 11¼.

21633-0 Paperbound $5.00

THE FITZWILLIAM VIRGINAL BOOK, edited by J. Fuller Maitland and W. B. Squire. Full modern printing of famous early 17th-century ms. volume of 300 works by Morley, Byrd, Bull, Gibbons, etc. For piano or other modern keyboard instrument; easy to read format. xxxvi + 938pp. 8⅜ x 11.

21068-5, 21069-3 Two volumes, Paperbound $10.00

KEYBOARD MUSIC, Johann Sebastian Bach. Bach Gesellschaft edition. A rich selection of Bach's masterpieces for the harpsichord: the six English Suites, six French Suites, the six Partitas (Clavierübung part I), the Goldberg Variations (Clavierübung part IV), the fifteen Two-Part Inventions and the fifteen Three-Part Sinfonias. Clearly reproduced on large sheets with ample margins; eminently playable. vi + 312pp. 8⅛ x 11.

22360-4 Paperbound $5.00

THE MUSIC OF BACH: AN INTRODUCTION, Charles Sanford Terry. A fine, nontechnical introduction to Bach's music, both instrumental and vocal. Covers organ music, chamber music, passion music, other types. Analyzes themes, developments, innovations. x + 114pp.

21075-8 Paperbound $1.50

BEETHOVEN AND HIS NINE SYMPHONIES, Sir George Grove. Noted British musicologist provides best history, analysis, commentary on symphonies. Very thorough, rigorously accurate; necessary to both advanced student and amateur music lover. 436 musical passages. vii + 407 pp.

20334-4 Paperbound $2.75

JOHANN SEBASTIAN BACH, Philipp Spitta. One of the great classics of musicology, this definitive analysis of Bach's music (and life) has never been surpassed. Lucid, nontechnical analyses of hundreds of pieces (30 pages devoted to St. Matthew Passion, 26 to B Minor Mass). Also includes major analysis of 18th-century music. 450 musical examples. 40-page musical supplement. Total of xx + 1799pp.
(EUK) 22278-0, 22279-9 Two volumes, Clothbound $17.50

MOZART AND HIS PIANO CONCERTOS, Cuthbert Girdlestone. The only full-length study of an important area of Mozart's creativity. Provides detailed analyses of all 23 concertos, traces inspirational sources. 417 musical examples. Second edition. 509pp.
21271-8 Paperbound $3.50

THE PERFECT WAGNERITE: A COMMENTARY ON THE NIBLUNG'S RING, George Bernard Shaw. Brilliant and still relevant criticism in remarkable essays on Wagner's Ring cycle, Shaw's ideas on political and social ideology behind the plots, role of Leitmotifs, vocal requisites, etc. Prefaces. xxi + 136pp.
(USO) 21707-8 Paperbound $1.50

DON GIOVANNI, W. A. Mozart. Complete libretto, modern English translation; biographies of composer and librettist; accounts of early performances and critical reaction. Lavishly illustrated. All the material you need to understand and appreciate this great work. Dover Opera Guide and Libretto Series; translated and introduced by Ellen Bleiler. 92 illustrations. 209pp.
21134-7 Paperbound $2.00

BASIC ELECTRICITY, U. S. Bureau of Naval Personel. Originally a training course, best non-technical coverage of basic theory of electricity and its applications. Fundamental concepts, batteries, circuits, conductors and wiring techniques, AC and DC, inductance and capacitance, generators, motors, transformers, magnetic amplifiers, synchros, servomechanisms, etc. Also covers blue-prints, electrical diagrams, etc. Many questions, with answers. 349 illustrations. x + 448pp. $6\frac{1}{2}$ x $9\frac{1}{4}$.
20973-3 Paperbound $3.50

REPRODUCTION OF SOUND, Edgar Villchur. Thorough coverage for laymen of high fidelity systems, reproducing systems in general, needles, amplifiers, preamps, loudspeakers, feedback, explaining physical background. "A rare talent for making technicalities vividly comprehensible," R. Darrell, *High Fidelity*. 69 figures. iv + 92pp.
21515-6 Paperbound $1.25

HEAR ME TALKIN' TO YA: THE STORY OF JAZZ AS TOLD BY THE MEN WHO MADE IT, Nat Shapiro and Nat Hentoff. Louis Armstrong, Fats Waller, Jo Jones, Clarence Williams, Billy Holiday, Duke Ellington, Jelly Roll Morton and dozens of other jazz greats tell how it was in Chicago's South Side, New Orleans, depression Harlem and the modern West Coast as jazz was born and grew. xvi + 429pp.
21726-4 Paperbound $3.00

FABLES OF AESOP, translated by Sir Roger L'Estrange. A reproduction of the very rare 1931 Paris edition; a selection of the most interesting fables, together with 50 imaginative drawings by Alexander Calder. v + 128pp. $6\frac{1}{2}$x$9\frac{1}{4}$.
21780-9 Paperbound $1.50

AGAINST THE GRAIN (A REBOURS), Joris K. Huysmans. Filled with weird images, evidences of a bizarre imagination, exotic experiments with hallucinatory drugs, rich tastes and smells and the diversions of its sybarite hero Duc Jean des Esseintes, this classic novel pushed 19th-century literary decadence to its limits. Full unabridged edition. Do not confuse this with abridged editions generally sold. Introduction by Havelock Ellis. xlix + 206pp. 22190-3 Paperbound $2.00

VARIORUM SHAKESPEARE: HAMLET. Edited by Horace H. Furness; a landmark of American scholarship. Exhaustive footnotes and appendices treat all doubtful words and phrases, as well as suggested critical emendations throughout the play's history. First volume contains editor's own text, collated with all Quartos and Folios. Second volume contains full first Quarto, translations of Shakespeare's sources (Belleforest, and Saxo Grammaticus), Der Bestrafte Brudermord, and many essays on critical and historical points of interest by major authorities of past and present. Includes details of staging and costuming over the years. By far the best edition available for serious students of Shakespeare. Total of xx + 905pp.
21004-9, 21005-7, 2 volumes, Paperbound $7.00

A LIFE OF WILLIAM SHAKESPEARE, Sir Sidney Lee. This is the standard life of Shakespeare, summarizing everything known about Shakespeare and his plays. Incredibly rich in material, broad in coverage, clear and judicious, it has served thousands as the best introduction to Shakespeare. 1931 edition. 9 plates. xxix + 792pp. (USO) 21967-4 Paperbound $3.75

MASTERS OF THE DRAMA, John Gassner. Most comprehensive history of the drama in print, covering every tradition from Greeks to modern Europe and America, including India, Far East, etc. Covers more than 800 dramatists, 2000 plays, with biographical material, plot summaries, theatre history, criticism, etc. "Best of its kind in English," *New Republic*. 77 illustrations. xxii + 890pp.
20100-7 Clothbound $8.50

THE EVOLUTION OF THE ENGLISH LANGUAGE, George McKnight. The growth of English, from the 14th century to the present. Unusual, non-technical account presents basic information in very interesting form: sound shifts, change in grammar and syntax, vocabulary growth, similar topics. Abundantly illustrated with quotations. Formerly *Modern English in the Making*. xii + 590pp.
21932-1 Paperbound $3.50

AN ETYMOLOGICAL DICTIONARY OF MODERN ENGLISH, Ernest Weekley. Fullest, richest work of its sort, by foremost British lexicographer. Detailed word histories, including many colloquial and archaic words; extensive quotations. Do not confuse this with the Concise Etymological Dictionary, which is much abridged. Total of xxvii + 830pp. 6½ x 9¼.
21873-2, 21874-0 Two volumes, Paperbound $7.90

FLATLAND: A ROMANCE OF MANY DIMENSIONS, E. A. Abbott. Classic of science-fiction explores ramifications of life in a two-dimensional world, and what happens when a three-dimensional being intrudes. Amusing reading, but also useful as introduction to thought about hyperspace. Introduction by Banesh Hoffmann. 16 illustrations. xx + 103pp. 20001-9 Paperbound $1.00

POEMS OF ANNE BRADSTREET, edited with an introduction by Robert Hutchinson. A new selection of poems by America's first poet and perhaps the first significant woman poet in the English language. 48 poems display her development in works of considerable variety—love poems, domestic poems, religious meditations, formal elegies, "quaternions," etc. Notes, bibliography. viii + 222pp.

22160-1 Paperbound $2.50

THREE GOTHIC NOVELS: THE CASTLE OF OTRANTO BY HORACE WALPOLE; VATHEK BY WILLIAM BECKFORD; THE VAMPYRE BY JOHN POLIDORI, WITH FRAGMENT OF A NOVEL BY LORD BYRON, edited by E. F. Bleiler. The first Gothic novel, by Walpole; the finest Oriental tale in English, by Beckford; powerful Romantic supernatural story in versions by Polidori and Byron. All extremely important in history of literature; all still exciting, packed with supernatural thrills, ghosts, haunted castles, magic, etc. xl + 291pp.

21232-7 Paperbound $2.50

THE BEST TALES OF HOFFMANN, E. T. A. Hoffmann. 10 of Hoffmann's most important stories, in modern re-editings of standard translations: Nutcracker and the King of Mice, Signor Formica, Automata, The Sandman, Rath Krespel, The Golden Flowerpot, Master Martin the Cooper, The Mines of Falun, The King's Betrothed, A New Year's Eve Adventure. 7 illustrations by Hoffmann. Edited by E. F. Bleiler. xxxix + 419pp. 21793-0 Paperbound $3.00

GHOST AND HORROR STORIES OF AMBROSE BIERCE, Ambrose Bierce. 23 strikingly modern stories of the horrors latent in the human mind: The Eyes of the Panther, The Damned Thing, An Occurrence at Owl Creek Bridge, An Inhabitant of Carcosa, etc., plus the dream-essay, Visions of the Night. Edited by E. F. Bleiler. xxii + 199pp. 20767-6 Paperbound $1.50

BEST GHOST STORIES OF J. S. LEFANU, J. Sheridan LeFanu. Finest stories by Victorian master often considered greatest supernatural writer of all. Carmilla, Green Tea, The Haunted Baronet, The Familiar, and 12 others. Most never before available in the U. S. A. Edited by E. F. Bleiler. 8 illustrations from Victorian publications. xvii + 467pp. 20415-4 Paperbound $3.00

MATHEMATICAL FOUNDATIONS OF INFORMATION THEORY, A. I. Khinchin. Comprehensive introduction to work of Shannon, McMillan, Feinstein and Khinchin, placing these investigations on a rigorous mathematical basis. Covers entropy concept in probability theory, uniqueness theorem, Shannon's inequality, ergodic sources, the E property, martingale concept, noise, Feinstein's fundamental lemma, Shanon's first and second theorems. Translated by R. A. Silverman and M. D. Friedman. iii + 120pp. 60434-9 Paperbound $2.00

SEVEN SCIENCE FICTION NOVELS, H. G. Wells. The standard collection of the great novels. Complete, unabridged. *First Men in the Moon, Island of Dr. Moreau, War of the Worlds, Food of the Gods, Invisible Man, Time Machine, In the Days of the Comet.* Not only science fiction fans, but every educated person owes it to himself to read these novels. 1015pp. (USO) 20264-X Clothbound $6.00

LAST AND FIRST MEN AND STAR MAKER, TWO SCIENCE FICTION NOVELS, Olaf Stapledon. Greatest future histories in science fiction. In the first, human intelligence is the "hero," through strange paths of evolution, interplanetary invasions, incredible technologies, near extinctions and reemergences. Star Maker describes the quest of a band of star rovers for intelligence itself, through time and space: weird inhuman civilizations, crustacean minds, symbiotic worlds, etc. Complete, unabridged. v + 438pp.   (USO) 21962-3 Paperbound $2.50

THREE PROPHETIC NOVELS, H. G. WELLS. Stages of a consistently planned future for mankind. *When the Sleeper Wakes,* and *A Story of the Days to Come,* anticipate *Brave New World* and *1984,* in the 21st Century; *The Time Machine,* only complete version in print, shows farther future and the end of mankind. All show Wells's greatest gifts as storyteller and novelist. Edited by E. F. Bleiler. x + 335pp.   (USO) 20605-X Paperbound $2.50

THE DEVIL'S DICTIONARY, Ambrose Bierce. America's own Oscar Wilde— Ambrose Bierce—offers his barbed iconoclastic wisdom in over 1,000 definitions hailed by H. L. Mencken as "some of the most gorgeous witticisms in the English language." 145pp.   20487-1 Paperbound $1.25

MAX AND MORITZ, Wilhelm Busch. Great children's classic, father of comic strip, of two bad boys, Max and Moritz. Also Ker and Plunk (Plisch und Plumm), Cat and Mouse, Deceitful Henry, Ice-Peter, The Boy and the Pipe, and five other pieces. Original German, with English translation. Edited by H. Arthur Klein; translations by various hands and H. Arthur Klein. vi + 216pp.
20181-3 Paperbound $2.00

PIGS IS PIGS AND OTHER FAVORITES, Ellis Parker Butler. The title story is one of the best humor short stories, as Mike Flannery obfuscates biology and English. Also included, That Pup of Murchison's, The Great American Pie Company, and Perkins of Portland. 14 illustrations. v + 109pp.   21532-6 Paperbound $1.25

THE PETERKIN PAPERS, Lucretia P. Hale. It takes genius to be as stupidly mad as the Peterkins, as they decide to become wise, celebrate the "Fourth," keep a cow, and otherwise strain the resources of the Lady from Philadelphia. Basic book of American humor. 153 illustrations. 219pp.   20794-3 Paperbound $1.50

PERRAULT'S FAIRY TALES, translated by A. E. Johnson and S. R. Littlewood, with 34 full-page illustrations by Gustave Doré. All the original Perrault stories— Cinderella, Sleeping Beauty, Bluebeard, Little Red Riding Hood, Puss in Boots, Tom Thumb, etc.—with their witty verse morals and the magnificent illustrations of Doré. One of the five or six great books of European fairy tales. viii + 117pp. 8⅛ x 11.   22311-6 Paperbound $2.00

OLD HUNGARIAN FAIRY TALES, Baroness Orczy. Favorites translated and adapted by author of the *Scarlet Pimpernel.* Eight fairy tales include "The Suitors of Princess Fire-Fly," "The Twin Hunchbacks," "Mr. Cuttlefish's Love Story," and "The Enchanted Cat." This little volume of magic and adventure will captivate children as it has for generations. 90 drawings by Montagu Barstow. 96pp.
22293-4 Paperbound $1.95

THE RED FAIRY BOOK, Andrew Lang. Lang's color fairy books have long been children's favorites. This volume includes Rapunzel, Jack and the Bean-stalk and 35 other stories, familiar and unfamiliar. 4 plates, 93 illustrations x + 367pp.
21673-X Paperbound $2.50

THE BLUE FAIRY BOOK, Andrew Lang. Lang's tales come from all countries and all times. Here are 37 tales from Grimm, the Arabian Nights, Greek Mythology, and other fascinating sources. 8 plates, 130 illustrations. xi + 390pp.
21437-0 Paperbound $2.50

HOUSEHOLD STORIES BY THE BROTHERS GRIMM. Classic English-language edition of the well-known tales — Rumpelstiltskin, Snow White, Hansel and Gretel, The Twelve Brothers, Faithful John, Rapunzel, Tom Thumb (52 stories in all). Translated into simple, straightforward English by Lucy Crane. Ornamented with headpieces, vignettes, elaborate decorative initials and a dozen full-page illustrations by Walter Crane. x + 269pp.
21080-4 Paperbound **$2.00**

THE MERRY ADVENTURES OF ROBIN HOOD, Howard Pyle. The finest modern versions of the traditional ballads and tales about the great English outlaw. Howard Pyle's complete prose version, with every word, every illustration of the first edition. Do not confuse this facsimile of the original (1883) with modern editions that change text or illustrations. 23 plates plus many page decorations. xxii + 296pp.
22043-5 Paperbound $2.50

THE STORY OF KING ARTHUR AND HIS KNIGHTS, Howard Pyle. The finest children's version of the life of King Arthur; brilliantly retold by Pyle, with 48 of his most imaginative illustrations. xviii + 313pp. 6⅛ x 9¼.
21445-1 Paperbound $2.50

THE WONDERFUL WIZARD OF OZ, L. Frank Baum. America's finest children's book in facsimile of first edition with all Denslow illustrations in full color. The edition a child should have. Introduction by Martin Gardner. 23 color plates, scores of drawings. iv + 267pp.
20691-2 Paperbound $2.50

THE MARVELOUS LAND OF OZ, L. Frank Baum. The second Oz book, every bit as imaginative as the Wizard. The hero is a boy named Tip, but the Scarecrow and the Tin Woodman are back, as is the Oz magic. 16 color plates, 120 drawings by John R. Neill. 287pp.
20692-0 Paperbound $2.50

THE MAGICAL MONARCH OF MO, L. Frank Baum. Remarkable adventures in a land even stranger than Oz. The best of Baum's books not in the Oz series. 15 color plates and dozens of drawings by Frank Verbeck. xviii + 237pp.
21892-9 Paperbound $2.25

THE BAD CHILD'S BOOK OF BEASTS, MORE BEASTS FOR WORSE CHILDREN, A MORAL ALPHABET, Hilaire Belloc. Three complete humor classics in one volume. Be kind to the frog, and do not call him names . . . and 28 other whimsical animals. Familiar favorites and some not so well known. Illustrated by Basil Blackwell. 156pp.
(USO) 20749-8 Paperbound $1.50

EAST O' THE SUN AND WEST O' THE MOON, George W. Dasent. Considered the best of all translations of these Norwegian folk tales, this collection has been enjoyed by generations of children (and folklorists too). Includes True and Untrue, Why the Sea is Salt, East O' the Sun and West O' the Moon, Why the Bear is Stumpy-Tailed, Boots and the Troll, The Cock and the Hen, Rich Peter the Pedlar, and 52 more. The only edition with all 59 tales. 77 illustrations by Erik Werenskiold and Theodor Kittelsen. xv + 418pp. 22521-6 Paperbound $3.50

GOOPS AND HOW TO BE THEM, Gelett Burgess. Classic of tongue-in-cheek humor, masquerading as etiquette book. 87 verses, twice as many cartoons, show mischievous Goops as they demonstrate to children virtues of table manners, neatness, courtesy, etc. Favorite for generations. viii + 88pp. 6½ x 9¼. 22233-0 Paperbound $1.25

ALICE'S ADVENTURES UNDER GROUND, Lewis Carroll. The first version, quite different from the final Alice in Wonderland, printed out by Carroll himself with his own illustrations. Complete facsimile of the "million dollar" manuscript Carroll gave to Alice Liddell in 1864. Introduction by Martin Gardner. viii + 96pp. Title and dedication pages in color. 21482-6 Paperbound $1.25

THE BROWNIES, THEIR BOOK, Palmer Cox. Small as mice, cunning as foxes, exuberant and full of mischief, the Brownies go to the zoo, toy shop, seashore, circus, etc., in 24 verse adventures and 266 illustrations. Long a favorite, since their first appearance in St. Nicholas Magazine. xi + 144pp. 6⅝ x 9¼. 21265-3 Paperbound $1.75

SONGS OF CHILDHOOD, Walter De La Mare. Published (under the pseudonym Walter Ramal) when De La Mare was only 29, this charming collection has long been a favorite children's book. A facsimile of the first edition in paper, the 47 poems capture the simplicity of the nursery rhyme and the ballad, including such lyrics as I Met Eve, Tartary, The Silver Penny. vii + 106pp. (USO) 21972-0 Paperbound $1.25

THE COMPLETE NONSENSE OF EDWARD LEAR, Edward Lear. The finest 19th-century humorist-cartoonist in full: all nonsense limericks, zany alphabets, Owl and Pussycat, songs, nonsense botany, and more than 500 illustrations by Lear himself. Edited by Holbrook Jackson. xxix + 287pp. (USO) 20167-8 Paperbound $2.00

BILLY WHISKERS: THE AUTOBIOGRAPHY OF A GOAT, Frances Trego Montgomery. A favorite of children since the early 20th century, here are the escapades of that rambunctious, irresistible and mischievous goat—Billy Whiskers. Much in the spirit of Peck's Bad Boy, this is a book that children never tire of reading or hearing. All the original familiar illustrations by W. H. Fry are included: 6 color plates, 18 black and white drawings. 159pp. 22345-0 Paperbound $2.00

MOTHER GOOSE MELODIES. Faithful republication of the fabulously rare Munroe and Francis "copyright 1833" Boston edition—the most important Mother Goose collection, usually referred to as the "original." Familiar rhymes plus many rare ones, with wonderful old woodcut illustrations. Edited by E. F. Bleiler. 128pp. 4½ x 6⅜. 22577-1 Paperbound $1.00

Two Little Savages; Being the Adventures of Two Boys Who Lived as Indians and What They Learned, Ernest Thompson Seton. Great classic of nature and boyhood provides a vast range of woodlore in most palatable form, a genuinely entertaining story. Two farm boys build a teepee in woods and live in it for a month, working out Indian solutions to living problems, star lore, birds and animals, plants, etc. 293 illustrations. vii + 286pp.

20985-7 Paperbound $2.50

Peter Piper's Practical Principles of Plain & Perfect Pronunciation. Alliterative jingles and tongue-twisters of surprising charm, that made their first appearance in America about 1830. Republished in full with the spirited woodcut illustrations from this earliest American edition. 32pp. 4½ x 6⅜.

22560-7 Paperbound $1.00

Science Experiments and Amusements for Children, Charles Vivian. 73 easy experiments, requiring only materials found at home or easily available, such as candles, coins, steel wool, etc.; illustrate basic phenomena like vacuum, simple chemical reaction, etc. All safe. Modern, well-planned. Formerly *Science Games for Children*. 102 photos, numerous drawings. 96pp. 6⅛ x 9¼.

21856-2 Paperbound $1.25

An Introduction to Chess Moves and Tactics Simply Explained, Leonard Barden. Informal intermediate introduction, quite strong in explaining reasons for moves. Covers basic material, tactics, important openings, traps, positional play in middle game, end game. Attempts to isolate patterns and recurrent configurations. Formerly *Chess*. 58 figures. 102pp. (USO) 21210-6 Paperbound $1.25

Lasker's Manual of Chess, Dr. Emanuel Lasker. Lasker was not only one of the five great World Champions, he was also one of the ablest expositors, theorists, and analysts. In many ways, his Manual, permeated with his philosophy of battle, filled with keen insights, is one of the greatest works ever written on chess. Filled with analyzed games by the great players. A single-volume library that will profit almost any chess player, beginner or master. 308 diagrams. xli x 349pp.

20640-8 Paperbound $2.75

The Master Book of Mathematical Recreations, Fred Schuh. In opinion of many the finest work ever prepared on mathematical puzzles, stunts, recreations; exhaustively thorough explanations of mathematics involved, analysis of effects, citation of puzzles and games. Mathematics involved is elementary. Translated by F. Göbel. 194 figures. xxiv + 430pp. 22134-2 Paperbound $3.50

Mathematics, Magic and Mystery, Martin Gardner. Puzzle editor for Scientific American explains mathematics behind various mystifying tricks: card tricks, stage "mind reading," coin and match tricks, counting out games, geometric dissections, etc. Probability sets, theory of numbers clearly explained. Also provides more than 400 tricks, guaranteed to work, that you can do. 135 illustrations. xii + 176pp.

20335-2 Paperbound $1.75

MATHEMATICAL PUZZLES FOR BEGINNERS AND ENTHUSIASTS, Geoffrey Mott-Smith. 189 puzzles from easy to difficult—involving arithmetic, logic, algebra, properties of digits, probability, etc.—for enjoyment and mental stimulus. Explanation of mathematical principles behind the puzzles. 135 illustrations. viii + 248pp.

20198-8 Paperbound $1.75

PAPER FOLDING FOR BEGINNERS, William D. Murray and Francis J. Rigney. Easiest book on the market, clearest instructions on making interesting, beautiful origami. Sail boats, cups, roosters, frogs that move legs, bonbon boxes, standing birds, etc. 40 projects; more than 275 diagrams and photographs. 94pp.

20713-7 Paperbound $1.00

TRICKS AND GAMES ON THE POOL TABLE, Fred Herrmann. 79 tricks and games—some solitaires, some for two or more players, some competitive games—to entertain you between formal games. Mystifying shots and throws, unusual caroms, tricks involving such props as cork, coins, a hat, etc. Formerly *Fun on the Pool Table*. 77 figures. 95pp.

21814-7 Paperbound $1.00

HAND SHADOWS TO BE THROWN UPON THE WALL: A SERIES OF NOVEL AND AMUSING FIGURES FORMED BY THE HAND, Henry Bursill. Delightful picturebook from great-grandfather's day shows how to make 18 different hand shadows: a bird that flies, duck that quacks, dog that wags his tail, camel, goose, deer, boy, turtle, etc. Only book of its sort. vi + 33pp. 6½ x 9¼. 21779-5 Paperbound $1.00

WHITTLING AND WOODCARVING, E. J. Tangerman. 18th printing of best book on market. "If you can cut a potato you can carve" toys and puzzles, chains, chessmen, caricatures, masks, frames, woodcut blocks, surface patterns, much more. Information on tools, woods, techniques. Also goes into serious wood sculpture from Middle Ages to present, East and West. 464 photos, figures. x + 293pp.

20965-2 Paperbound $2.00

HISTORY OF PHILOSOPHY, Julián Marias. Possibly the clearest, most easily followed, best planned, most useful one-volume history of philosophy on the market; neither skimpy nor overfull. Full details on system of every major philosopher and dozens of less important thinkers from pre-Socratics up to Existentialism and later. Strong on many European figures usually omitted. Has gone through dozens of editions in Europe. 1966 edition, translated by Stanley Appelbaum and Clarence Strowbridge. xviii + 505pp. 21739-6 Paperbound $3.50

YOGA: A SCIENTIFIC EVALUATION, Kovoor T. Behanan. Scientific but non-technical study of physiological results of yoga exercises; done under auspices of Yale U. Relations to Indian thought, to psychoanalysis, etc. 16 photos. xxiii + 270pp.

20505-3 Paperbound $2.50

Prices subject to change without notice.

Available at your book dealer or write for free catalogue to Dept. GI, Dover Publications, Inc., 180 Varick St., N. Y., N. Y. 10014. Dover publishes more than 150 books each year on science, elementary and advanced mathematics, biology, music, art, literary history, social sciences and other areas.